AF479188

Focusing on the steel industry during the postcommunist transition from 1989 through 2009, Aleksandra Sznajder Lee traces the transformation of flagship state enterprises in the Czech Republic, Poland, Romania, and Slovakia into subsidiaries of large, transnational corporations. By analyzing this transformation at the three levels of enterprise, sector, and national-international nexus, she identifies the players—from international investors and European Union institutions to national labor unions and local industry managers—in the political economy of reform. Even in the midst of the transition to a capitalist, democratic system, Sznajder Lee finds, the state plays a key role in mediating between domestic vested interests and external pressures from international financial markets and institutions, on the one hand, and regional institutions on the other. Whereas state power may be employed to require domestic firms to operate as capitalists in the international market, it may also be used to shield enterprises from market pressures in order to promote the political and personal preferences of the elite.

This study has broad implications for the political economy of reform because it illuminates the political determinants of privatization and internationalization and the resources used to resist them. In addition, Sznajder Lee sheds new light on why some countries are more likely than others to be subject to external constraints, such as IMF conditionality, and how some allegedly pro-market reformers manage to maintain public ownership over certain industry sectors.

Aleksandra Sznajder Lee is Associate Professor of Political Science at the University of Richmond.

TRANSNATIONAL CAPITALISM IN EAST CENTRAL EUROPE'S HEAVY INDUSTRY

From Flagship Enterprises to Subsidiaries

Aleksandra Sznajder Lee

University of Michigan Press
Ann Arbor

Copyright © by Aleksandra Sznajder Lee 2016
All rights reserved

This book may not be reproduced, in whole or in part, including illustrations, in any form (beyond that copying permitted by Sections 107 and 108 of the U.S. Copyright Law and except by reviewers for the public press), without written permission from the publisher.

Published in the United States of America by the
University of Michigan Press
Manufactured in the United States of America
♾ Printed on acid-free paper

2019   2018   2017   2016   4   3   2   1

A CIP catalog record for this book is available from the British Library.

Library of Congress Cataloging-in-Publication Data

Names: Sznajder Lee, Aleksandra, author.
Title: Transnational capitalism in East Central Europe's heavy industry : from flagship
   enterprises to subsidiaries / Aleksandra Sznajder Lee.
Description: Ann Arbor : University of Michigan Press, [2016] | Includes bibliographical
   references and index.
Identifiers: LCCN 2016001990 | ISBN 9780472119875 (hardback) | ISBN 9780472121915
   (e-book)
Subjects: LCSH: Steel industry and trade—Europe, Eastern—Finance. | Capitalism—
   Europe, Eastern. | Investments, Foreign—Europe, Eastern. | Post-communism—
   Europe, Eastern. | BISAC: POLITICAL SCIENCE / Economic Conditions.
Classification: LCC HD9525.E852 S93 2016 | DDC 338.4/309437—dc23
LC record available at http://lccn.loc.gov/2016001990

For my parents,
Joanna and Roman

Contents

Abbreviations

APAPS	Authority for Privatization and Administration of State Ownership (*Autoritatea pentru Privatizarea şi Administrarea Participaţiilor Statului*)
AVAS	Authority for State Assets Recovery (*Autoritatea pentru Valorificarea Activelor Statului*)
AWS	Solidarity Electoral Action (*Akcja Wyborcza "Solidarność"*)
CMEA	Council for Mutual Economic Assistance
CZK	Czech Koruna
ČSSD	Czech Social Democratic Party (*Česká Strana Sociálně Demokratická*)
EBRD	European Bank for Reconstruction and Development
EC	European Community
ECE	East Central Europe
EU	European Union
FDI	Foreign Direct Investment
FESAL	Financial and Enterprise Sector Adjustment Loan
FNM	National Property Fund (*Fond Národního Majetku*, Czech Republic; *Fond Národného Majetku*, Slovak Republic)
FPS	State Ownership Fund (*Fondul Proprietăţii de Stat*)
FSN	National Salvation Front (*Frontul Salvării Naţionale*)
GDP	Gross Domestic Product
HZDS	Movement for a Democratic Slovakia (*Hnutie za Demokratické Slovensko*)
IFC	International Finance Corporation
IFI	International Financial Institution
IMF	International Monetary Fund

ISD	Industrial Union of Donbas
KOZ	Slovak Confederation of Trade Unions (*Konfederácia Odborových Zväzov*)
MEBO	Management-Employee Buyout
NATO	North Atlantic Treaty Organization
NIK	Supreme Audit Office (*Najwyższa Izba Kontroli*)
ODA	Civic Democratic Alliance (*Občanská Demokratická Aliance*)
ODS	Civic Democratic Party (*Občanská Demokratická Strana*)
OPZZ	All-Poland Trade Union Alliance (*Ogólnopolskie Porozumienie Związków Zawodowych*)
OS KOVO	Czech Metalworkers' Federation KOVO (*Odborový Svaz KOVO*)
OZ KOVO	Slovak Metalworkers' Federation KOVO (*Odborový Zväz KOVO*)
PAV	Public Against Violence (*Verejnosť Proti Násiliu*)
PDSR	Party of Social Democracy of Romania (*Partidul Democraţiei Sociale din România*)
PHARE	Poland and Hungary Assistance for the Restructuring of the Economy
PHS	Polish Steelworks (*Polskie Huty Stali*)
PLN	Polish Złoty
PSAL	Public Sector Adjustment Loan
PSD	Social Democratic Party (*Partidul Social Democrat*)
PSL	Polish People's Party (*Polskie Stronnictwo Ludowe*)
ROL	Romanian Lei
SKK	Slovak Koruna
SLD	Democratic Left Alliance (*Sojusz Lewicy Demokratycznej*)
SOE	State-Owned Enterprise
USSR	Union of Soviet Socialist Republics
UW	Freedom Union (*Unia Wolności*)
VSŽ	East Slovak Steelworks (*Východoslovenské Železiarne*)
ZOHO	Works of Ostrowiec—Ostrowiec Steelworks (*Zakłady Ostrowieckie Huta Ostrowiec*)

Acknowledgments

The completion of this book has been a long journey. During that time, I benefited from the generosity of numerous individuals who helped in various capacities to bring this project to fruition. The origins of this book reach back to my dissertation, and I remain deeply indebted to my dissertation committee, David R. Cameron, Anna Grzymała-Busse, Susan Rose-Ackerman, and Iván Szelényi, for all the advice, inspiration, constructive criticism, and encouragement they have given me along the way. I am especially grateful to Anna Grzymała-Busse for her suggestions for turning the dissertation into a book.

I remain in awe of the generosity of many people who went out of their way, literally and figuratively, to assist me during the research process. I am also greatly indebted to the numerous interviewees who patiently answered my questions, sometimes on more than one occasion, and who shared their knowledge, experience, and insights. I am also immensely grateful to those interviewees who trusted me with their stories and who helped me understand the human dimension of postcommunist transition.

In the initial stages of the project, I greatly profited from discussions with Magda Pustoła, Paweł Ruszkowski, Agnieszka Sołtys, and Rafał Towalski. I also thank my friends and relatives in Warsaw, whose support and good humor were critical as the project was getting off the ground. My warm thanks go to the late Mrs. Jadwiga Węgrzecka, who hosted me during my research in Poland. Her joy in life in the face of physical suffering and the trauma of being a survivor of the Majdanek concentration camp was an incredible lesson in human resilience.

My research in the Czech Republic was greatly facilitated by contacts I obtained from Dagmar Ašerova, Gerald McDermott, and Milada Vachudova. Veronika Rábová kindly assisted me with correspondence in Czech. In Slovakia, I remain especially indebted to Marta Kahancová and her family, Ľubica Kollová, and Jan Baca. In Romania, my warmest thanks go to the late Aurel Radi, Laurenţiu Ispir, and Călin Ţânţăreanu. I also thank Monica Drăgan and Mark Wenig. I am grateful to Călin Ţânţăreanu and his family for their hospitality and for making me feel at home in Bucharest. I also thank Irina Cătălin for her assistance with Bucharest-based research.

Throughout the research process and writing of this book, I have been generously supported by numerous institutions. I thank the National Science Foundation for the Graduate Research Fellowship, which helped to finance part of my fieldwork and enabled me to study Czech and Romanian, which were indispensable for my field research. I am grateful to the Yale Center for International and Area Studies, which supported various stages of this project through the European Union Studies Grant, the Globalization and Self-Determination Grant, and the Dissertation Fellowship, as well as to Yale University for the Dissertation Fellowship. The American Council of Learned Societies Dissertation Fellowship in East European Studies enabled me to concentrate solely on data analysis during the post-fieldwork year. The following year, the Department of Political Science at Yale University offered me a part-time teaching position as I was finishing the dissertation.

The Postdoctoral Fellowship at the Harriman Institute at Columbia University provided the perfect opportunity to rethink the framework of the project and to start reworking it into a book. I thank David Stark for welcoming me to the Networks and Institutions in Postcommunism project and my colleagues Roger Schoenman, Eugene Raikhel, and Balazs Vedres for their insights and camaraderie.

At the University of Richmond, I found a very collegial environment in which to grow as a scholar and teacher. I am particularly grateful to Andrea Simpson for her encouragement. The University of Richmond's School of Arts and Sciences Faculty Research Committee and the Dean's Office have generously supported my summers to pursue additional research for this book and other projects. I am also indebted to the Woodrow Wilson International Center for Scholars for awarding me the Title VIII Research Scholar Grant. The spring 2013 residence at the Wilson Center created a truly special and stimulating environment in which to finalize the manuscript. I also thank Keith Weber for his research assistance at the Wilson Center.

For their feedback on various aspects of the manuscript, and in different capacities, I am indebted to Oksan Bayulgen, Rachel Epstein, Timothy Frye, Scott Gehlbach, Stephen Hanson, Juliet Johnson, Mitchell Orenstein, Jennifer Pribble, Graeme Robertson, Roger Schoenman, David Stark, Silvana Tarlea, Vera Trappmann, Milada Vachudova, Sarah Wilson Sokhey, and Vineeta Yadav. I also thank Jessica Fortin for sharing her dataset on state capacity. My thanks go to Springer Science+Business Media for allowing me to reprint excerpts from my article "Between Apprehension and Support: Social Dialogue, Democracy, and Industrial Restructuring in Central and Eastern Europe," which appeared in *Studies in Comparative International Development* 45 (2010): 30–56.

At the University of Michigan Press, I thank Melody Herr for her interest and support for this project. I am grateful for her patience and good humor in guiding me through the publication process. I also extend my thanks to Mary Hashman and the rest of the editorial and production teams for their work on the manuscript. Last, I thank the three anonymous reviewers for their feedback and excellent suggestions.

Foremost, however, I want to thank my husband, Chinyelu Lee, who has been putting up with this book project throughout our marriage, and made numerous sacrifices to give me the time and space to complete it. He patiently, and thoroughly, read all the versions of the manuscript, and his sharp intellect and suggestions to make the argument crisper and the writing clearer, have greatly improved this book. On top of his wise counsel, his love and encouragement have given me the strength to keep going in the midst of life's uncertainties, upheavals, and numerous moves.

I also want to thank both our families for their incredible support, and for tolerating my preoccupation with this project, to the detriment of spending quality time together. I especially thank my little boy, Frederick, for his patience with my absences as I was putting finishing touches on the book. Frederick's arrival in the late stage of this project brought incredible joy and wonder to my life and has given me a sense of perspective. When the book was in production, it was baby Joanna's turn to fill our lives with yet more amazement and happiness. Last, I want to thank my parents, who have been a constant source of inspiration and support. It is to them that I dedicate this book, with tremendous gratitude for their selfless love, and for being there for me—and now for us—every step of the way.

ONE

Introduction

From Flagship Enterprises to Subsidiaries

The steel sector is a concrete exemplification of all the difficulties
connected to democratization and economic reform in Poland.
    —Former Polish Deputy Minister of Economy

Introduction

Twenty years after the fall of communism in East Central Europe (ECE),
the centrally planned economies in the region have given way to a capitalist
system marked by the strong presence of foreign direct investment (FDI).
Increased transnationalism, as evidenced by increasing trade flows and
exchanges of capital, with inward investment consistently far exceeding
outward investment, is an outcome of globalization in ECE countries. As a
result, these countries are dependent on the decisions of foreign investors
who base their investment calculus and business plan on the global interests of their corporations. The powerful role of foreign investors in the
region has even led to the identification of the ECE economies as "dependent market economies."[1]

Heavy industry, and the steel industry more specifically, was among the
numerous sectors in the ECE economies dominated by foreign investors.
The transformation of the steel industry was highly symbolic and represented one of the most striking features of the transition. The behemoths
of yesteryear, the once-proud standard bearers of the communist industrial prowess, became modest and significantly scaled-down subsidiaries of

large multinational corporations fighting for meager profits on a hugely competitive world market. This book explains that treacherous journey.

Heavy industry restructuring—defined as actions by management intended to bring about greater efficiency within a company—is challenging, irrespective of geographical region. In ECE, the difficulty of restructuring and privatizing the steel sector was compounded by several factors. First, the industry enjoyed a very privileged position during communism. Second, it inherited pervasive overcapacity, which was a legacy of measuring economic success and prestige in terms of production volume. Third, the postcommunist transition coincided with dramatic changes in the steel industry globally. The transition began after more than a decade of fundamental steel-sector restructuring and capacity cuts in the West, and it coincided with a period of enterprise mergers and fierce competition from new, powerful market players from newly industrializing countries. Given these obstacles, the emergence of transnational capitalism begs the question of how these communist-era flagship enterprises, filled with purportedly nefarious vested interests, not only survived the transition but became profitable subsidiaries of leading multinational corporations. This common outcome is all the more intriguing in that it holds equally for the countries with the best and the worst political and economic reform records.

This book examines the emergence of transnational capitalism in the steel sector, a critical sector of the postcommunist economy, in the four biggest steel-producing countries in ECE: Poland, Czech Republic, Romania, and Slovakia. The analysis focuses on a long-neglected actor in the transition, the state, and on the tension it confronted between domestic vested interests and external pressures. I show that these countries followed different pathways to a common outcome of transnational capitalism and that state capacity played a crucial role in determining which pathway each country followed. Surprisingly, my findings demonstrate that relatively high state capacity is a double-edged sword. Although institutional strength and sophistication can be employed to discipline firms into engaging in market-oriented behavior, it may also enable political actors to shield enterprises from market pressures and promote political and personal preferences that are potentially inefficient. This finding has clear implications for designing economic reform packages not just in postcommunist countries but also in other parts of the world.

The Puzzle

The steel sector was a political hot potato. On one hand, because steelworks were often located in geographically concentrated, mono-industrial towns whose survival depended on them, a great deal hinged on the fate of the steel industry. Moreover, the governments in the region faced workers and managers who had become habituated to privilege under communism. Allowing the steelworks to go under, no matter how economically efficient, was tantamount to political suicide. On the other hand, the ECE governments faced a clear endgame: due to the provisions of the Europe Agreements they had signed with the European Community (EC) in the early 1990s, they were expected to cease state assistance, preferably by 1996/1997 or by the time of accession, at the latest. In other words, after about ten years, the state could no longer assist the steel sector if the candidate country wished to become a member of the European Union (EU).

The EU was not bluffing for two reasons. First, the EU underwent an extremely painful and costly restructuring process of its own sector during the 1980s and 1990s. Second, the EU steel producers found themselves under increasing pressure from global competitors originating in emerging markets, most notably, Ispat International (predecessor of ArcelorMittal). The EU, therefore, had a keen interest in the steel industry, and its member states were extremely wary of any potential largesse being actively or passively bestowed upon steelmakers by candidate states.[2] Beyond the restrictions on state aid, however, the EU played a more indirect role as well; membership became a prized political goal that helped strengthen the other external pressures vis-à-vis the transition governments.

Thus, the key challenge for companies in EU candidate states was access both to the investment capital necessary for restructuring and to a vehicle to help them integrate into the international production networks. The governments' choices for facilitating this process were limited by the paltry financial resources at their disposal and by the lack of a domestic capitalist class with sufficient wealth. Thus, restructuring required strategic foreign investors. As the cases discussed in the subsequent chapters illustrate, being a foreign buyer was no guarantee of quality or of having sufficient financial resources to undertake the restructuring task. Foreign adventurers were more than willing to make a quick buck at the expense of the companies they were claiming to save, with dire consequences for the workers. By contrast, strategic foreign investors had both the intent and the wherewithal to restructure companies. (For the sake of brevity, I refer to "strategic foreign investors" simply as "foreign investors," with the understanding that *investor* implies an ability to generate funds and to lead the restructuring process.)

Sales to foreign investors were certainly not easy; discussion of such sales prompted allegations of trying to sell the "family silver," and the overgrown mills were not exactly hot commodities. Furthermore, separating the working assets from nonworking assets, not to mention company housing units, day-care facilities, local medical clinics, and cultural and recreation centers, to name a few, demanded substantial state involvement. All in all, it was a buyer's market that offered several advantages to prospective investors. Most importantly, the region had an inexpensive and well-qualified workforce with substantial local market opportunities in construction, infrastructure upgrading, and manufacturing, especially in automobile and appliance production. In addition, the region's proximity to Western Europe was attractive due to both relatively low transportation costs and the impending free trade of steel products with the EU.[3]

These investment opportunities became more obvious as the transition process unfolded and economic outlooks became more optimistic. However, the courtship continued throughout the transition process. By the early 1990s, Czech steel producers had already received substantial interest in forming joint ventures from reputable foreign steel producers, such as Krupp, Mannesman, Thyssen, Voest-Alpine, and Usinor Sacilor. In Poland, the Italian Lucchini Group actually purchased the relatively small Warsaw Steelworks in 1992. In the mid-to-late 1990s, several well-known foreign investors, such as Voest-Alpine, Hoogovens, British Steel, and, later, Corus, engaged in far-reaching, albeit failed, privatization negotiations with the government.

Interest by foreign investors in the steel sectors of these two countries is consistent with the received wisdom that FDI tends to flow to countries that have a positive reform record.[4] The reason is straightforward: a better reform record signals greater political stability and results in better legal infrastructure and a predictable legal and administrative climate. When one compares the ratings of economic reform of the four countries in 1999, the Czech Republic emerges as the leader, closely followed by Poland, then by Slovakia, and finally by Romania.[5] As expected, investors' money was correlated with the reform record in each. Between 1989 and 1999, the comparably more populous Poland had attracted a total of over $20 billion in FDI, followed by the Czech Republic with nearly $15 billion. By contrast, Romania attracted $5.6 billion and Slovakia $2.1 billion.[6]

Given the initial foreign interest in the steel sectors of the Czech Republic and Poland and the political and economic difficulties of restructuring the industry, one would have expected the biggest steelworks in these two countries to attract foreign capital sooner than either Slovakia or Romania. After all, the Czech Republic and Poland, as the top reformers, were already attracting more FDI and would have been expected to have the ability to overcome the resistance to sales to foreign investors by vested interests, such as managers, unions, or state agents, interested in blocking reform. Second, the Czech Republic and Poland had more developed market institutions, which were needed to run such sophisticated industrial operations effectively.

However, in the steel sector, Slovakia and Romania became the pathbreakers in selling their biggest steel producers to international corporations. Slovakia sold its largest steelworks, Východoslovenské Železiarne (Eastern Slovak Steelworks [VSŽ]), to U.S. Steel in 2000 while Romania sold its major producer, Sidex Galaţi, to LNM Holdings, now ArcelorMittal, in 2001.

Lagging behind these two countries, considered at the time to be the "laggards" of transition, the Czech Republic and Poland eventually followed suit. The Czechs sold Nova Hut' to LNM Holdings in 2002, and Poland sold Polskie Huty Stali (Polish Steelworks) in 2003. Moreover, unlike in the Slovak and Romanian sales, EU pressure was dominant and direct in the Polish and Czech privatizations.

In Slovakia, the EU was conspicuously absent from the privatization discussions and considerations, even though EU accession was a central goal of the reform-minded post-Mečiar administration. In Romania, a civil servant at the European Commission lauded the Sidex privatization as a "huge achievement" by the government, which "has been clear as to the goals and desire to bring about change."[7] However, at the time of privatization, the closure of EU negotiations was a prospect very much in the future, and it was the World Bank that exerted the pressure to privatize Sidex. By contrast, the privatizations in Poland and the Czech Republic would likely have taken even longer, if they had taken place at all, had it not been for the pressure from the EU in the run-up to the closure of the accession negotiations.

To finish the accession negotiations with the European Commission—itself under pressure from EU steel producers to cut the EU candidate countries no slack—the accession country governments had no realistic alternative to negotiating the permissible amount of state aid with the European Commission. Given the accumulated enterprise debts, at that point, state aid had to be retroactive in nature, granted only once, and contingent upon restructuring and privatization measures assuring sectoral viability on the free market without subsequent aid. The other option available to candidate countries was to provide no aid to the sector whatsoever, which would have meant bankruptcy—and possible liquidation—of their biggest steel producers, an ordeal no government wanted to face. It would have also produced a political drama in which "Brussels" wanted no role, and certainly not that of the culprit. Hence, it was in both sides' interest to reach an agreement prior to accession.

The EU was adamant about the need to solve the state-aid question in the run-up to the closure of accession negotiations. The Czechs were being prodded by European Commission officials to develop a solution to their steel-sector woes. As one Commission official sternly told the Czech government, "This is your problem. If you want to solve it, you better get cracking . . . but time is running very short. . . . If you don't make up your mind over the next few weeks, it will be too late [to grant state aid]."[8] In Poland, the government engaged in elaborate brinkmanship to delay privatization, and the lack of a solution for the steel sector endangered the timely closure of Poland's EU accession negotiations. By the Polish civil servants' own admission, EU entry was "a pistol held to our head" as the government decided how to deal with the privatization of the steel sector in the summer of 2003.[9]

Eventually, the largest steelworks in all four countries were sold to foreign investors, exactly as the dependent capitalist model posits. This common outcome, however, obscures the differences in the trajectories the four countries followed in arriving at this point. As table 1 indicates, by the mid-1990s, the Czech Republic and Slovakia had begun creating a class of domestic capitalists whereas Poland and Romania maintained state owner-ship.[10] By mid-2005, the sector ownership structures in all four countries were dominated by foreign investors.

Thus, the reform trajectory of the steel sector in each country was marked by differences not only in the type of external pressure that was dominant in the sale to foreign investors but also in the initial policies adopted in the aftermath of the collapse of communism.

TABLE 1. Change in the Steel-Sector Ownership Structure, 1996–2005: Number of Enterprises

Country	1996			2005			Diff. (2005–1996)		
	SOE	DPO	FI	SOE	DPO	FI	SOE	DPO	FI
Czech Republic[a]	1	3	0	0	1	2	–1	–2	+2
Poland[b]	7	2	1	1	0	8	–6	–2	+7
Slovakia[c]	0	2	0	0	1	1	—	–1	+1
Romania	11	0	0	0	0	11	–11	—	+11

SOE, state-owned enterprise; DPO, domestic private ownership; FI, foreign investors.

[a]Vítkovice (VS), remained under state ownership in 1996, although a privatization contract was signed with the management and the state devolved responsibility for its oversight to the management. One of the companies, Poldi, was liquidated prior to 2005.

[b]One of the Polish companies, Baildon Steelworks, was liquidated prior to 2005.

[c]VSŽ (foreign-owned) is responsible for more than 93% of Slovak steel production.

Simply observing the similar outcomes reveals little about the transformation of the steel sectors of these four countries because the *process* through which the observed outcome was attained could be attributed to any number of causes. For example, the observed convergence in the steel sector could result from differences in partisan politics and reform proclivities, the economic significance of the enterprises, variations in labor and managerial organization and pressure, or disparity in country wealth. One could also reach for demand-side arguments and focus on enterprise attractiveness to investors. Finally, one could turn to external pressures, which usually entail the requirements of EU membership. The same outcome could also result from different combinations of any of these potentially causal factors. Thus, understanding the mechanics of the reform process helps to answer the questions of what domestic and external actors want, when, and under what conditions—all of which are important for tailoring future reform policies to local conditions. From the point of view of external actors, examining the reform process can help rectify the much-criticized, one-size-fits-all reform prescriptions and correct misleading assumptions about the preferences of domestic actors.

The task at hand, therefore, is to identify the determinants of the political decision to sell the enterprises to foreign investors in the steel sector. Since the mechanisms responsible for the outcomes may entail complex interaction effects that could easily be missed in statistical analysis, methodologically, these determinants are best isolated through process tracing the restructuring and privatization trajectories of each country.

Summary of the Argument and Definitions

In this book, which traces the process of reform between 1989 and 2009, I explain the convergence on transnational capitalism and show that different causal mechanisms were at play in the four countries. These diverse pathways resulted from the interactions of domestic institutions and external pressures. In a nutshell, the differing levels of state capacity associated with domestic institutions determined the extent and fiscal consequences of restructuring. These, in turn, shaped the converging trajectories of each country by determining which of the various external pressures, such as international financial institutions (IFIs), international financial markets, or the EU, proved *dominant* in which country.

I define "state capacity" as the ability of formal state institutions to implement policy and enforce legal sanctions.[11] Thus, state capacity rests on the infrastructural power of the state, defined by Michael Mann as the capacity of the state to penetrate civil society and implement political decisions throughout the territory. It is especially important in capitalist societies, where the state apparatus needs to regulate the economic processes.[12] Even though the communist states resembled leviathans that needed to undergo significant adjustment to the tasks of capitalist systems, they nonetheless possessed different levels of capacity at the outset of the transition process.[13] As the subsequent chapter shows, communist legacies left the transitioning states with bureaucracies that differed as far as training and access to technocratic resources were concerned. Thus, the different institutional endowment at the outset of transition, in part, explains the diversity of economic performance throughout the postcommunist world.

Communist legacies were key to understanding the disparate models employed in the challenging reform context of the steel industry. Facing the exigencies of economic transition, politicians needed to reform the steel sector. Initially, however, they secured their political and personal interests by opting for politically safer incomplete reform, epitomized by continued state ownership or privatization to nonstrategic, generally domestic, investors. As chapter 2 explains in greater detail, and as chapter 3 demonstrates empirically, the initial policy choice depended on the governing coalition's commitment to privatization and on the strength and preferences of managers and unions. All three variables were rooted in the political context shaped by the communist experience of the given country. The implementation of the initial policy choice at the enterprise and sectoral levels varied based on the capacity of a given state, and it determined the degree of restructuring and the resulting market adjustment that took place, including the consequences for the public purse.

At the enterprise level, developed in chapter 4, the analysis focuses on the relationship between state actors and the managers of the individual enterprises, with labor playing a secondary role. Here, the question is whether state actors were able to rein in managerial investment ambitions and rationalize production while checking their own impulse to seek rent. At the level of the sector, explored in chapter 5, the analysis turns to the institutional framework of social dialogue and the implications this has for restructuring.

Figure 1 summarizes the causal chain developed in chapters 4, 5, and 6, as it shows the relationship between the level of state capacity, the resulting domestic features of restructuring, and the dominant reasons for a sale to foreign investors. Given the overarching goal of "return to Europe," the figure also summarizes the role EU membership considerations played in the decision to sell the enterprises to foreign owners.

As figure 1 indicates, states with relatively low capacity, such as Romania,

State Capacity	Domestic Features of Restructuring	Dominant Reason for Sale to Foreign Investors	Role of EU Pressure in Sale to Foreign Investors
Low	State unable to introduce hard budget constraint, curb rent-seeking and managerial prestige maximizing; Lack of institutional capacity for quasi-market intervention; Captured unions and weak and dysfunctional social dialogue.	***International Financial Institutions*** Difficulty maintaining macroeconomic stability gives the IFISs leverage over the government and forces targeted sales.	***Indirect*** EU membership as primary political goal strengthens the hand of IFIs who need to "certify" the functioning market economy status for EU accession.
Medium	State able to curb some rent-seeking behavior (medium soft budget constraint) and managerial prestige maximizing; Lack of institutional capacity and funds for quasi-market intervention; Captured unions and weak and dysfunctional social dialogue.	***International Financial Markets*** Difficulty repaying loans to foreign lenders and severely limited state ability to bail out enterprises leads to a sell-to-FDI-or-perish scenario.	***Indirect*** EU membership as primary political goal strengthens the bargaining position of international market players vis-à-vis the government.
High	State able to introduce relatively hard budget constraint, curb significant rent seeking, and contain some managerial prestige-maximizing; State-led restructuring episodes and intervention to prop up failing enterprises using quasi-market means; Autonomous unions and functioning social dialogue.	***European Union*** Ban on state aid to the sector eliminates the use of quasi-market tools of intervention and forces sales for long-term survival.	***Direct*** EU greatly restricts the options available to the government due to the strict state aid provisions. ***Indirect*** Government desire for EU membership strengthens the bargaining power of foreign investors.

Fig 1. State Capacity, Domestic Features of Restructuring, and the Dominant Reasons for Sale to Foreign Investors

were unable to curb rent-seeking behavior and managerial prestige maximizing. Social dialogue was poorly institutionalized and dysfunctional and marked by captured, rather than autonomous, unions. Endemic rent seeking made it difficult to maintain macroeconomic stability, which triggered the pressure from the IFIs to sell to foreign investors. IFI conditionality was strengthened further by the EU's requirement that Romania be considered a functioning market economy by the IFIs as a prerequisite to membership.

Medium-capacity states, like Slovakia, were more successful in their restructuring endeavors, as they were able to harden the budget constraint somewhat by limiting some rent-seeking behavior and managerial prestige maximizing. In Slovakia, social dialogue was also poorly institutionalized and dysfunctional and marked by captured, rather than autonomous, unions. Moreover, the Slovak state did not have the trust of the international lenders, and it also lacked quasi-market intervention instruments for propping up failing enterprises. As a result, although it was able to deflect the IFI pressures, it could not withstand the pressures of the international financial markets and banks that pressed for sales to foreign investors able to pay back the loans taken out on the international financial markets. The EU played an indirect role as well, as it provided an overarching reform trajectory for the government, but it did not need to intervene directly to press for privatization to foreign owners. The international banks had already done so.

States with higher capacity, such as the Czech Republic and Poland, were better positioned to restructure, and they did so, in part, by constraining excessive rent-seeking distributional coalitions and some managerial prestige maximizing that threatened reform. At the sectoral level, both states also had autonomous unions and a functioning social dialogue. Critically for their ability to deflect external pressures, they were able to maintain macroeconomic stability, and they had the quasi-market tools to intervene in ailing (and failing) enterprises. Paradoxically, higher state capacity became a double-edged sword that obstructed attempts to reform the sensitive steel sector, because it enabled states to maintain politically convenient status quos by staving off external pressures through the deployment of targeted market-based tools. These tools included debt workouts and parastatal entities that supported the embattled enterprises. In other words, the higher capacity of the Czech and Polish states allowed their governments to deflect the IFI and international financial market pressures to which their Slovak and Romanian counterparts had succumbed. The initial reform efforts were incomplete and would not be completed until these states were constrained by the requirements of EU accession, and specifically, the EU ban on state aid. Thus, the EU played a direct role in the decision to sell to foreign investors: the requirements of accession gave the governments no viable alternative to seeking strategic investors to ensure long-term survival. At the same time, the EU also played an indirect role in the process, as it strengthened the bargaining power of foreign investors interested in acquiring the enterprises.

Implications

This book draws on and contributes to a number of scholarly literatures. First, it adds to the literature on the political economy of reform—specifically, to the studies on the role of the state in economic reform, industrial restructuring, and industrial relations. Second, because of its focus on how the state mediated domestic pressures, on the one hand, and external ones, on the other, in the context of postcommunist economic reform, the book brings together and contributes to three distinct literatures: postcommunist transition, varieties of capitalism, and European integration studies.

First, although the role of the state and the importance of state capacity for development have been increasingly recognized by the IFIs, most notably by the World Bank,[14] I demonstrate that the relationship between higher state capacity and reform is more complex than these entities and the scholarly literature tend to recognize. The institutional sophistication indicative of higher state capacity is a double-edged sword that can be used to discipline firms into engaging in market-oriented behavior or, by contrast, to shield enterprises from market pressures and otherwise promote the inefficient preferences of political actors.

The dual nature of state capacity travels to other settings in both developing and developed countries. It sheds new light on why some countries are more likely to succumb to external constraints, such as IMF conditionality, than others. Differences in state capacity can also explain the variations in degree of economic reform among advanced market economies, such as the most recent bout of reforms in Greece and Spain, on the one hand, and the lack of reform in France, on the other. It also clarifies seeming contradictions, such as why some avid market reformers, such as Chile, manage to maintain public ownership over certain sectors of industry, as exemplified by the Codelco copper mining company.[15] In short, I illuminate the political determinants of economic reform more broadly, and privatization specifically, including the resources used to resist it, by focusing on the deployment of state resources by politicians.[16]

Second, the analysis also clarifies the potential range of developmental consequences resulting from privatization. At a basic level, the present study of the initial policy choice corroborates Hector Schamis's argument that far from being a silver bullet, privatization may be a gateway to enrichment for the governing coalition's allies.[17] However, unlike Schamis's approach to the state as the dependent variable, this study treats the state as an intervening variable that leads to privatization and, more generally, to reform outcomes. At the same time, this study addresses the criticism of treating the state as a unitary actor by explicitly recognizing the complex and often contradictory network of state agencies and institutions.[18]

Third, the book's systematic and central focus on the role of state capacity in mediating industrial restructuring fills an important void in the literature on the political economy of postcommunist transition. For a long time, this particular literature has ignored the state as an important actor, and existing studies of the state have tended to treat it as the dependent variable.[19] As my account makes clear, the state plays a central role in restructuring due to its interaction with the domestic actors. In other words, how well the state is able to rein in rent seeking and managerial ambitions and build an institutionalized relationship with the trade unions also determines its susceptibility to different types of external pressures. The sources of state capacity are deeply rooted in these countries' communist and precommunist legacies; thus, they are thoroughly domestic.[20] The analysis of the implementation of the initial policy choice in the four countries makes clear that the level of state capacity rather than ownership type—whether domestic private ownership or continued state ownership—accounts for better or worse restructuring outcomes.

Fourth, the book supports two propositions concerning effective restructuring in the context of market reform. First, neither abandoning the sector to the nascent market forces nor using the preexisting industrial networks for restructuring will foster significant restructuring without a transparent policy and sufficient state capacity to intervene and rein in managerial investment ambitions.[21] Second, contrary to the arguments for insulating decision makers from social actors, I show that, far from hindering the restructuring process, engaging unions in sectoral-level social dialogue is conducive to restructuring and can lead to unions becoming the agents of restructuring.[22]

Fifth, the book bridges the literatures on the varieties of capitalism and EU integration studies by illuminating the role of the EU and other external pressures in the convergence on transnational capitalism in the region. The type of capitalism emerging in the transition economies became a subject of considerable scholarly debate in the field of political economy of postcommunism. The proliferation of categories into which the countries are divided stems from the realization that the categories of liberal and coordinated market economies, originally developed by Peter Hall and David Soskice for the most advanced industrialized countries, are of limited applicability to the institutionally fluid postcommunist economies.[23] However, whether the categories center on the main institutional coordinating mechanism, insertion into the world production networks, or social forces and domestic institutions, there is a growing recognition of the central role played by foreign capital in the region.[24] Seeking to add to the Hall and Soskice framework, Nölke and Vliegenthart explicitly refer to the economies of the Central European region as "dependent market economies," in which external dependency is the principal coordinating mechanism in the economy. Thus, foreign capital plays a paramount role.[25]

This book addresses the understudied political process through which the outcome of transnational capitalism was achieved.[26] The few existing accounts that have tackled the complex process leading to the emergence of transnational capitalism have tended to emphasize either the active role of state elites and institutions in attracting FDI or, at least, their complicity.[27] The present study moves beyond these accounts by demonstrating that the entry of foreign investors into the steel sector not only lacked the active support of the elites but was resisted in an effort to retain the status quo. Rather, it focuses on the coercive pressures exerted by international actors in a sector in which the domestic actors did not unfold the welcome mat, even in those states, such as Poland, that have historically been relatively foreign-capital friendly in other sectors, such as banking. Thus, this book explains why and when the process of convergence holds even if the elites are not enthusiastic about internationalization and try to resist it.

This book adds to the EU studies literature through its examination of the EU's role in the emergence of transnational capitalism in the region.[28] The EU's transformative role in ECE has generated an impressive literature and sometimes contradictory conclusions, many of which depend on the sector and policy examined.[29] Most often, the EU effect is the focus of analysis, and other external pressures are treated as complements in the liberalization/market transformation project.[30] However, the relationship among these external pressures, and the conditions under which each of them plays a significant role, have not been given sufficient attention, and this is precisely the gap that this book addresses. Certainly, the EU set the parameters of reform for all accession countries in the wide-ranging *acquis communautaire*, or the body of all of EU's rules, regulations, treaty obligations, and court rulings.[31] However, in the case of this challenging sector, the EU had to exert direct pressure on the transition and enlargement leaders (Poland and the Czech Republic) to force compliance with EU regulations.[32] Even though the EU was avowedly neutral as to ownership type, the requirements of EU membership left the countries little leeway in choosing to sell to foreign investors. At the same time, due to the differences in domestic institutions, in the case of the transition "laggards" (Romania and Slovakia), the EU's *direct* pressure was preempted by other external pressures: IFIs and the financial markets.

Case Selection and Data

The book is comparative on several dimensions: it process traces the restructuring and privatization of a critical and difficult-to-reform sector across three levels of analysis in four countries over twenty years. The nested research design, progressing from the enterprise level to the sector level and on to the national-international nexus, creates an opportunity for making across- and within-case comparisons and for examining the salience and interactions of domestic and external variables over time. Such a research design complements prior work by linking the macro and micro levels. By examining a single sector, the analysis is broad enough to embed the sector within the overarching country-reform trajectories and narrow enough to engage in a systematic comparison at the level of the enterprise.

Steel Sector

Steel, a particularly sensitive sector in communist countries, is the ideal sector for examining the relationship between successive governments, the state, and managerial and labor interests during the transition process. The steel sector under communism operated in an ideologically loaded context, and a country's performance was measured by its annual steel output. The steelworkers were seen as the epitome of the proletariat, and their work was remunerated handsomely compared with other industrial sectors. It is, therefore, not surprising that at the outset of transition, the steel sector was expected to be filled with vested interests inimical to reform.[33] The heightened challenge of reforming sectors characterized by high capital intensity and overcapacity was only exacerbated by the steel sector's formerly privileged status and great symbolic value as a communist-nationalist project; it was in the steel mills that the new communist man was to be forged.

The construction of the Lenin Steelworks in 1949 at the outskirts of Kraków, Poland, was accompanied by the building of a model socialist-realist city, Nowa Huta, literally "New Foundry." In Romania, the construction of Sidex, the country's largest steelworks, in the eastern Romanian city of Galați, became central to the 1960 feud between the Romanian communist leader, Gheorghe Gheorghiu-Dej, and the Soviet leader, Nikita Khrushchev, over Romania's role in the Council for Mutual Economic Assistance (CMEA). With Gheorghiu-Dej rejecting the vision of Romania as the communist bloc's breadbasket, Sidex came to symbolize Romania's national communism.[34] Similarly, in Slovakia, the 1958 decision to build Eastern Slovak Steelworks (VSŽ) as part of the Second Five-Year Plan, was essential for meeting the developmental objective of putting the Czech and Slovak lands on an equal economic footing; VSŽ became Slovakia's flagship enterprise.[35] Even in the Czech lands, with their long history of steelmaking, the 1952 construction of Nová Huť Klementa Gottwalda (Klement Gottwald New Steelworks), named after the Czechoslovak Stalinist leader, was intended to represent a new era in Czech industrial development.[36] These legacies, combined with the sector's economic importance and global market pressures, shed light on the political economy of reform and the region's convergence on transnational capitalism.

Beyond its ideological and political legacies, the steel sector held an important position in the economies of the Czech Republic, Poland,

Romania, and Slovakia. At the outset of transition, the share of total manufacturing industry (in current prices) held by the basic metals and fabricated metal products branch, of which the steel sector is the core, ranged from the 19.0% in Slovakia to 14.7% in Romania. In addition, the steel sector was similarly positioned in the economies of the four countries—a key comparative point because, following the insights of Shafer's sectoral analysis, dominant sectors can shape the restructuring of the national economy and affect the state itself. Sectors marked by high capital intensity, high economies of scale, and high production and asset/factor inflexibility are particularly influential.[37] In Slovakia, the metals branch was the biggest industrial sector; in the Czech Republic (17.2%), Poland (16.2%), and Romania, it was the second-largest branch.[38]

The branch was also an important employer. At the outset of transition, the basic metals and fabricated metal products branch was the biggest industrial employer in the Czech Republic, with a 17.6% share of total employment in manufacturing. In Romania (12.7%), the branch was the third-largest industrial employer, and the metals branch was the fourth-largest industrial employer in both Poland (11.5%) and Slovakia (9.9%).[39] Thus, the fate of the steel sector had important implications for the economy and the labor force in all four countries.

In this analysis, I use the biggest companies in the individual countries to identify different causal pathways, to present a theory of convergence, and to test initial hypotheses concerning external effects. I then test the claims about the role of state capacity in restructuring against the evidence garnered in the other big and medium-size enterprises in the steel sector of these countries. In the final chapter, I demonstrate that the other countries in the region also fit the pattern flowing from the theoretical expectations identified in this chapter. Thus, I test the theory of convergence within and across country cases, using the enterprises as units of analysis. I compare sectoral-level labor dynamics and managerial competition at the country level.

Country Cases

The choice of country cases follows the most similar research design. Given that the outcome to be explained is the difference in the converging trajectories, the countries selected are similar on several crucial variables. First, they have all faced the task of postcommunist transition and the simultaneous drastic political, economic, and social change it entails. Second, as the previous section has shown, the four selected country cases had steel sectors of similar political and economic domestic stature and salience. Third, none of the countries relied on natural resource wealth for export earnings. In 1999, during the mid-transition period, fuel exports as a percentage of merchandise exports equaled 2.8% in the Czech Republic, 4.7% in Slovakia, and 4.9% in Poland and Romania. By contrast, in Russia, this number stood at 41.8% in 1999.[40]

Finally, as noted earlier, all four countries took part in the EU accession process. Given the sensitivity of the steel sector in the EU, these states needed to reach the same standard as already established EU member states—namely, the viability on the market without state aid. Because the four countries represented the biggest steel producers among the EU accession states, the EU would have been expected to take a close interest in all four countries' production potential and restructuring process. Due to the importance of the EU in setting the broader parameters of reforms and because of its close attention to the developments in the steel sector, the universe of cases for this study is limited to the EU applicants.

The one key variable on which these countries differ, as the next chapter will show, is the level of state capacity. As the similarities in the ultimate trajectories of Poland and the Czech Republic demonstrate, it was the level of state capacity, rather than the differences in the initial policy choices, that drove the process of convergence.

Data

The following chapters process trace restructuring and privatization of the enterprises in the steel sector in the four countries over the entire transition period, 1989 to 2009. I used a wealth of local sources in Czech, Polish, Slovak, and Romanian, including government documents, policy papers, and publications by labor unions and by employer and industrial associations. I relied on hundreds of newspaper accounts of restructuring and privatization events from over fifty local newspaper sources, including those drawn from enterprise newspapers located in labor union archives. In addition, I conducted more than 125 open-ended interviews, listed in Appendix A, with various actors involved in the restructuring and privatization process in each country.

Chapter 2 provides the theoretical framework for the empirical analysis presented in subsequent chapters. It examines in greater detail the political economy of restructuring and privatization at the three levels of analysis and discusses the central role of state capacity in the reform process.

TWO

State Capacity and the Political Economy of Industrial Restructuring

Privatization is like an apple on a tree—it takes time to ripen.
—Civil servant at the Polish Ministry of the Economy

Steel-sector reform was a touchy and challenging task fraught with economic traps and political perils that successive governments hoped to avoid. The former Polish deputy minister of economy even explicitly stated that the steel sector exemplified all the difficulties connected to democratization and economic reform in Poland.[1] The same could be said of all the countries discussed in this book. The country-level responses to the challenge posed by restructuring differed. Some, such as Poland and Romania, retained state ownership, while the Czech Republic and Slovakia embarked on privatization to domestic owners. In both cases, however, the initial policy choices represented incomplete reform that would generally not be able to ensure the long-term survival of the enterprises. These country-level policy differences are the subject of chapter 3. This chapter will discuss the model of the political economy of reform in the steel sector, and the process of convergence on transnational capitalism. Thus, the chapter presents the entirety of the causal mechanism, from the initial policy choice to policy implementation to extrication from incomplete reform.

The chapter first presents the end goal of restructuring and the high bar that was set by the requirements for EU membership. It then examines the challenges of restructuring. The discussion subsequently turns to the politics of the initial policy choice and to the role of state capacity in industrial restructuring. The chapter examines state capacity in the context of communist legacies and presents indicators of state capacity during the transition period. It then addresses the role state capacity played in industrial restructuring at the enterprise and sectoral levels, before turning to the third level of analysis—the national-international nexus—and to the question of sustainability of incomplete reform in the face of different external pressures. As this chapter illustrates, although the accession states faced a clear set of criteria in the steel sector, and ultimately converged on the same outcome, the roads that led to the fulfillment of these criteria differed.

The End Goal of Restructuring

Just as the experiences of the four countries under the communist regime differed significantly, so did the politics of their transition out of communism. The various means by which each country cast off the communist system—be it through negotiation (Poland), the regime's outright implosion (Czechoslovakia), or preemptive "revolution" by a segment of the old regime (Romania)—gave way to differently structured democratic competition.[2] In the case of Romania, it is even difficult to call the political system democratic in the aftermath of Nicolae Ceaușescu's fall. The political dynamics in the four countries—following the split of Czechoslovakia into the Czech Republic and Slovakia in 1993—also reflected different economic reform preferences. The political dynamics and policy choices in the four countries will be explored in depth in the following chapters; for now, it suffices to say that the countries shared a fundamental goal: the "return to Europe." In other words, these countries stressed their desire to become part of the European Community (EC), or the soon-to-be European Union. Their path began with the signing of the Europe Agreements in the early 1990s.[3]

The Europe Agreements were free-trade agreements that simultaneously provided for harmonization of EU and national legislation and, de facto, became the first step toward EU membership. These agreements, which bound the transition countries closer to the EU and reoriented their economic and political focus to Western Europe, were carefully crafted to protect the EC's economic interests, especially in areas that were considered to be sensitive, such as agriculture, textiles, and steel. The steel sector was particularly fragile, having undergone massive production cuts and painful restructuring, albeit with generous state support, and the EC members were particularly keen to make sure that their eastern neighbors did not disturb the precarious status quo. Therefore, they incorporated stiff regulations concerning state aid to the sector that were encapsulated in Protocol 2 of the Europe Agreements.

Protocol 2 focused on mutual liberalization of trade and the conditions for state aid to the sector. On the prior issue, the EC agreed to an asymmetric stepwise liberalization of trade in steel products. The EC committed to eliminating its duties on steel products originating in the associate countries by 1997.[4] For the associate countries, the tariff reduction schedule differed by country and product, but trade with the EU was to be liberalized between the end of 1997 and 2000.[5]

The key issue, however, was state aid to the sector and the related question of total production capacity. The two were linked because maintaining unused production capacity lowered the already low profit margin in the sector, making the enterprises less able to operate without direct or indirect state aid. Protocol 2, therefore, laid out the criteria for granting state aid to the sector. Public aid was permissible for restructuring purposes only, and it had to meet the following conditions (as determined by the EU prior to the provision of aid): public aid had to be structured so as to lead to the viability of the benefiting firms, under normal market conditions, at the end of the restructuring period; it could last no longer than five years; the amount and intensity of the aid were strictly limited to what was absolutely necessary to restore viability and were progressively reduced; and the accompanying restructuring program was linked to a global rationalizing and reduction of capacity in the countries concerned.

Although state aid could initially last for five years only—until the end of 1996 in Poland, the Czech Republic, and Slovakia, and the end of 1997 in Romania—the European Commission's policy evolved to accept the possibility of extending the grace period.[6] The Commission agreed to extend the grace period until the date of accession and even to acquiesce to state aid retroactively, provided that the country in question presented a "credible and realistic" sectoral restructuring plan that was acceptable to the Commission.[7] In no case could state aid be granted after accession to the EU, or December 2003, whichever came first.[8] Protocol 2 became the basis for EU accession negotiations regarding the so-called chapter, or the body of EU rules—known as the *acquis communautaire*—dealing with competition. As will become apparent in the discussion of the Czech and Polish cases in chapter 6, these negotiations became acrimonious, and the existing EU member states paid close attention to the resolution of differences between the EU and the accession hopefuls. In short, by signing the Europe Agreement, the countries committed themselves to attaining the medium- to long-run economic viability of the enterprises in the steel sector *without state aid* or to adopting policy solutions that credibly promised that viability would be attained. Viability was to be assessed based on clearly defined economic criteria.[9] The following section discusses the numerous challenges these countries faced as they embarked on the arduous restructuring process.

The Restructuring Challenge in Practice

Industrial restructuring can assume various interrelated forms. At the most general level, it refers to the actions taken by management to achieve viability or, at minimum, greater efficiency by their firm. Restructuring can be broken down into three types: *employment, organizational*, and *technological*. Employment restructuring refers to the downsizing of the labor force and is often tied to organizational restructuring, which entails the closing of obsolete production lines or facilities with excess capacity, or eliminating nonproductive assets, such as cafeterias and child-care facilities. By contrast, technological restructuring involves capital investment in new technologies, facilities, and product lines.[10]

The three types of restructuring are complementary but face different kinds of challenges. Employment and organizational restructuring, which are particularly sensitive for social and political reasons because they lead to job losses, were particularly relevant in postcommunist countries. Prior to transition, the steel sector was driven by the demand of the Soviet military-industrial complex. Thus, the initial challenge was in reorienting trade with the Soviet Union toward Western markets, which were already struggling with overcapacity. For example, in 1989, a third of Polish iron and steel exports were destined for the USSR. By 1991, the number had dropped to less than 1%.[11] In 1988, half of all Czechoslovak exports were destined for the CMEA partners. The number had dwindled to 15% by 1992.[12]

The challenge of trade reorientation was exacerbated by the initial collapse in domestic demand for steel, resulting from permanent changes in the industrial structure, and from the deep recession of the early transition period. Czech and Slovak steel consumption did not reach 70% of the record 1987 consumption level until 2004, the year both countries entered the EU.[13] In the same year, Polish consumption reached 62% of the record

1987 level. The comparable percentage for Romania was 34%, reflecting both the belated nature of Romania's economic recovery and the degree of structural change that took place.[14] During the post-EU-accession boom, which also coincided with the record year for global steel markets—2007—the combined apparent consumption of steel equaled 94% of the 1987 levels in the Czech Republic and Slovakia, 88% in Poland, and 53% in Romania.[15] Thus, especially during the transition-era economic crisis, it was clear that organizational restructuring, including capacity closures, was necessary.

Capacity closures were inextricably linked to employment restructuring. Overemployment was an endemic problem in the communist-era steel sector; in 1989, productivity was 139 tons of crude steel output per employee in Slovakia, 103 tons in Poland, 83 in the Czech lands, and 65 in Romania.[16] By comparison, in 2003, labor productivity in the EU-15 was equal to 600 tons per employee and had been so since the early 1990s because the bulk of the EC's restructuring occurred in the 1980s.[17] Politically, employment restructuring was the most contentious aspect of restructuring, especially in the case of steel and other heavy industry sectors because it contributed to regionally concentrated unemployment.

Technological restructuring was sensitive for both political and economic reasons. Technological restructuring involves enormous financial investment, and these multimillion-dollar ventures required either a strategic investor or state guarantees for loans. To the extent that technological investments give one enterprise a competitive advantage over another, the choice of which state-owned enterprise (SOE) should receive state support for investment often becomes a contentious political decision. The scope of the modernization challenge faced by the communist-era industry can be illustrated by two figures related to the production of crude steel: the use of the obsolete open-hearth furnaces and the percentage of total steel produced using the more efficient continuous casting system.[18] In 1989, 27% of Romanian steel was produced using the open-hearth method, and the number stood at nearly 36% in Poland and 40% in Czechoslovakia (mostly in the Czech lands). By contrast, the method had been virtually eliminated in the EC by 1989.[19] As for continuous casting, it was used for about 90% of steel produced in the EC in 1989. However, only 2% of steel produced in the Czech lands, 8% in Poland, 25% in Slovakia, and 34% in Romania used continuous casting.[20] Thus, the technological gulf between these countries and Western Europe was enormous at the basic level of steel production; the gap was even more pronounced with steel finishing, where much of the value-added is produced.

It was clear that far-reaching restructuring needed to occur. The EU did not mandate the means through which viability was to be attained, such as privatization, but given the amount of investment necessary for restructuring and the dearth of domestic capital, sale to a *strategic* investor, presumably foreign, was the most logical solution for cash-strapped enterprises. Thus, attempts at full reform—that is, concrete steps taken to ensure enterprise viability on the free market—should have involved a vigorous restructuring program coupled with a credible search for strategic investors.[21]

Nonetheless, viability on the free market was not the immediate priority of decision makers. Rather, they opted for politically convenient, incomplete reforms, such as continued state ownership, concomitant with stopgap reform measures or sales to nonstrategic investors. These choices and their political determinants are the subject of chapter 3. The following section presents the theory of the politics of heavy industry reform in transition countries.

The Politics of Initial Policy Choice

The initial policy choice in heavy industry was overwhelmingly determined by domestic-level variables. During the initial years of transition, international financial institutions (IFIs), promoting the Washington Consensus solutions of stabilization, liberalization, and privatization, did not target specific industrial sectors, preferring to focus on the macroeconomic sphere. Although the World Bank financed sectoral studies meant to assist these countries in developing sectoral restructuring policies, there was no external pressure to act on these proposals. Thus, the policies adopted in the individual countries were the product of internal political dynamics.

This book, adopting a rationalist approach, works from the premise that politicians optimize their reform plans so as to reconcile policy preferences with reelection considerations. Even otherwise reform-minded governments opportunistically defer reforms that involve political risk and harm their popularity. In cases where the status quo directly benefits the governing coalition or its allies, the case for reform is even weaker. In the steel sector, both avoiding political risk and preserving a beneficial status quo were powerful disincentives to engaging in full reform.

Political risk resulted from the presence of relatively strong organized labor and managerial pressures (compared with other sectors) that, during the initial stages of reform, increased the government's political costs in regard to radical change. The beneficial status quo was due to structural characteristics of the sector that created numerous opportunities to reward political allies via patronage and rent seeking. These opportunities included, for example, enabling political supporters to obtain a stake in the enterprises, setting up intermediaries to sell overpriced inputs and to buy underpriced products from the enterprises (transfer pricing), and facilitating barter-based, interenterprise debt trade. Consequently, confronted with these two disincentives to full reform, incomplete reform—that is, ensuring plant survival only in the short term and leaving the status quo largely intact—became the preferred solution for government actors.[22]

As figure 2 illustrates, initial policy choice is determined by the governing coalition embedded within a particular domestic context.[23] Coalitions are distinguishable by their level of commitment to privatization, which is a function of the preferences of the coalition partners *and* their cohesion regarding privatization.[24] The domestic context is characterized by the relative strength and preferences of managerial and labor pressures.[25] Due to overall strong disincentives to engage in full reform, the preferred policy choice is a variant of the following incomplete reforms: continued state ownership with stopgap reform measures at best, or privatization to nonstrategic buyers. Over time, the payoffs from the status quo, resulting from the initial policy choice, reinforce the desirability of this policy compared with the possibility of full reform.

Sustainability of Initial Policy Choice: State Capacity and Restructuring

Defining State Capacity

The implementation of a chosen policy choice depends on the level of state capacity. Although state capacity can be applied in a variety of contexts, here it refers to the infrastructural power of the state in the economic domain—that is, the power of the state apparatus to penetrate the economy and society to regulate and oversee the functioning of economic processes, collect taxes, and ensure that market abuse does not take place.[26] Table 2 summarizes the attributes of states with low and high levels of capacity. "The state," however, is not meant to imply a unitary actor.[27] Rather, it refers to the sum total of actions of the set of formal institutions that implement policy and enforce legal sanctions. Thus, various institutions can disagree and even work at cross-purposes. To the extent that these dis-

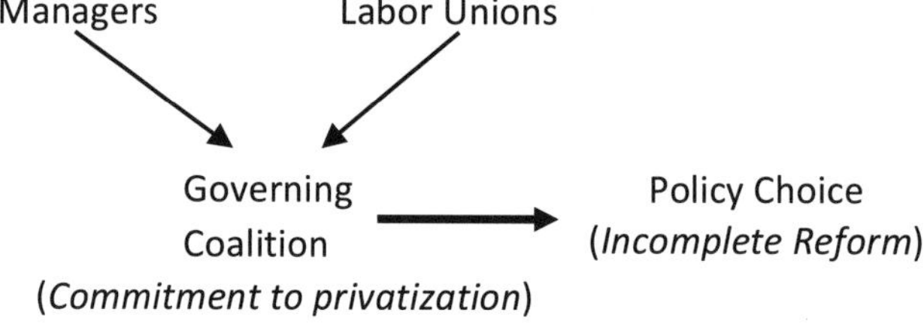

Fig 2. Determinants of Initial Policy Choice

agreements impede oversight and regulation of economic processes, they lower the state's capacity.

State capacity affects a government's ability to support, inter alia, market adjustment, a positive business environment, and privatization, and to intervene selectively in enterprises. In terms of market adjustment, high state capacity signifies the state's ability to impose relatively hard budget constraints on enterprises. It also suggests the ability to create an environment that is conducive to market processes found in advanced market economies. In the case of privatization, this means vetting the buyers to assess their abil-

TABLE 2. Attributes of States with Low and High Levels of Capacity

High State Capacity	Low State Capacity
State apparatus able to impose hard budget constraints on enterprises:	State unable to establish hard budget constraint:
Collects taxes and utility bills of SOEs	Foregoes taxes and utility bills of SOEs
Does not tolerate tax arrears	Tolerates tax arrears
Creates a real threat of bankruptcy	Fails to create a real threat of bankruptcy
State able to vet buyers to ensure their compatibility with hard budget constraints	State unable to vet buyers
	High incidence of leveraged buy-outs and unreliable investors
Low incidence of leveraged buy-outs	
Search for strategic investors	Unclear privatization contracts
State seen as legitimate business partner by foreign lenders	State lacks credibility vis-à-vis foreign lenders
High sovereign risk ratings	Low sovereign risk ratings
	State lacks market-based tools to intervene selectively in failing enterprises
State has market-based tools to intervene selectively in failing enterprises	
State actors facilitate debt workouts between lenders and debtor enterprises	
State offers other market-based vehicles to prop up failing enterprises	

ity to operate under hard budget constraints. High state capacity also signals to foreign lenders that a state is a reliable business partner.

As for selective intervention, high state capacity indicates the availability of an institutional framework for facilitating debt workouts between lenders and debtor enterprises. It also means state-led reorganizing of the assets of SOEs as well as intervention using the intermediation of other state-owned entities to assist enterprises in need. After all, an indiscriminate imposition of hard budget constraints in transition economies may result in a chain reaction of bankruptcies and liquidations, which, in addition to entailing disastrous social consequences, could be detrimental to economic performance.[28] The state, often as the majority owner or a significant shareholder, may want to temporarily intervene to help resolve the enterprises' problems while promoting restructuring. The key is to intervene in a manner that prevents the creation of a moral hazard whereby the enterprises obtaining state assistance do not alter their inefficient behavior in anticipation of future bailouts.

When dealing with politically powerful enterprises, high state capacity may be a double-edged sword because state intervention may reflect the desire to maintain the benefits accruing from the status quo of incomplete reform. In other words, the government may be willing and may have the tools to intervene by providing stop-gap measures sufficient to address the enterprise's problems in the short run but not in the medium to long run, which may prove ultimately counterproductive. By contrast, a state with low capacity lacks the ability to harden budget constraints and to promote restructuring. Consequently, it keeps on foregoing taxes, tolerates arrears to other SOEs, including utility companies, and is not able to create a credible threat of bankruptcy. As a result, it does not mandate significant adjustment to market conditions.

Communist Legacies and State Capacity

Even though formal state structures underwent change during the transition, there was a remarkable continuity in state capacity between the outset of transition and the date of EU accession. According to Jessica Fortin, "[T]he initial level of state capacity is the strongest determinant of subsequent levels of state capacity."[29] Initial state capacity, in turn, was influenced by the communist legacies, which affected the level of technocratic resources at the disposal of state actors, the capacity of social actors to organize in opposition to government initiatives, and the relationship between the political and economic spheres.

Herbert Kitschelt and colleagues' typology provides a comprehensive examination of the communist legacies and places the four countries in separate categories spanning the entire spectrum of communist regimes. The Czech communist regime represented a pure type of "bureaucratic-authoritarian regime"; the Polish regime was a hybrid of "bureaucratic-authoritarian" and "national-accommodative" types, though the authors' discussion suggests there was much greater affinity to the latter type.[30] On the other end, the Romanian regime was labeled as "patrimonial communism," and the Slovak regime was described as a hybrid of "patrimonial" and "national-accommodative" communism.[31]

In the case of the bureaucratic-authoritarian regime, exemplified by the Czech Republic, a rule-guided bureaucratic machine was governed by a planning technocracy and a disciplined, hierarchically stratified Communist Party. The high levels of formal bureaucratization meant that rules, rather than discretion, marked the relationship between the communist bureaucracy (the state) and the state's economic agents (managers), which led to less corruption. Moreover, party rule was enforced by repression, rather than co-optation of society. Consequently, there was little room for bargaining between the state and the economic and social agents, such as the enterprise-level managers and workers.[32] As for the technocratic skills of the orthodox communist bureaucracy, the authorities tolerated some independent economic thought, especially at the end of the 1980s. This was evident in the creation of the Prognostic Institute, which became the neoliberal hub in the Czech lands.[33]

At the opposite end of the spectrum was the patrimonial communist system, which relied "on vertical chains of personal dependence between leaders in the state and party apparatus and their entourage, buttressed by extensive patronage and clientelist networks."[34] The level of rational-bureaucratic institutionalization was low because the small ruling clique penetrated the state apparatus through nepotistic appointments and high levels of corruption. Moreover, social actors were repressed, making it difficult for them to organize in the future. Romania was an exemplar of patrimonial communism.

The national-accommodative communism ideal type was located between the two extremes. This system had intermediate levels of bureaucratic professionalization, and the more developed formal-rational bureaucratic governance (compared with patrimonial communism) partially separated party rule and technical state administration. This partial separation permitted a greater degree of corruption than was possible in bureaucratic-authoritarian regimes. At the same time, though, national-accommodative regimes relied heavily on societal co-optation, rather than repression, to quell dissent and ensure compliance. The national-accommodative system was most closely exemplified by Poland, which was located between this type and the bureaucratic-authoritarian regime. What distinguished Poland from the national-accommodative communism ideal type was the high level of technocratic capacity, which flowed from the communist regime's openness to economic reform and interaction with Western market institutions.[35] By contrast, the Slovak case was located between the national-accommodative and the patrimonial communist systems, and its reservoir of technocratic capacity was limited because its economic policy-making was dominated by Prague. Thus, based on the communist legacies, one would expect the highest level of state capacity in the Czech Republic, followed in decreasing order of state capacity by Poland, Slovakia, and Romania.[36]

Measuring State Capacity

It is challenging to measure state capacity directly, especially because it can vary by sector. After all, sectoral actors acting collectively can force or coax state agents into preferential treatment. Consequently, the market-adjustment measures that operate at the broader level of the economy could take a laxer form at the level of the specific sector.[37] However, given that the sectoral characteristics are similar among the four country cases, it is reasonable to expect that the relationship between overall state capacity and the capacity these states exhibited in the steel sector would be highly correlated across these states. Furthermore, using economy-wide indicators of state capacity as a proxy for state capacity in the steel sector has the advantage of removing any suspicion of circular reasoning that could arise should outcomes in the steel sector be used to infer the level of state capacity.

The ability to implement decisions, to collect taxes, and to regulate the economy is difficult to capture by one measure. Jessica Fortin developed a comprehensive index of state capacity in the economic domain, which encompasses the entire transition period in the postcommunist region, until 2006, and is an aggregation of standardized scores for five indicators of the quality of public goods provision.[38] These are taxing capacity, progress in infrastructure reform, levels of corruption, quality of property rights, and ratio of noncurrency money to total money supply ("contract intensive money").[39] In line with the legacies argument presented in this book, the initial level of state capacity, using Fortin's index, is the strongest predictor of state capacity at the end of the transition period and the two are very highly correlated (Pearson's *r* of .87). I use the averages for only the 1990–2000 time period, rather than 1990–2006, in order to capture the measurement of state capacity in real time. In 2000, the first important privatization to a strategic foreign investor took place in the steel sector in Slovakia, and thus it is a good reference point for comparison.[40]

Fortin's state capacity index measures the extent to which the state is able to provide public goods, but the preceding discussion also suggests that another indicator of state capacity is whether a state is perceived as a reliable business partner by external market players. This is captured by sovereign risk ratings, which indicate the degree to which the international lenders see the postcommunist states as trustworthy borrowers.

As discussed above, controlling for the sectoral characteristics should yield a similar correlation between the economy-wide and steel-sector-relevant state capacities in the country cases. A way of testing this assumption and of examining state capacity in the sector is to look at the level of state aid granted to the sector. After all, state aid is a proxy for the softness of the budget constraint. This is especially important in the transition context, where state aid rarely took the form of sector-specific investments. Rather, it resulted overwhelmingly from tax arrears and debts to state-owned utility companies. More often than not, state aid was a sign of inadequate adjustment to nascent market conditions. As subsequent chapters illustrate, where effective, state involvement tended to limit itself to the reorganization of assets using quasi-market means and modest levels of state aid.

Table 3 lists the indicators of state capacity in the four countries. The three measures are almost perfectly correlated with each other, and they

TABLE 3. Indicators of State Capacity

Country	State Capacity Index[a]	Euromoney Country Risk Rankings 2000[b]	Average State Aid $ per Ton of Steel Produced[c]
Czech Republic	97.5	63.1	5.4
Poland	92.2	63.6	5.8
Slovakia	78.5	53.0	9.1
Romania	–1.4	36.6	21.1

[a]Calculated by the author based on data provided by Jessica Fortin, featured in Jessica Fortin, "A Tool to Evaluate State Capacity in Post-Communist States, 1989–2006," *European Journal of Political Research* 49 (2010): 654–86. I multiplied the standardized score by 100 for ease of comparison.

[b]Keri Geiger, "Waiting for the Dust to Settle," *Euromoney* 377 (2000): 214–24.

[c]Author's calculation, based on levels of state aid agreed with the EU and actually granted, and on the sector's production output, 1993–2004 (see Appendix B).

indicate that the overall level of state capacity is proportional to the state capacity at the sectoral level.[41] Moreover, internal measures of state capacity, such as the ability to provide public goods, are essentially the same as external measures—that is, international lender perception of the state as a reliable financial partner.[42]

The comparative statistics in table 3 indicate that the Czech Republic had the highest level of state capacity, followed closely by Poland. Slovakia had the intermediate position, while Romania trailed on all three counts. Case studies presented in the subsequent chapters provide further qualitative evidence of these differences, as evidenced by the institutionalization of debt workouts and support by parastatal companies. Having addressed the origins of differences in state capacity and the measurement thereof, the discussion now turns to the means by which state capacity is exercised during the implementation process, at both the enterprise and sectoral levels.

State Capacity and Restructuring: Considering the Levels of Analysis

As figure 3 illustrates, the implementation of initial policy choice entails the exercise of state capacity and leads to a restructuring/privatization outcome characterized by a specific trade-off between economic efficiency and rent seeking.[43] This outcome produces a feedback effect, which takes the form of financial pressure on the government and affects the sustainability of the previous policy choice.

Government policies operate at two interrelated levels of analysis, where different economic actors predominate: the enterprise level and the sectoral level. The enterprise level is dominated by managers whereas the sectoral level, largely due to managerial unwillingness to act collectively, is dominated by labor unions. During the initial transition period, the lack of sectoral policies made the enterprise the locus of the restructuring effort. It was only when the inadequacy of the enterprise-centered perspective became obvious that governments with a relatively high level of state capacity began creating sectoral policies.

Enterprise Level

At the enterprise level, the crucial task for economic reformers was to entice the enterprises to adjust to the new market conditions by engaging in restructuring. Enterprises are embedded in networks of mutual dependence; they should be considered in relation to each other during the process of restructuring, and not as atomistic units.[44] Such was certainly the conclusion of the World Bank– and EC-financed sectoral studies, which

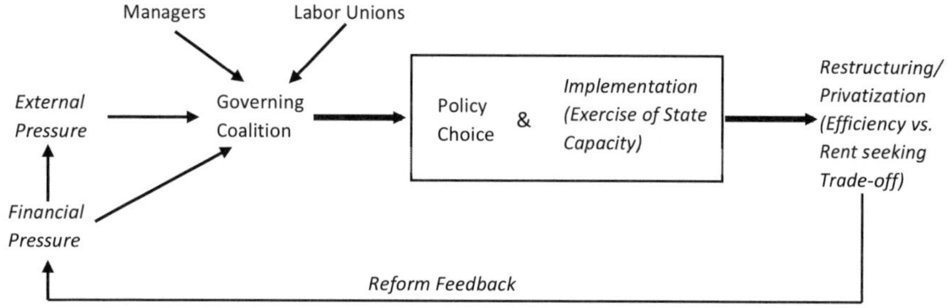

Fig 3. General Model of the Political Economy of Reform in the Steel Sector

were carried out by Western consultants in the early 1990s. These studies highlighted the restructuring priorities while pointing out the need to engage in interenterprise cooperation in the restructuring process.[45] Suggested cooperation was intended to take advantage of the existing steel-plant complementarities and strengths while preventing redundant investments. The enterprises could merge, form holding companies, or independently engage in far-reaching marketing collaborations, with the exact type of cooperation to be determined by plant complementarities. These early reports warned of uncoordinated investment, which would lead to overcapacity and redundancy, combined with insufficient reliance on economies of scale. The ultimate danger of these inefficiencies was that they could precipitate the sector's collapse by leaving it unable to stand up to well-organized competition in the wake of the full liberalization of trade in steel products. Yet, early in the transition process, sectoral coordination was resisted both by the reluctant governments and by the even more reluctant enterprise managers. Consequently, the enterprise became the locus of restructuring and the sectoral approach was neglected. I now turn to the discussion of the theoretical underpinnings of enterprise-level restructuring.

Restructuring success is measured by whether the state is able to force or entice specific enterprises to adjust to market conditions. Yet, the effective hardening of the budget constraint, which would stimulate enterprise restructuring, involves not only the deployment of technocratic resources by the state but also a separation between the political and economic spheres, along with the ability of state actors to stand up to pressure by sectoral actors. The separation of political and economic spheres and the ability to resist sectoral actors are difficult to achieve, for two reasons. First, the enterprises provide an attractive and tempting source of rent seeking and patronage. Second, the government and state agents are subject to strong managerial pressures to support the enterprises in ways that will fulfill the managerial vision of the restructuring endgame.

The managers, especially those at the prominent steel mills, had their own vision of restructuring, which I call *managerial prestige maximization*. Before explaining the theory of managerial prestige maximization, however, I first address the costs and benefits of cooperation from the managerial perspective. Although the benefit of cooperation is a competitive steel sector in the medium to long run—the socially optimal outcome—the costs of cooperation are substantial for the individual plants in the short term. First of all, the enterprises need to coordinate reductions in steel production capacity to optimize economies of scale. This means retaining more modern capacities in some plants while closing obsolete ones in others, which makes the enterprises involved mutually dependent.

Coordinated capacity cuts also entail substantial layoffs. Moreover, managers need to decide on the distribution of new investments, relying on economies of scale and avoiding duplication.[46] Cooperation in this respect means foregoing some of the investments that an enterprise's manager would have otherwise wanted to make. Finally, any mergers or even holding-company-type arrangements also mean the loss of an individual plant's identity and the loss of the sole leadership position by the top manager. Given these challenges and costs, instead of cooperating, managers engage in prestige maximization and actively eschew collaboration.

Prestige is understood to be the power and social status derived from being the head of a given enterprise. It depends on two components: position of the individual within the enterprise and position of the enterprise in the sector. As far as the individual aspect of prestige is concerned, it means that to the extent that their enterprise's survival is not threatened, managers strive to remain the central decision makers and therefore are unwilling to share power with other managers, let alone become subordinate to someone else. This logic is in line with the sociological critique that holds that economic goals are accompanied by noneconomic ones, such as status, power, or approval.[47] Individual prestige increases along with the enterprise's prestige in the sector.[48] The enterprise's prestige is derived from several components: production capacity, technological investment and modernization, and employment.

This prestige-maximizing strategy is highly popular with the workforce of the enterprise. After all, resistance to layoffs, maintaining production capacity, and upgrading capital stock within the enterprise are seen as key to maintaining jobs and creating new ones. Efforts to rationalize production and employment are therefore resisted by the workforce. They can also be opposed by other members of top management, who, by promising to carry out restructuring in a less socially painful manner, can use conflict between the workforce and a restructuring-oriented top manager as a means of personal advancement.

Thus, in the short run, in addition to increasing personal and the enterprise's prestige in the sector, the manager secures his position within the enterprise by avoiding steps that would make him unpopular. Such a strategy is all the more rational if one compares the certainty of high short-term costs with the uncertainty of medium- to long-term benefits. After all, the medium to long run is shrouded in uncertainty because the manager does not know how hard the budget constraint really is. Consequently, it is rational to be overly optimistic about the long-term implications of a preferable short-term strategy.

Summing up the discussion concerning the managerial approach to restructuring, I identify the following propositions:

> P1: Managers resist "defensive" restructuring measures, such as layoffs and capacity closures, unless a crisis situation develops within the enterprise.
>
> P2: Managers attempt to engage in technological restructuring; that is, they seek opportunities to upgrade and technologically enhance the capital stock of their enterprises, irrespective of market capacity.
>
> P3: Managers eschew cooperation with the managers of other enterprises and resist sectoral consolidation measures.
>
> P4: Managers demand protection by the state, in the form of bailouts and financial support.

These propositions stand in contrast to the network approach to restructuring. David Stark and László Bruszt have argued that the preexisting communist-era enterprise networks should be viewed as both the units to be restructured and as the entities undertaking the restructuring process. Their argument is straightforward: restructuring decisions should take into account the interenterprise ties of mutual dependence and cooperation, and because network actors possess the most information on the nature and relationship of the assets at their disposal, they are in the best position to make restructuring decisions.[49] The steel industry, with its limited number of key enterprises and managers who know each other as well as the assets in each enterprise, should be the perfect candidate for networks to exert their transformative power. Yet, as my account shows, rather than use network resources to devise restructuring strategies and act collectively, the managers adopted, instead, a prestige-maximizing strategy, whereby they resisted capacity cuts and aimed at maximizing technological investment in their enterprises. The bottom line is that while the managers are embedded in interdependent networks, they resist using them as the basis for engaging in coordinated sectoral restructuring.

The prestige-maximizing attitude fits into the larger family of principal-agent problems, well known in advanced capitalist economies.[50] Effective corporate governance may remove many of the agents' incentives to act contrary to the interests of their principals, but in the postcommunist world, this constraint on managerial behavior has been strengthening only gradually. Given state ownership of the sector at the outset of transition, the higher the level of state capacity, the greater the ability of state actors to rein in managerial ambitions.

Yet, exerting state capacity at the enterprise level involves far more than controlling the managers' investment desires. An indicator of a high level of state capacity is the ability of the state to intervene in situations where the enterprise is experiencing problems in an effort to bring about market adjustment. Rather than mere bailouts, such intervention takes the form of restructuring-contingent debt workouts, requiring sophisticated institutional support and coordination from state actors. An extreme form of such a state intervention involves significant state redeployment of enterprise assets by bringing in managerial "dream teams": rescue managers connected to specialized state agencies that engage in comprehensive organizational and financial restructuring. However, it is crucial to keep in mind that such interventions may simply be aimed at keeping the enterprise afloat and maintaining the status quo largely unchanged, rather than reaching the restructuring gold standard—that is, an enterprise capable of operating on international markets without state aid.

High levels of state capacity are also required during the privatization process. First and foremost, it is the state's responsibility to pick buyers who are capable of undertaking the exacting restructuring process. Second, the state often needs to deploy significant resources to make the enterprise more attractive for the potential investors, effectively engaging in asset and employment restructuring prior to privatization. Therefore, the state's role as a monitor, mediator, and negotiator with network actors is unique during restructuring.[51]

Sectoral Level

Despite the initial inability or reluctance to devise restructuring policies at the level of the sector, the financial feedback effect quickly created demand for sectoral-level solutions. These eventually took two forms: state reorganization of assets and sectoral policies of employment restructuring. To the extent that the state redeployment of assets at the sectoral level took place, it was carried out as a result of external pressures that will be discussed in the next section. As for employment restructuring measures, these were closely tied to the institutionalization of social dialogue, and their success depended on state capacity. In other words, the story of sectoral-level restructuring is largely the story of organized labor and its interaction with state institutions. In this section, I discuss labor's role in restructuring, including union attitudes to restructuring, the role of social dialogue, and the necessity of state institutional support for both social dialogue and the implementation of concluded agreements. My key claim is that social dialogue, especially at the sectoral level, is conducive to restructuring, but it requires a highly capable state.

The role of social dialogue in transition, as in economic reform more broadly, has been contentious. With the benefit of hindsight, the relative quiescence of labor was a surprising development in the former workers' states.[52] Although unions are either paltry or lacking altogether in the service sector, and even though they are severely weakened in many industrial sectors,[53] in the former flagship industries, whose employees were privileged under the former regime, unions have remained relatively powerful in comparison. In fact, these sectors witnessed several significant strikes, and the politicians often used the rhetoric of fostering "social peace" to postpone far-reaching reform in these sectors.

At the outset of transition, the privileged workers in the heavy industry, especially coal and steel, were seen to be the potential losers of transition and the greatest opposition to radical change of the status quo. They were expected to spearhead the popular backlash against reform after the initial "time of extraordinary politics" was over.[54] For this reason, the neoliberal reformers in ECE, as in the developing regions, following the executive insulation thesis, advised isolating the prospective "losers" in the transition process from the technocrats in charge of making the painful, but necessary, decisions.[55] After all, according to the neoliberals, in enterprises experiencing financial problems "workers are interested in grabbing whatever income they can before they are forced to find alternative employment. . . . In general, the best bet of the current insiders is to maximize their own current income, come what may."[56] As a corollary to income maximization, the insiders were also perceived to want to defend their privileged position inside the enterprise. Hence, the unions were not expected to want to give up control over their enterprises to outsiders, including foreign investors.[57]

The critics of the executive insulation thesis, on the other hand, argue that embedding the executive in a network of social institutions aimed at consultation leads to increased transparency because it makes the decision makers more accountable to the public. A stake in the decision-making *process* also extends the time horizons of the representatives of labor and of civil society more generally, as well as of the decision makers themselves, with the emerging policies becoming more cohesive. Rather than undermine the reform process, these critics contended, consultation makes it more resilient.[58]

Both, the proponents of executive insulation and their critics implicitly assume that the government is pro-reform and argue for either isolating or engaging the unions, respectively. Yet, the examination of restructuring of sensitive sectors begs precisely the question of what the union preferences are in the *absence* of pro-reform prodding from the government. After all, as stated earlier, the governments in power are not necessarily pushing for full reform and often enjoy the benefits of maintaining incomplete reform.

When examining the preferences of organized labor regarding reform, there are two dimensions that need to be considered: the structure of social dialogue and the nature of unions' relationship with the state (degree of autonomy). The structural dimension refers to whether organized social dialogue exists at the sectoral—or just at the enterprise—level. At the enterprise level, one union, or several, represents worker interests in negotiations with enterprise managers. At the sectoral level, umbrella union organizations delegate individuals who represent the interests of labor in bipartite negotiations with the employers or tripartite negotiations with both employers and state representatives.[59] The distinction is whether the government is insulated from the unions or whether it engages them institutionally. In the absence of sectoral social dialogue, enterprise-level politics predominate.

The second dimension considers whether the unions are autonomous vis-à-vis the government or are captured by it. Union capture would indicate a blurring between the political and economic/social spheres and a presence of patrimonial relations, which is characteristic of a low level of state capacity. Generally, individual captured unions do not preclude the presence of a separate, autonomous sectoral organization, albeit the latter is weaker than it would be otherwise. Although one could entertain the possibility that the government would attempt to capture the unions in order to push the reform process forward, it is much more plausible that the reason for union capture would be to ensure their complicity in rent-seeking activities benefiting those in government and the management.

Faced with the reform challenge, the task of the unions is to preserve employment and wages, in that order. Thus, unions oppose radical measures that would deprive their members of a livelihood. At the same time, they are likely to support measures that ensure the future of the enterprise and its workers, on the condition that the workers would be compensated for any job losses and that future developments would be relatively predictable. The organization of social dialogue at the sectoral level has several advantages. Not only does it elongate the time horizons of the main actors involved, but it also provides a sectoral, rather than enterprise-specific, perspective on the restructuring challenge. This has two effects.

First, following the logic used by Mancur Olson when discussing the salutary effects of encompassing organizations, it enables the participants to recognize their proposals' effects on the rest of society, and, perhaps more importantly for the case at hand, on other enterprises in the sector. This, in turn, tempers rent-seeking temptations. According to Olson, encompassing organizations, unlike very narrow distributional coalitions, "encompass a substantial portion of the societies of which they are a part."[60] In contrast to distributional coalitions, they "care about the excess burden [to the rest of society] arising from distributional policies favorable to its members and . . . strive to make the excess burden as small as possible."[61]

However, Olson's discussion of what constitutes an encompassing organization makes it difficult to operationalize the concept, as it is highly context specific.[62] As Olson admits, it is the *incentives* facing encompassing organizations that matter the most for his discussion, rather than their choices in particular circumstances.[63] The present discussion borrows the idea of incentives.

Second, in addition to providing incentives to take sectoral interests into account, social dialogue at the sectoral level supplies crucial information, without which the unions could miscalculate the feasibility of maximizing both workplaces and wages. Without the sectoral perspective, union attitudes to reform are determined by the local context and unions' calculations as to what course of action is feasible for maximizing their goals. Factors that influence this calculation are enterprise size and importance to the local economy, as well as the perception of a hard budget constraint. Thus, enterprise-level unions that are part of sectoral organizations are less likely to become pawns in the managerial game of prestige maximizing.

The preceding discussion begs the question of how union capture takes place. After all, ceteris paribus, if the unions want to ensure the workers' future, they should be reluctant to engage in activities that could endanger the enterprise's welfare. Thus, going beyond Olsonian distributional coalitions, in the presence of inducements from the government, found in countries where the political-economic/social relations approach patrimonialism—often places with relatively low levels of state capacity—individual unions may be inclined to enter into clientelistic relations with the government. While harmful to the economy, the sector, and the enterprise in the long term, the short-term (or even medium-term) benefits for the union might be significant in preserving both jobs and wages. The government and state actors, in exchange, would gain the individual union's support and a free hand when engaging in rent-seeking activities within the enterprise.

Summing up union preferences regarding restructuring, along with the critics of the executive insulation approach, this account suggests that autonomous union organizations are more likely to support reform when organized social dialogue exists at the sectoral level. Moreover, they continue to support reform even when the government support is lacking. Thus, faced with the financial feedback effect from enterprise-level restructuring, autonomous unions organized at the sectoral level push for comprehensive solutions at the sectoral level. By contrast, captured unions become an obstacle to economic reform. When the union organizations are autonomous, yet organized social dialogue is lacking at the sectoral level, the behavior of the enterprise-level unions is more difficult to predict as it is grounded in the particularities of enterprise-level politics, which include the perception of the relative hardness of the budget constraint.

The preceding discussion lends itself to the formulation of the following propositions:

> P5: Autonomous unions support (complete) reform when organized social dialogue exists at the sectoral level.
>
> P5a: Autonomous unions support (complete) reform when organized social dialogue exists at the sectoral level, even when the government is not spearheading reform.
>
> P6: Captured unions oppose (complete) reform regardless of whether social dialogue is organized at the sectoral level.

The discussion thus far has concentrated on union preferences toward restructuring as mediated by the structure of social dialogue. As chapter 5 shows, employment restructuring agreements became the pinnacle of domestically driven sectoral policy, and sectoral-level union organizations pushed for such policies to be adopted. However, only countries with a high level of state capacity—the Czech Republic and Poland—adopted and implemented such programs. Moreover, the quality of state institutions is also reflected in the practice of social consultation during privatization: where the quality of state institutions is higher, the social partners are much more likely to be thoroughly consulted prior to privatization. Thus, not only does the level of state capacity affect the structure and functioning of social dialogue at the sectoral level, it also determines the extent to which adopted sectoral agreements are implemented and consequently have transformative power in the sector.

State Capacity and External Pressures

As figure 3 illustrates, the implementation of the preferred policy choice at the enterprise and, later, the sectoral levels, produces a financial feedback effect. To the extent that the resulting financial pressures spur the government and state agents to engage in greater restructuring, which eases these financial pressures, they may lead to the preservation of the ownership-type status quo. However, they can also trigger external pressures for policy change. External pressures are crucial for convergence on the eventual structure of ownership because, absent the political will to do so, domestic market/financial pressures resulting from reform feedback are subject to political vagaries and lobbying and are therefore too weak to affect the status quo. External pressures can be divided into the categories of globalization and Europeanization/EU accession. Globalization pressures consist of two subcategories: top-down, or institutional pressures emanating from the IFIs; and bottom-up international financial market pressures.

Figure 4 illustrates the process of extrication from incomplete reform, as it shows the relationship between different dominant external pressures and various levels of state capacity, marked by specific trade-offs between economic efficiency and rent seeking. In countries where state capacity is relatively low, the government has difficulties hardening the budget constraints and ensuring macroeconomic stability, and it is implicated in large amounts of rent seeking. These countries become dependent on IFI loans to make up for the revenue shortfall. As creditors, the IFIs strive to ensure return on their investment and become the principal drivers of full reform/privatization to foreign investors who will ensure loan repayment. Thus, they preempt other pressures that are present in these economies, such as those of the EU, as the drivers of the privatization process. Although the initial IFI involvement largely operated at the macro level, as the transition process unfolded, the IFIs took an increasingly micro-level approach, aimed at privatizing and restructuring specific enterprises.

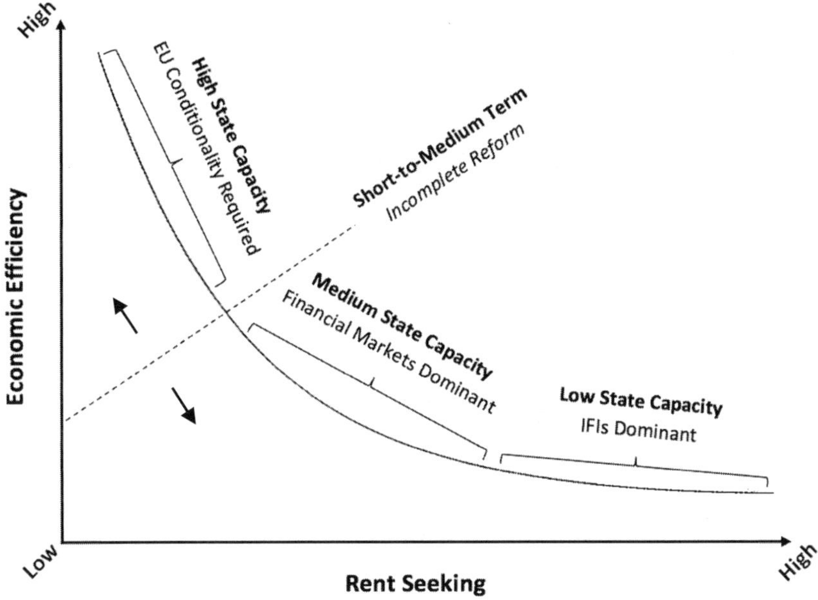

Fig 4. Modes of Extrication from Incomplete Reform

As the level of state capacity increases, moving up the curve and reaching the medium level, the state can ensure macroeconomic stability, thereby allowing it to stave off IFI pressures. However, it is unable to protect enterprises from the bottom-up globalization pressures of the international financial markets. Specifically, if a given enterprise becomes an active actor on international capital markets, it takes on the obligations of this participation, including any debt repayment. Should an enterprise become unable to repay its debts, it can rely on the state for a bailout or hope for a debt workout, or it can face bankruptcy.[64] If the state is unable to bail the enterprise out or organize a debt workout, its only solution, other than bankruptcy, is to try to transfer the enterprise to an investor who will settle with the creditors.

States with relatively high capacity can manage to avert both IFI and financial market pressures. They do so both by generating internal financing for the enterprises' use and by forcing the enterprises to undertake market adjustment measures—that is, forcing them to restructure. In such states, incomplete reform may, counter-intuitively, be maintained longer, as they are able to dampen external pressures for full reform. Rather than oversee thorough restructuring, the state apparatus would repeatedly mediate between creditors and debtor enterprises, leading to a series of partial, short-term adjustments. Thus, although there might be sufficient restructuring to stave off the pressures of the international financial markets or prevent IFI involvement, incomplete reform still threatens the long- and, perhaps, medium-term survival of the enterprises in these countries.

This dual nature of relatively high state capacity, protective, but also inducing market adjustment, is crucial. The question is not just whether the state has sufficient funds to sink into a failing enterprise, but what it does to foster market-oriented restructuring and how its efforts are perceived by creditors. In states with relatively high capacity, which are able to stave off globalization pressures, incomplete reform is directly challenged by the EU accession obligations. Here, the relevant issue is not the magnitude of the financial pressure but the fact that the enterprises are unable to operate without state aid. To grant any state aid would be to violate the rules of EU competition policy; therefore, a strategic investor becomes a guarantee of meeting EU requirements.

Whereas the IFIs clearly favor privatization, the EU is nominally indifferent as to the method through which viability is achieved. Nonetheless, given the restrictions on state aid and the amount of investment necessary to complete restructuring, privatization to strategic investors becomes the preferred option. Thus, where the state is capable of stimulating partial market adjustment while not triggering external market or IFI pressures, the EU can pull the last lever of coercion to induce full reform: prohibition of state aid. It is noteworthy that the external pressures are not exclusive; rather, they all act upon the transition countries. However, states with higher capacity are better able to resist the IFI and international financial market pressures than their counterparts with lower capacity.

According to the state capacity indicators, Romania, with clearly the lowest capacity should be most susceptible to the pressures of the IFIs. Slovakia, with its intermediate position should be more vulnerable to the pressures of the international financial markets, to the extent that these come earlier than pressures from the EU. Finally, given their relatively high level of state capacity, the Czech Republic and Poland would be expected to maintain the status quo until the EU mandates a resolution to the state aid issue.

Conclusion

This chapter laid out the model of the political economy of reform in the steel sector and presented the causal mechanism linking the initial policy choice to the eventual outcome of transnational capitalism in the region. The chapter discussed the origins, attributes, and measurement of state capacity, which was the central intervening variable in the restructuring process. The model of reform presented the ways in which state capacity mediated domestic and external pressures that acted on the government in power.

The following chapters focus on the discrete parts of the causal mechanism presented here. Chapters 4, 5, and 6 illustrate how state capacity mediates the outcomes at the three levels of analysis—enterprise, sector, and national/international—respectively, in the four countries. Chapter 3 places the steel-sector policy developments in the broader context of the political economy of reform in these countries, as it discusses the initial policy choices made by the four countries.

THREE

Putting Reform in Context

Politics of Initial Policy Choice

[T]he managements are there, in that area and they [have the local knowledge]—the scope of intervention, the scope of knowledge of the civil servants is simply too small to steer this well.

—Civil servant at the Polish Ministry of Economy

[T]here was a cry in parliament: "In my [steel]works, nobody will order [me around]!"

—Expert at the Organization of Romanian Steel Producers

In the early years of transition, the challenges facing the steel sector were daunting: overcapacity, overemployment, technological obsolescence, the need to redirect trade from east to west, and rapidly approaching full trade liberalization. To assist with restructuring, the World Bank and the European Community financed studies proposing solutions for the sector in the individual countries that would have involved their governments in far-reaching redeployments of assets. However, because of managerial pressure and, in most cases, the ideological predilections of the governments, these sectoral plans remained on paper. Rather, the governments took on reform strategies with lower political risks that tracked their broader reform pathways. In other words, the initial policy choices combined the ideological preferences of the government with pragmatic political considerations. In Poland and Romania, the government maintained state ownership whereas in the Czech Republic and Slovakia, the government opted for privatization to nonstrategic domestic owners. In all four countries, however, the government devolved responsibility for the restructuring process to the managers of the individual enterprises. This chapter explores the politics of these initial policy choices, and embeds them in the broader reform pathways the four countries followed. First, however, it briefly introduces the steel industry and discusses the structural features of the sector in each country. It then discusses the recommendations of the initial sectoral studies before turning to the politics of the initial policy choice in Poland, Czech Republic, Slovakia, and Romania.

Structural Features of the Steel Sectors

As the following section illustrates, the steel sectors in the four countries had structural differences, which manifested themselves in the ratios of flat products to long products and in regional concentration. However, despite these differences, all four steel sectors played a key role in their regional and local economies, and were emblematic of the communist modernization project.

Romanian Steel Industry

Romania entered the transition period with seven integrated steelworks and a steelmaking tradition dating back to the 1771 construction of Reşiţa Steelworks. Before the communist takeover, steel production, like most of industry, was largely confined to Transylvania. In addition to Reşiţa Steelworks, there was Socumet Oţelu Roşu, established in 1796; Siderurgica Hunedoara, established in 1882; and Câmpia Turzii Wire Factory (Industria Sârmei Câmpia Turzii), established in 1920.

After the communists took power in 1947, heavy industrialization—of which the steel sector was the centerpiece—became the main thrust of the Romanian development and modernization strategy. Romania, breaking with the policy of the Soviet Union, which envisioned Romania as the breadbasket of the Soviet bloc, built the enormous Sidex steelworks in the Danube port city of Galaţi in 1960. In the 1970s, with the help of Western credit, more investment followed, resulting in the construction of the Specialty Steels Enterprise at Târgovişte (Combinatul de Oţeluri Speciale Târgovişte [COS Târgovişte]) in 1973, and Siderca Călăraşi in 1976. The latter project was never completed, and by 1989, Siderca was only producing 164,000 tons of steel per year, compared to an envisioned capacity of 3.6 million tons.[1]

Specializing in flat products, Sidex Galaţi came to dominate the Romanian steel sector. Its 1989 output of nearly 7.7 million tons of crude steel represented over half of Romania's total steel production that year. Throughout the transition period, the importance of Sidex grew, and by 2003, it was responsible for about 80% of Romanian steel production; this was largely due to the near collapse of Siderurgica Hunedoara.[2] With the exception of Sidex, the six integrated steelworks specialized in either long products (Siderurgica Hunedoara, COS Târgovişte, Industria Sârmei Câmpia Turzii, Reşiţa Steelworks, Socomet Oţelu Roşu, and Siderca Călăraşi) or specialty steels (COS Târgovişte). In 1989, flat products represented 55% of total production, which compared favorably with both Poland and the Czech Republic. A higher ratio of flat products to long products is preferable because flat products—used in automobile and household-goods production—have much higher value added than long products, such as rails and sections.

As figure 5 illustrates, the Romanian steel sector is geographically less concentrated than its Polish and Czech counterparts. Southwestern Transylvania, where the steel-producing towns of Reşiţa, Oţelu Roşu, and Hunedoara are located, is the exception. However, as far as employment restructuring is concerned, any social advantages that could result from the greater dispersal of the industry are offset by the abundance of mono-industrial towns, such as Reşiţa, Oţelu Roşu, Hunedoara, Câmpia Turzii, and Roman. Finally, Sidex dominated the economy and politics of Galaţi. In 1989, the Romanian steel sector employed about 200,000 workers, about 40,000 of whom worked in Sidex alone.[3]

Slovak Steel Industry

Slovakia's steel sector was shaped by the 1958 decision by the communist authorities to build Eastern Slovak Steelworks (Východoslovenské Železiarne [VSŽ]), in the city of Košice in eastern Slovakia (figure 5). The investment in the flat-products manufacturer—most notably, of thin plates—was driven by the goal of industrializing Slovakia to the same degree as the Czech lands of Bohemia and Moravia. The investment overshadowed the existing central Slovak plant, Podbrezová Steelworks (Železiarne Podbrezová). The latter plant, dating back to the 1840s, eventually specialized in pipe manufacturing, and its crude

Fig 5. Major Steel Producers in the Czech Republic, Poland, Romania, and Slovakia. (Adapted from Google Maps Engine using GIMP.)

steel output in 1990 amounted to about 300,000 tons of crude steel, about 6% of total production. The rest of Slovak steel was produced by VSŽ.

The authorities achieved their goal of industrializing Slovakia; by 1989, Czechoslovakia could boast one ton of steel produced per capita, with the Czech Republic's ten million inhabitants producing 10.6 million tons of steel, and the five million Slovaks producing 4.9 million tons. Yet, because of the structural features of its sector, Slovak production was more efficient; the entire steel sector employed about 35,000 people, compared to the Czech Republic's 128,000.[4] As many as 26,000 worked at VSŽ and about 5,500 at Podbrezová Steelworks; the rest of the workers were employed by smaller plants, such as the ferroalloy producer OFZ, a.s., Istebné. Slovakia's focus on the more profitable flat products stood in stark contrast with the Czech production profile, which entailed only 31% flat products and was dominated by long products, such as rails and construction steel.[5]

Czech Steel Industry

At the outset of the transition period, steel production in the Czech lands topped one ton per capita, and the sector employed 128,000 people.[6] By the time of the communist takeover in 1948, the Czech lands already had three significant steel producers: Třinec Steelworks (Třinecké Železárny), which specialized in long products; Poldi Steelworks, which produced specialty and construction steel; and Vítkovice Steelworks, which was an integrated steel plant combining an engineering facility with steelworks specializing in thick plate production. There were also several smaller steel producing and processing enterprises, but their importance vis-à-vis the big producers was marginal. These included Hrádek Steelworks, Chomutov Steelworks, and Veselí Steelworks.

Soon after the establishment of the communist regime, in 1952, the existing enterprises were overshadowed by the child of the First Five-Year Plan, Nová Huť Klementa Gottwalda, or Klement Gottwald New Steelworks (Nová Huť). Similarly to Třinecké Steelworks, this darling of the new regime specialized in long products.[7] Figure 5 illustrates that the Czech steel industry, like its Polish counterpart, was regionally concentrated, with the "big three"—Třinecké Steelworks, Nová Huť, and Vítkovice—all situated in the North Moravia region, close to the Polish border. Nová Huť and Vítkovice were located in the city of Ostrava, and Třinecké Steelworks was in Třinec, about 30 kilometers away.[8] Poldi Steelworks, located in Kladno, 25 kilometers from Prague, was the exception.

Polish Steel Industry

At the time of transition, the Polish steel sector was fragmented; its two largest steelworks, Tadeusz Sendzimir Steelworks (Huta im. T. Sendzimira) and Katowice Steelworks (Huta Katowice) competed for the title of the leading enterprise in the sector. In 1990, the sector encompassed twenty-five enterprises, of which fifteen were involved in steel processing and production, four were predominantly pipe producers, three produced machines and equipment, and three manufactured other steel products.[9] In 1989, the sector employed about 147,000 workers and produced 15.1 million tons.

As figure 5 illustrates, the industry is concentrated in the Upper Silesian region of southern Poland. The ten enterprises discussed in this section represent nearly all of the country's steel production. Historically, the Silesian region had been the industrial heartland, and numerous small steel mills, such as Florian, Cedler, Baildon, Łabędy and Zawierce, dotted its landscape prior to the communist takeover.[10] Nonetheless, the biggest enterprises that came to dominate the sector were constructed as part of the communist industrialization drive, notably the Tadeusz Sendzimir Steelworks, formerly known as the Lenin Steelworks (Huta im. Lenina), and Katowice Steelworks.

The construction of Sendzimir Steelworks began in 1949 on the outskirts of Kraków. The integrated mill, specializing in flat products, was completed in 1954 and became Poland's biggest enterprise. Some of the other notable investments in the early 1950s include Warsaw Steelworks (Huta Warszawa) and Częstochowa Steelworks (Huta Częstochowa). The construction of Częstochowa Steelworks began in 1951, and the enterprise specialized in thick plate production; the construction of Warsaw Steelworks started in 1952, and production was focused on specialty steels and long products. Neither enterprise threatened Sendzimir Steelworks' supreme position in the sector.

In 1972, at the beginning of Edward Gierek's Western credit-financed investment boom, Sendzimir Steelworks' powerful position was challenged by a decision to build a gigantic steelworks about 40 miles away in Dąbrowa Górnicza, in the heart of Silesia. The new rival, Katowice Steelworks, was a white elephant project and the product of intensive lobbying efforts by the Silesian industrial lobby. Technologically, the new enterprise's crude steel processing facilities were very modern, but its final-product manufacturing section was never completed.[11] Although the flat-products manufacturing never got underway, the mill, in addition to semifinished goods, manufactured long products, such as rails, and became one of the key CMEA crude steel producers.

In part because Katowice Steelworks never developed the capacity to manufacture flat products, Poland had a disadvantageous ratio of long products to flat products. To compare, the share of the greater-value-added flat products in the total hot-rolled production in the EU stood at approximately 60% in mid-1990s; in Poland, this share stood at approximately 30%, similar to that of the Czech Republic.[12]

Sectoral Studies and Restructuring Recommendations

The structural differences among the sectors notwithstanding, it was clear that the reform process would be both painful and difficult because of challenges all four sectors shared: overcapacity, overemployment, the need for technological upgrading, and the upcoming trade liberalization. To make matters worse, trade with the former Soviet bloc collapsed, as did domestic consumption, so the countries simultaneously needed to reorient their trade elsewhere. To assist with restructuring, the World Bank and the European Community financed studies assessing the individual country sectors and proposing solutions for comprehensive restructuring. These programs shared several characteristics. First, they provided a sober assessment of the weaknesses of the different sectors. Second, they projected the countries' steel-production needs in light of the transition to a market economy and identified the necessary adjustment measures their sectors should undertake: crucially, capacity and employment cuts. Third, they highlighted the need to involve the state in a far-reaching redeployment of assets during restructuring to prevent redundant investments that would result in overcapacity.

Initial Restructuring Proposals in Romania

In 1990, solicited by the Romanian government, foreign experts were already advocating the creation of a single holding structure that would facilitate restructuring by coordinating capacity closures and investment across the sector. The Ministry of Industry and Trade subsequently created two "holding-type" structures, which grouped enterprises in the sector together and were intended to provide a strategic direction for the enterprises to take. These "holding-type" structures, however, were proved proven to be ineffectual as they were endowed with neither decision-making power nor financial resources.[13]

As in Poland and Czechoslovakia, the Romanian government subsequently commissioned a study, funded by the European Community's PHARE program.[14] The study was carried out in 1992 by French experts, from Sofres Conseil, Sofresid, and Cedres, who had earlier advised the French government on the restructuring of the French steel sector.[15] The study forecast that steel production in Romania would decrease to about 8.7 million tons by 2002 and recommended that the total production *capacity* decrease to 11.1 million tons by that year.[16]

The study results, presented in 1993, stressed the need for state control of the restructuring process. The report echoed recommendations, made by foreign consultants in 1990, to create a single holding structure within the sector to facilitate strategic decision making. In addition to coordinating investments and capacity closures, the holding company was supposed to monitor the financial performance of the enterprises and adjudicate any disputes among them.[17]

Moreover, the program recommended by the experts from Sofres Conseil, Sofresid, and Cedres envisaged concentration of production around four centers, which would specialize in flat products (Sidex), long products (Siderurgica Hunedoara), special steels (COS Târgovişte–Oţelinox), and seamless pipes (Romtub). The creation of these production centers would involve mergers of constituent enterprises. Although Sidex was to remain an enterprise in its own right, Siderurgica Hunedoara, Siderca Călăraşi, and CS Reşiţa were to be merged. Similarly, COS Târgovişte was to be merged with two smaller enterprises.[18] Finally, four producers of different kinds of seamless pipe were to be merged as well.[19] The mergers were intended to ensure optimal coordination of capacity closures and specialization of production within individual plants while simultaneously lowering administrative costs.[20] As far as employment restructuring was concerned, the initial forecasts envisioned a reduction of total employment in the steelmaking sector to 45,000 workers by the year 2002.[21]

Initial Restructuring Proposals in Czechoslovakia

Czechoslovakia also commissioned a study of the steel sector, which was paid for using the PHARE funds and carried out by a consortium of Western consulting companies led by Roland Berger Strategy Consultants. The final report, presented in mid-1992, projected a lowering of production capacity from 14 million to 9 million tons and a reduction in the number of employees in the sector to about 45,000 or 50,000 by the year 2000.[22] Despite the recognized importance of privatization, the study delegated the responsibility for implementing and supervising the restructuring to the state, requiring an expected 2.8 billion dollars' worth (CSK 80 billion) of technological investment by the state. Crucially, the report pointed to the need for close cooperation among the enterprises via the formation of product groups, concentrated around VSŽ for flat products, Poldi for special steel, and Třinecké for long products. The study devised a holding-type organizational structure for the sector, as it warned against the dangers of observed centrifugal tendencies within the Czech steel industry.[23]

Initial Restructuring Proposals in Poland

In March of 1991, the Polish minister of industry and trade commissioned a consortium of Canadian consulting companies to carry out a sectoral study to serve as the basis for future policy toward the sector.[24] The results of the "Canadian program," as it became known in the industry, were presented in the summer of 1992. The program pointed out the obvious ills: obsolete and environmentally unfriendly technologies, an unfavorable balance between flat and long products in the overall production structure, and the fragmentation and concomitant lack of specialization in many of the steelworks.[25]

Using moderate growth forecasts, the consulting firms calculated that Polish steel-producing power should be reduced from more than 15 million tons per year to 11.7 million tons and that the finished goods capacity should be capped at 8 to 8.5 million tons per year.[26] Moreover, the production profile was to be changed to increase the share of the higher-value-added flat products—especially sheet metal—used in the automobile and appliance industries. The program devised a detailed plan for how to restructure each of the plants by a target year of 2000.[27]

Most importantly, the fragmented sector was to be consolidated. A proposed merger of Katowice and Sendzimir Steelworks intended to make optimal use of their complementary production strengths and eliminate potential rivalry in the rather small domestic market. Although Katowice's crude steel production was its forte, without completing investments that would guarantee higher-value-added production—the development of flat products—it would have been unable to survive in the long run.[28] In the case of Sendzimir Steelworks, its obsolete crude steel production facilities were slated for closure by the year 2000, and its cold-rolled product facilities were to be renovated and upgraded.[29] Sendzimir would have specialized exclusively in flat products, but it would have lost its status as an integrated works, because it would have had to rely on Katowice for crude steel input for further processing.

A second proposed merger was to join Częstochowa Steelworks and Zawiercie Steelworks. The Częstochowa plant was supposed to close down its crude steel production facilities and rely on Zawiercie, which would have specialized in crude steel production. Other steel mills were to be grouped together and were to cooperate according to production profiles, enabling them to optimize the use of existing assets and to coordinate needed investments, sales, and marketing strategies, leading to increased competitiveness. As many as seven plants were to be closed.[30] Employment restructuring would be harsh: the total workforce was to be reduced from about 140,000 to no more than 43,000 workers.[31]

The costs of restructuring were estimated at $4.45 billion, out of which $1.65 billion was intended for basic investments, $300 million for social programs connected to employment restructuring, $550 million for increased working capital, $1.8 billion for renovations, and $150 million for other smaller investments. The study called on the government to take decisive action in overseeing its implementation. Not only was an inter-ministerial group supposed to devise an implementation strategy, but the government was seen as a crucial partner, responsible for solving social problems connected to employment restructuring and for guaranteeing some of the needed investment prior to sales to foreign investors.[32] Kato-wice Steelworks and Sendzimir Steelworks were to be immediately placed in a holding structure and a common management established. Importantly, the Polish state was not supposed to endorse any further investments until an implementation schedule was worked out.

Reform Pathways, Initial Policy Choice, and the Emergence of Incomplete Reform

Even though all three studies called for a coordinated sectoral approach, which would capitalize on the strengths of the different enterprises and prevent redundant investments and overcapacity, the call for the state to get involved in a far-reaching redeployment of assets went unheeded in the four countries. Rather, the governments de facto, and sometimes even formally, devolved restructuring responsibility to the managers of the individual enterprises. The exact form of this devolution—whether under the auspices of state ownership (Poland and Romania) or domestic privatization (Czech Republic and Slovakia)—was embedded in the broader transition pathways of these countries. These initial policy choices reflected the governing coalition's commitment to privatization, which was a function of the preferences of the coalition partners and their cohesion regarding privatization. Table 4 summarizes the political determinants of policy choice in these countries during the early to middle years of transition. Because political and economic reforms began in Poland, that country will serve as the point of departure for assessing the steel sector in the broader national-reform context.

Poland's Reform Pathway

In Poland, the initial policy choice in the steel sector, almost by default, was to maintain state ownership. It resulted from ideologically diverse coalitions that could not agree on privatization policy. Following its negotiated exit from communism, Poland began the transition process with an ideologically diverse Solidarity coalition. The ideological cleavages became obvious shortly after the surprising victory of Solidarity in the first semi-free elections of 1989, and it became clear that the strongest, if not the sole, factor holding the fragile post-Solidarity coalitions together was the common legacy of anti-communist opposition and continued opposition to the communist successor party. The neoliberal contingent became very strong in the Solidarity movement in the late 1980s and was distinct from the nationalist/social conservative camp, which represented statist tendencies and a far more reserved approach to radical economic reform. The social liberal faction was also strong, and it was oriented toward fast integration with the West. Finally, there was a significant worker self-management movement that had a vastly different ideological outlook from that of the neoliberal contingent.[33] The highly fragmented nature of the first postcommunist parliament and the ideological composition of the resulting governing coalitions translated into multifaceted approaches to economic reform.[34]

Pressured by rising inflation, in December 1989, the coalition made a show of unity by supporting the initial shock-therapy stabilization plan, named the Balcerowicz Plan after Poland's minister of finance. Privatization of SOEs, however, proved to be far more contentious. In fact, the Balcerowicz Plan aimed at macroeconomic stabilization, and it left privatization for the second stage of reform.[35] Privatization became the fodder for protracted political battles between the various factions. The neoliberals were split between proponents of rapid mass privatization and supporters of capital privatization via sales to foreign investors, delaying mass privatization until capital market institutions—namely, the stock market—were established.[36] The conservatives were wary of privatization, especially to foreign capital, and preferred to keep "strategic" sectors in state hands. Finally, the worker self-management movement emphasized the need to transfer ownership to the workers of industrial enterprises, thus bringing

TABLE 4. Political Determinants of Policy Choice: Early to Mid-Transition Years

Country	Coalition Support for Privatization	Preferred Privatization Route for the Steel Sector	Dominant Policy Choice
Poland	Mixed support	Foreign investors	State ownership
Czech Republic	Strong support	Domestic owners	Privatization to domestic owners
Slovakia	Strong support	Domestic owners	Privatization to domestic owners
Romania	Weak support	None	State ownership

to fruition the historic 1980–1981 Solidarity demands for greater worker rights and self-management within the enterprises.[37]

The law regulating large-scale privatization, adopted in July 1990, reflected the rifts within the Solidarity camp and aimed to reconcile the competing proposals. It envisioned three privatization routes: capital privatization, management-employee buyouts (MEBOs) or leases, and mass privatization. The latter was left to subsequent legislation that was not adopted until 1996 and pertained to only 530 enterprises.[38] Having opened the way for MEBOs, the 1990 law propelled privatization to enterprise insiders, which became the dominant means of privatizing in the early stages of transition.[39] However, MEBOs had little relevance for large industrial enterprises, which required sizable investments and were much more conducive to capital privatization.

Capital privatization encountered the typical problem of postcommunist societies: the paucity of domestic capital. This factor, combined with the lack of strong domestic capitalist traditions, led many economic elites to deem foreign ownership to be the best alternative to state ownership where privatization of large industrial enterprises was concerned. Still, the idea of sales to foreigners was not easily accepted by the social conservative/nationalist faction of Solidarity, for whom continued state ownership of significant portions of "strategic" industries was the preferred option.

In contrast to the Czech and Slovak Republics, creating a class of "domestic capitalists" using large-scale privatization never became an important political project or salient idea in Poland. In addition to the obstacles to capital privatization previously discussed, domestic capitalists were popularly associated with the *nomenklatura*—that is, with individuals holding positions requiring the approval of the communist authorities. After all, the first notable privatizations were the late-communist-era "*nomenklatura* privatizations," in which communist apparatchiks translated their political connections into tangible economic benefits, often by asset-stripping SOEs.[40] Even though the procedure was largely stopped at the end of 1989, this was enough to arouse popular suspicion that domestic capital was originating from close and illicit ties with the communist establishment.[41] At the same time, a micro version of domestic capitalism thrived, making use of the possibilities afforded by the small-scale privatization law, which rapidly transformed the landscape of retail shops and small enterprises.[42]

Because of these divisions, Polish reformers, instead of making privatization the priority, emphasized the construction of market institutions. Consequently, initial efforts to privatize the steel industry were postponed, though foreign ownership was seen as the preferred future option for the sector. The exception that proved the rule—and a harbinger of privatization policy—was the 1992 sale of the relatively small Warsaw Steelworks to the Italian Lucchini Group. Because potential sales to foreign investors raised significant opposition from nationalists and domestic investors were never treated as a serious alternative, the default option was continued state ownership, a status quo pleasing to the conservative/nationalist plank of the center-right Solidarity coalitions.

Unable to agree on an overarching approach to privatization, successive governments of Solidarity provenance, as well as those dominated by the communist successor party, placed the privatization of heavy industry on the policy back burner. In April 1993, Janusz Lewandowski, the neoliberal minister of ownership transformation (minister of privatization) explicitly stated that "coalmines, the steel industry, and ports . . . will not be privatized in the near future."[43]

All in all, maintaining state ownership was a politically safer option: it saved the government from criticism by its own allies, not to mention the opposition. Moreover, it preempted the potential resistance of sectoral actors. Soon, these payoffs were augmented by the discovery of rents and patronage opportunities that accrued to the governing coalitions.[44] Unlike with privatization to foreign investors, the opposition could avail itself of these opportunities once it won power. Thus, the status quo of state ownership became self-perpetuating.

Nomination of supporters to the supervisory boards of enterprises was one such benefit. Moreover, state ownership created substantial rents for industry insiders and the political figures connected to them. Rent seeking came in three forms. The first was the widespread use of politically connected firms that would contract with the SOEs to provide consulting and repair services at inflated prices, and that became intermediaries, selling overpriced iron ore, scrap metal, and other production inputs. According to a former deputy minister of economy, the firms that did business with the steelworks and won these contracts changed depending on the political parties in office.[45]

Another, related, form of rent seeking was the involvement of politically connected intermediaries in inter-enterprise debt trade. Politically connected companies would purchase an enterprise's liabilities in barter-based exchanges among numerous enterprises and procure rent by taking a cut of each transaction. The third form of rent seeking was the notification of certain business partners of an impending debt workout and offer to repay the debt on more favorable terms in exchange for a kickback.[46]

Even though privatization of the sector had entered the political agenda by 1996, it was not accorded priority. At the time, a coalition composed of the Democratic Left Alliance (Sojusz Lewicy Demokratycznej [SLD]) and the Polish People's Party (Polskie Stronnictwo Ludowe [PSL]) was in power, and foreign entry was perceived to be necessary to put the enterprises on a sound financial footing and to finish the necessary technological investments.[47] Sendzimir Steelworks' strategic position notwithstanding, in 1996, the minister of industry and trade announced that he did not oppose selling a majority of Sendzimir shares to a foreign investor, provided that it would benefit the company and that there would be a "good, solid partner." He also asserted that "we have no ideological prejudices [against privatization to foreign investors]."[48] However, no concrete initiatives followed.

In fact, industry insiders admitted in interviews that in mid-1990s there was neither a perceived need nor a desire to privatize the enterprises, especially Katowice and Sendzimir, in part because they were seen to be functioning relatively well.[49] This lack of political will to privatize was seconded by the 2003 assessment of the Supreme Audit Office, which stated that the successive governments had lacked a clear vision of privatization policy:

> According to the Office, the basic reason for the lack of privatization of the biggest steel mills was inadequate priority assigned to privatization in the subsequent Restructuring programs as well as delayed—by several years—action on the part of subsequent Ministers responsible for privatizing the steel mills.[50]

In the fall of 1997, the SLD-PSL coalition—much criticized by the post-Solidarity parties for stalling the privatization process—lost the parliamentary elections in favor of the post-Solidarity coalition between the Solidarity Electoral Action (Akcja Wyborcza Solidarność [AWS]) and the Freedom Union (Unia Wolności [UW]).[51] The new government, however, was marked by policy continuity. Soon after the elections, Janusz Steinhoff, the new minister of economy, pronounced that the steel industry, along with several other sectors, was "in dire need of restructuring programs, which *should* end with privatization."[52] In his new capacity as minister of state treasury, Emil Wąsacz claimed that Polish steelworks *should* consolidate quickly and form a big steel conglomerate to be privatized with the participation of foreign capital.[53] Nonetheless, he left both sectoral consolidation and the privatization initiative to the individual managements.

The successive governments were also not keen on decisively intervening in the sector to reorganize assets, thus leaving the restructuring process in the hands of the managers. Even though the "Canadian program" was accepted as a "directional strategy," the restructuring and implementation schedule it advocated was never devised.[54] Importantly, Katowice and Sendzimir continued to be under separate management. All in all, individual steel mills maintained the uncoordinated market-adjustment measures they had begun prior to the program's adoption.[55]

In fact, in response to strong managerial pressure, the government had already committed itself to giving six loan guarantees for restructuring in various plants.[56] Henryka Bochniarz, the former minister of industry and trade (August 1991 to January 1992), admitted that the government stepped in with assistance where managerial and trade union activities attracted media attention.[57] According to a Sendzimir Steelworks insider and former parliamentarian, "[T]he governments were not determined and were afraid to touch the subject [of sectoral consolidation], even though it was a costly subject. . . . Everyone created this situation that 'somehow we'll manage.'"[58]

One could attempt to attribute the delegating-to-management strategy to the ideological predilections of Solidarity-camp governments, rather than political expediency. Indeed, the former center-right AWS minister of economy stressed that for the neoliberals, consolidation of the sector went against the principle of market competition.[59] However, the period of relatively radical neoliberalism had ended by 1991.[60] When the center-left SLD-PSL coalition came to power in the fall of 1993, it did not bring much to the table as far as sectoral restructuring policy was concerned, despite its prior criticism of its predecessors' policies. Rather, the government continued to delegate restructuring to the managerial strata, even as it attempted to create sectoral programs as part of its "[i]ndustrial policy program for the years 1995–1997," which had clearly statist overtones.

In 1995, the Ministry of Industry and Trade criticized the sector's fragmentation, which was to be rectified through the creation of a holding structure, as laid out in the "Industrial policy program for the years 1995–1997."[61] Given that the managements of Katowice Steelworks and Sendzimir Steelworks adopted restructuring strategies that provided for additional continuous casting lines in each plant, the need for a merger or holding was less pressing.[62] Rather, the discussion switched to consolidating production around product groups.

However, although it favored sectoral consolidation, the ministry was not going to "impose" any solutions on the steelworks.[63] In 1996, an inter-ministerial team created to coordinate sectoral restructuring activities prepared four subsectoral programs for the long, flat, special steel, and pipe products, but these were never adopted as government policy.[64] Rather, official pronouncements took on normative overtones. A 1997 statement by a senior civil servant at the Ministry of Economy (successor to the Ministry of Industry and Trade) is illustrative:

> Steelworks should consolidate [production] around long [in Katowice] and flat [in Sendzimir] production cycle. We do not, however, intend to force anyone into such associations. I presume that today the directors and managements of some of the steelworks understand very well the benefits, which these [associations] have to offer.[65]

Likewise, when the AWS-UW coalition was in power, Emil Wąsacz, the minister of state treasury, tried to persuade the managements of Katowice Steelworks and Sendzimir Steelworks to create a merger. However, he repeated the words of his predecessors when he underscored that he "would not merge the steelworks by force."[66]

According to Wąsacz, successive governments delegated the responsibility for restructuring to the individual managers without trying to promote a strategy for the sector as a whole.[67] This conclusion was also suggested by a high-ranking civil servant at the Ministry of Economy, who stated point blank that "[t]he managements are there, in that area and they [have the local knowledge]—the scope of intervention, the scope of knowledge of the civil servants is simply too small to steer this well."[68] Ironically, the view of not wanting to engage in "central planning" could also be heard from the Ministry of Economy civil servants, who had overseen the sector as far back as the 1980s.[69]

Czech Republic's Reform Pathway

In the Czech Republic, the steel sector became part of a project of creating domestic capitalism undertaken by the cohesive and nominally neoliberal governing coalition. The conditions in the Czech Republic–part of Czechoslovakia at the time–after the communist regime collapsed were the most conducive to radical economic change in the region. The elite nature of the anti-communist opposition and its shallow roots in society led to a competition between two elite visions of transition within the Civic Forum, the anti-communist umbrella organization: the self-styled neoliberals, who advocated rapid economic change, and their social liberal counterparts, who favored a more gradual and interventionist approach. The vision of rapid economic and political change, advocated by Václav Klaus, the Czechoslovak Minister of Finance, and Vladimír Dlouhý, the federal Minister of Industry and Trade, prevailed and was endorsed in the 1990 election from which the Civic Forum emerged victorious over the communists. Václav Klaus quickly proceeded to break the Civic Forum apart and formed the Civic Democratic Party (Občanská Demokratická Strana [ODS]). ODS's power was further strengthened in the aftermath of the 1992 elections when it formed a coalition with the neoliberal Civic Democratic Alliance (Občanská Demokratická Aliance [ODA]), another splinter from the Civic Forum.[70]

Thus, the Czech lands entered the transition period with a cohesive governing coalition, which, in a feat of creative political entrepreneurship, managed to combine neoliberal rhetoric with economic nationalism that became known as the "Czech way" (*česká cesta*). Unlike its counterpart in Poland, the statist-oriented nationalist/conservative faction in the Czech lands was marginal because nationalists were attracted to the ODS by the slogans of domestic capitalism.[71]

In 1990, Klaus and Dlouhý's recipe for transition was similar in macroeconomic terms to that of Balcerowicz. Unlike in Poland, though, privatization became the central element of the reform agenda from the very beginning.[72] The slow progress of the Polish capital privatization gave Czech reformers an additional incentive to turn to the voucher method, which they closely tied to regime change, suggesting that it represented the only possible radical severing of ties with the past.

A large-scale privatization law was passed in February 1991, and even though it allowed a multiplicity of competing privatization plans to be submitted for any given enterprise, preference was given to plans relying on the voucher method. Mass privatization became a collective, participatory, national, market-building exercise—the "Czech way" of creating capitalism. Still, the Czech state took a relatively careful approach to large industrial enterprises, which were not supposed to be privatized exclusively using the voucher method. Rather, the state retained a majority stake in these enterprises initially, with only a minority share submitted for voucher privatization.[73]

The "Czech way" referred not just to the mass privatization program but also to the conscious effort to create Czech industrial-financial groups or a genuine Czech capitalist class, and it harked back to the Czech industrial tradition in the region.[74] Early in the transition period, building the Czech capitalist class became the preferred choice. Foreign investors were considered as possible participants in joint ventures with important Czech enterprises, including the major steel mills, but talks with prospective investors collapsed after the government refused to involve the state in pre-privatization (mostly financial) restructuring.[75] Because the two waves of voucher privatization were not intended to privatize large industrial enterprises, the government turned its attention to completing, at least formally, their privatization. Invoking a popular attitude against selling "family silver" to foreigners, the Klaus government started selling stakes in major industrial enterprises to domestic owners—usually derived from the managerial circles of these enterprises—in a series of leveraged buyouts.[76] The steel sector was a perfect illustration of these, ultimately, overwhelmingly failed efforts.

The formation of the Czech financial-industrial groups was politically convenient, because it both fulfilled ideological policy objectives and created obvious future allies of the governing party. Given two party-financing scandals surrounding privatization, discussed later, and the operation of the largest steel mills, there is evidence that the establishment of the class of domestic capitalists also had pecuniary benefits for the governing coalition. All in all, the preferred strategy was to turn to domestic investors, including the managers, as the future proprietors of the family silver.

The developments in the steel sector reflected the overarching pathway of the Czech transition. The 1992 Roland Berger study was never adopted as official policy, and the minister of privatization plainly stated that restructuring before privatization would be "contradicting our privatization philosophy."[77] Although the official rhetoric implied that restructuring should follow privatization, both restructuring and privatization followed pragmatic political considerations.[78] The reformers considered heavy industry to be a liability from the past, and, according to Vladimír Dlouhý, a former minister of industry and trade, privatization of the sector was not the highest priority in the initial years of transition, compared with privatization of the Czech economy as a whole.[79]

Unlike the program in Slovakia, discussed later, the famous Czech voucher privatization program only partially pertained to the largest steel enterprises, and only minority shares of some enterprises were put up for privatization. Among the Big Four steel producers, Poldi did not participate at all, Nová Huť and Vitkovice sold approximately 32% of their stock via vouchers, and Třinecké Steelworks sold 49%.

Dlouhý explained that in 1993, due to the delicate political and social climate of the North Moravia Ostrava region, he decided to pursue financial stabilization and modernization and to postpone the discussion of full privatization until later.[80] In short, the enterprises were to modernize on their own.

The policy of privatizing to domestic capitalists was not chosen by default. As noted earlier, the government passed up the opportunity to form joint ventures with foreign investors in the steel sector, as these would have required significant state involvement in the pre-privatization restructuring of assets. In fact, prior to the 1992 elections, which solidified the neoliberal contingent's dominance, there were plans for joint ventures with strategic foreign investors in the steel sector. The prospect initially attracted interest from the leading companies in the sector, such as Kaiser, Krupp, Thyssen, Mannesman, Voest-Alpine, and Usinor Sacilor.[81] These were particularly encouraged by Ján Vrba at the Federal Ministry of Industry, who wanted the government to become a financial and negotiating partner, and who was also a proponent of adopting the 1992 steel-sector restructuring plan.[82] With the lack of government involvement in facilitating these joint ventures following the 1992 elections, none of the core steel production activities were privatized with the participation of foreign investors.[83] Rather, the government turned its attention to domestic privatization.

Sales to domestic owners aimed at attaining two objectives: fulfilling the government's privatization program and gaining the support of the domestic industrial lobby that pushed to retain industry in "Czech hands." It also provided a way for the government to devolve responsibility for particular enterprises. The first notable privatization in the steel sector involved Poldi Ocel, the core steelmaking facilities of the Poldi holding company.[84] The acrimonious nature of this privatization led, in part, to the government's more cautious approach to Nová Huť and Vítkovice. Poldi did not participate in the voucher privatization and was heavily indebted. In 1993, the National Property Fund (Fond Národního Majetku [FNM]) refused to bail the enterprise out.[85] At that point, Dlouhý made a controversial claim that bankruptcy would offer a solution to Poldi's problems and that he would be willing to sell Poldi for a nominal sum of one Czech crown.[86] Instead, the FNM quickly offered a 66% stake in Poldi Ocel—the holding's steel-producing core—for sale via public tender, with price the determining factor in selecting the winner.[87]

In September 1993, little-known Bohemia Art was selected as the winner. The company submitted the highest bid: $60 million (CZK 1.75 billion) for Poldi shares and $377 million (CZK 11 billion) in prospective investments in the steelworks and the town of Kladno.[88] Led by Vladimír Stehlík, Bohemia Art was engaged in numerous unrelated activities, ranging from agriculture to newspaper publishing. It presented a woefully unrealistic business plan as well as a classic leveraged buyout, for which it relied on loans from domestic banks, most notably from Komerční Banka.

Rashly, the FNM granted Stehlík majority control of the enterprise after he offered to take over the holding company's debt to Komerční Banka. In hindsight, Dlouhý suggested that the sale to Bohemia Art had bought time for a steel producer that should have been slated for bankruptcy.[89] However, were the privatization to fail, it would not be the state that would be taking immediate responsibility for the steelworks' collapse, a less costly maneuver politically.

Bohemia Art's honeymoon period with the government ended relatively quickly when Stehlík missed the deadline for the payment for his shares early in 1996. In the meantime, the FNM received warnings that Poldi's working capital was being misused to finance Bohemia Art's loans and further acquisitions.[90] At that point, the government attempted to wrestle control from Stehlík, and, failing to do so, began complex bankruptcy proceedings, which ended in 1997.[91]

Whereas the privatization of Poldi was characterized by political expediency, the privatization of Třinecké Steelworks reflected the pecuniary benefits of creating a domestic capitalist class. In 1995, before the problems with Poldi began, the government turned its attention to the North Moravia steel giants, and to Třinecké Steelworks in particular. The company was the Czech Republic's second-largest steel producer and its seventh biggest enterprise, employing around 18,000 people in 1989. By 1995, this number had decreased to slightly over 12,000 workers.[92] Given Třinecké's relatively good standing, and even the government's proclamations that it did not need profound restructuring, the government decided to continue the privatization process by putting FNM's remaining 51% stake in Třinecké Steelworks up for sale using public tender.[93]

A Czech-Slovak company, Moravia Steel, owned in part and presided over by a former tennis star turned businessman, Milan Šrejber, won the tender in October 1995, offering the highest purchase price at $99.5 million (CZK 2.64 billion) and the biggest future investments, amounting to $644.3 million (CZK 17.1 billion).[94] The company was created specifically for the purpose of attempting to purchase the steelworks. As in the other domestic privatization cases, this purchase was a classic leveraged buyout, and, to pay for the shares, Moravia Steel took out a loan equal to the purchase price from a consortium of Czech banks led by Česká Spořitelna.[95]

What made the Třinecké Steelworks privatization distinctive was the political scandal that erupted in the aftermath of privatization. Around privatization time, donations in the amount of more than $500,000 (CZK 15 million) were given to the senior coalition partner, the ODS, supposedly by a Hungarian, long since deceased, as well as a citizen of Mauritius with no knowledge of ODS's existence. These contributions were traced back to Šrejber and to another person from Moravia Steel, creating suspicions that the company was chosen not on the merits of its proposed business plan, but because of its generosity to the ODS. Speculations arose that the outcome of the bidding process was determined in advance.[96] The scandal, combined with allegations of ODS's secret Swiss bank accounts, fueled public outrage and led to Klaus's resignation in November 1997.[97]

After the privatization of Třinecké Steelworks, the government turned its attention to two behemoths under state control: Nová Huť and Vítkovice. Governmental actions in these two privatization attempts were also controversial. First, the government pursued an unorthodox privatization method: it engaged in managerial buyouts of Nová Huť and Vítkovice, a method generally reserved for small and medium enterprises. At the time, the government did not even attempt to look for foreign investors to take over these enterprises, doubting that there would be substantial external investor interest.[98] Rather, seeing that the management was carrying out restructuring at the enterprise level, the government decided to sell some shares to the management as a further incentive.[99] The managerial privatization initiative met with the support of the managerial strata, which preferred that the steel mills remain in Czech hands.[100]

Having learned from the Poldi debacle, the FNM took a more careful approach this time, selling an initial 1% share to companies set up by the management: Petrcíle, s.r.o. in the case of Nová Huť, and Rafis Trading, s.r.o. in the case of Vítkovice. Both managements were given the option of purchasing further 10% to 15% stakes by December 31, 2001, but only once the conditions of the contract had been fulfilled. The exact conditions remained part of a confidential agreement, and, as was disclosed later, they entailed specific technological investments and a requirement to increase the value of the enterprises' shares.[101]

The government officially committed itself to supporting managerial restructuring measures in both enterprises. In the case of Nová Huť, it permitted Petrcíle to exercise shareholders' rights with respect to shares to be sold to Petrcíle in the future. Moreover, the FNM committed to not sell any shares—beyond those needed to reduce the enterprise's stake to 49%—before the end of 2001 unless both the FNM and Petrcíle agreed.[102] Furthermore, although in the case of Vítkovice, the FNM could cancel the contract for the subsequent shares in the enterprise if the management did not fulfill the requirements of technological restructuring, in the case of Petrcíle, it would have had to pay approximately $20 million (CZK 550 million) in penalties if it broke the contract before the end of 2001.[103] This provision tied the hands of the government until the end of 2001.

The circumstances of managerial privatization were also not scandal-free, as it appeared that Vítkovice had anonymously sponsored the ODA. Even though, according to Dlouhý, these activities were not illegal in the Czech Republic at the time, the 1998 allegations concerning this and other donations took a heavy toll on the party; its chairman resigned and its ranks thinned.[104]

Selling stakes to the managers, however, was not sufficient to make the enterprises private. Therefore, in an effort to reduce the state share to under 50%, the government committed to selling minority stakes in both enterprises on the capital market by June 30, 1997.[105] The privatization effort of Nová Huť was tied to the so-called minimill investment. Because Czechoslovakia's major flat products manufacturer—VSŽ—was located in Slovakia, following the "velvet divorce," Czech steel producers wanted to build a flat-products facility—a minimill —of their own. Due to its scope, the project necessitated state guarantees and all three major producers submitted rival proposals. The Czech government eventually decided to support Nová Huť's proposal, but in doing so, it strived to reconcile two objectives: free market ideology (no state involvement) and political pragmatism (necessity of investment guarantees). The government believed the best solution was not to underwrite the loan itself, but to solicit the assistance of the International Finance Corporation (IFC), the "private sector arm of the World Bank."[106]

According to its own rules, the IFC could only lend to private entities. With the government unable to sell the minority shares on the capital market, it engaged in what amounted to illusory privatization. The FNM transferred enough Nová Huť shares—18.25% to be precise—to the commercial bank Credit Suisse First Boston to reduce the state's share to 49% and make Nová Huť nominally private and therefore eligible for IFC lending. Credit Suisse was subsequently tasked with selling the shares on the market for the current market price. Should these efforts have failed by September 1999, the Czech government committed itself to reacquiring the shares.[107] The IFC was aware of this maneuver and the loan agreement with Nová Huť was actually signed before the official government resolution concerning the 18.25% stake was issued.[108] The $250 million loan was the biggest IFC investment in the Czech Republic.[109] The IFC also assisted Nová Huť in arranging a $100 million loan from a consortium of Czech banks. The IFC loan became the key to the restructuring strategy, estimated at $600 million, which was supposed to help the managers increase their share in the enterprise.

Thus, similarly to Poland, the Czech government devolved restructuring to the managements of the major enterprises, whether through privatization (Poldi, Třinecké), nominal privatization (Nová Huť) or continued state ownership (Vítkovice).

Slovakia's Reform Pathway

In Slovakia, the steel sector also became part of the project of creating domestic capitalism. The initial privatization via vouchers, pushed through by the neoliberal faction of an ideologically diverse coalition, was soon complemented by crony privatization undertaken by its populist former allies. In Slovakia, the anti-communist opposition was organized by Public Against Violence (PAV), the Czech Civic Forum's sister organization. Whereas the Civic Forum was divided between the neoliberals and gradualists/social liberals, PAV was more ideologically diverse; in addition to these factions, it contained a significant populist and nationalist faction whose importance grew over time.[110] In comparison with the Czech Republic, the neoliberal faction was substantially weaker, though it faithfully followed the ideas proposed by its Czech colleagues, a fact that caused discord in Slovak politics.[111]

After the 1990 elections, Czech economic thinking dominated the transition process in the entire Czechoslovakia. Although the federal economic ministerial posts were held by Czech neoliberals, they found faithful collaborators among their PAV colleagues. Because of the structural differences between the two economies, the same set of stabilization reforms had remarkably different effects. In particular, Slovakia's heavy industry, which relied on trade with the Soviet Union, was especially vulnerable to the collapse of the CMEA. Moreover, the political decision to cut Czechoslovak armaments production, two thirds of which was located in Slovakia, only exacerbated the existing hardships. Slovakia also lacked the "Prague factor," a major motor of job creation in the tourist industry, and remained a far less attractive region for potential foreign investment than its sister state to the west. Within two years of system change, the Slovak unemployment rate stood at 11.8% in 1991, as compared to 4.1% in the Czech Republic.[112]

Popular dissatisfaction with the course of economic reform gave considerable ammunition to the populist and nationalist political entrepreneurs within the PAV. The problem was exacerbated by PAV's neoliberal contingent, which was supportive of the Czech reformers who dismissed

Slovak economic hardship and refused to try to adjust policy to ameliorate specifically Slovak conditions.[113] The populist and nationalist contingent started to coalesce around the prime minister, Vladimír Mečiar. Thus, a fault line formed within PAV between an uneasy coalition of Slovak neo-liberals and conservatives, on the one hand, and the nationalist populists, on the other.

In April 1991, articulating his political agenda in an increasingly nation-alist language, Mečiar split from the PAV, formed his own party, Movement for a Democratic Slovakia (Hnutie za Demokratické Slovensko [HZDS]), and ran on a platform strongly critical of economic reform and advocat-ing its outright reversal. The reply of the remaining factions within PAV was to bring down the Mečiar government through a no-confidence vote. During the period that preceded the June 1992 elections, the neoliberal faction within the caretaker government formed by PAV pushed forward with its largely unaltered reform agenda, including the introduction of the first round of voucher privatization. This dismayed its socially conservative partner, who nonetheless wanted to distance itself from the HZDS.[114]

After the split of PAV, the neoliberal faction became more vocal as it made voucher privatization a symbol of reform against the populist-nationalist opposition. As in the Czech Republic, discussion concerning foreign participation in the privatization process was largely absent. How-ever, the Slovak reformers seemed to move in a more radical direction, as they did not limit the percentage of shares in large industrial enterprises that could be sold via voucher privatization to minority shares only. HZDS, comfortably in the opposition during the pre-electoral period, remained highly critical of the reform process, which proved to be a valuable strategy that led to HZDS victory in the 1992 elections.

HZDS's victory in 1992, combined with that of the ODS in the Czech Republic, made it much more difficult to reach consensus concerning com-mon constitutional arrangements with the Czech lands. Although Mečiar supported negotiating greater autonomy under a new constitutional arrangement, he did not openly back full sovereignty, which had weak, though growing, popular support. The conflicting approaches to economic reform subsequently made it much easier to reach a radical solution to the Czech and Slovak differences concerning constitutional arrangements. In a quick decision, the Czech and Slovak negotiating teams, led by Klaus and Mečiar, settled upon a "velvet divorce," citing irreconcilable differences.[115] For the two politicians, independence of the Czech Republic and Slovakia became the key to the unobstructed fulfillment of their respective political goals.[116]

Freed from Czech influence, Mečiar did not engage in radical policy reversals, but he initially slowed down the privatization process. In March 1994, the opposition, galvanized by HZDS's patrimonial and authoritarian tendencies, again orchestrated Mečiar's ouster. The broad coalition, which subsequently formed the government, united forces across the political spectrum. Under the premiership of Jozef Moravčík, the government cancelled several privatization decisions made by its predecessor that it found illegal and started to prepare the next stage of economic reform—that is, the second wave of voucher privatization. Despite HZDS's criticism, voucher privatization drew much public interest.[117] Notably, in subsequent elections, HZDS did not run explicitly on the platform of reversing voucher privatization.

Following its victory in the 1994 early fall elections, HZDS formed a coalition with two seemingly unlikely allies, the far-left Association of Workers of Slovakia and the far-right Slovak National Party. With the junior coalition partners vehemently opposed to reform, the coalition rhetoric became a mixture of illiberal nationalist populism, with clientelism the cementing force holding the coalition together. Ideological debates within the coalition became subordinated to the pragmatic goal of staying in power while seeking rent, complete with shrouding privatization policy in overtly nationalist rhetoric.[118] Although the HZDS coalition largely maintained its predecessors' macroeconomic policy, it made a clear departure regarding privatization. Contradicting his electoral platform, Mečiar announced the cancelation of the second round of voucher privatization. Claiming that "everything will be different and it will be better," Mečiar's coalition took to selling enterprises, at discounted prices, to political supporters who leveraged their loans.[119]

The divided, weak, and sidelined opposition had a difficult time formulating an alternative vision to the one presented by Mečiar and his allies, where patrimonial privatization policy was couched in language promoting domestic national capitalism. The government's stated goal in its 1995 program was to "create such ownership structures, in which a substantial share will consist of domestic businessmen and employees of privatized companies."[120] Thus, the fostering of domestic capitalism and excluding of foreign investors from the privatization process became part of the state-building project. The steel sector embodied these processes.

VSŽ was initially privatized using the voucher method. Attesting to the zeal of the Slovak neoliberals, as much as 97% of VSŽ shares were made available for purchase using the voucher method. This contrasted sharply with the more careful Czech approach, in which only minority shares of large industrial enterprises were put up for sale. In fact, the flagship VSŽ became the very first enterprise the minister of privatization put on the privatization list.[121] It was a risky move given that voucher privatization was a highly experimental program, and VSŽ was clearly the most important Slovak enterprise, responsible for the production of about 10% of Slovakia's GDP, 11% of its exports, and 25% of its foreign-currency inflow.[122] All in all, the enterprise was indirectly responsible for about 100,000 workplaces in the Košice region.

Just under 63% of VSŽ shares was sold via the voucher system, resulting in a fragmented ownership structure, and with only a 37.6% stake, the state retained a plurality of shares. The investment fund Harvard Capital and Consulting, which later became infamous for its deleterious activities in the Czech Republic, gained the second-largest share, 11.5%. The third-largest share, equal to 10.1%, was bought up by Cassoviainvest, an investment fund created by VSŽ management.[123]

The management and the Slovak government soon formed close ties as Mečiar's HZDS tightened its grip on Slovakia. After the June 1992 elections, Július Tóth, a former top-management member responsible for information technology, left the enterprise to become Slovakia's minister of finance. In March of 1994, he returned to VSŽ as president of the enterprise's supervisory board. By 1994, close ties took on clientelistic character. First, there were allegations that the state subsidized the enterprise while the coalition in power was receiving financial support from the enterprise.[124] Second, privatization became a tool for rewarding political allies.

The extent of clientelism was reflected by the "midnight privatizations" following the no-confidence vote in the first Mečiar government in independent Slovakia. On March 11, 1994, the day of the no-confidence vote, five VSŽ managers, including Alexander Rezeš, Ján Smerek, and Peter Hrinko, founded a company called Manager, a.s., which had only 3,100 dollars' worth of capital (SKK 100,000). On the very same day, the government promised Manager, a.s., a 9.53% stake in VSŽ for about $9.8 million (SKK 313.6 million). On the same day, Priemyselná Banka, a Košice bank that was later placed under a caretaker administration by the Slovak National Bank, promised a loan in the amount of $1.9 million (SKK 60 million) for the purchase of these shares. Three days later, the Slovak FNM signed the contract, selling more than 1.5 million shares having a book value of between $16 million and $23 million for $6 million, representing, at most, 38% of their market value.[125] Moreover, the terms of contract stipulated that the managers were to pay only 10% of the total price, with the remaining 90% to be paid by 1999.[126]

Július Tóth, the VSŽ-bred minister of finance and FNM chairman was instrumental in making the deal. Tóth was also known for granting financial favors to select companies. In the case of VSŽ, for example, he wrote off 50% of the enterprise's fines, in the amount of $113,750 (SKK 3.5 million), for the first quarter of 1993.[127] For its part, VSŽ helped to finance the HZDS's 1994 electoral campaign. As a reward, the enterprise obtained one ministerial post in the new government, which was allocated to Alexander Rezeš, VSŽ's vice president, who in December 1994 became the minister of transportation, postal services, and telecommunication in the new Mečiar government. During his ministerial tenure, Rezeš granted VSŽ a 13.5% to 20% reduction in transportation fees to be paid to the state-owned Slovak railways. This token of appreciation for VSŽ's contribution to HZDS's 1994 electoral campaign meant an annual subsidy of between $28.3 and $50 million.[128]

In addition to using VSŽ as a financial resource to fund political activities, HZDS also treated the enterprise as a prized source of support via party membership. At the end of 1995 and the beginning of 1996, one of the top managers engaged in outright recruitment of middle management to join the HZDS ranks. Middle management treated these "invitations" as offers it could not refuse.[129]

In July 1995, the FNM sold the remaining 15% stake in VSŽ to Ferrimex, at the same heavily discounted price. Ferrimex was a company set up by Hutník and ARDS. Hutník, a union-owned company that already had a 10% stake in VSŽ, owned a quarter of Ferrimex's shares. ARDS held the remaining 75% stake in Ferrimex and was owned by a group of VSŽ managers: Alexander Rezeš, (Ladislav) Drabik, and (Ján) Smerek, hence the acronym.[130] Ján Smerek's topping of the HZDS's electoral list during the 1998 elections highlighted the continuity of clientelistic relations between the Mečiar regime and VSŽ management.[131] Equally telling is the fact that the 1998 HZDS parliamentary elections campaign was managed by none other than Alexander Rezeš, who in April 1997 stepped down from his ministerial post and returned to serve as the chair of VSŽ's supervisory board.

In Slovakia, similarly to the Czech Republic, the government devolved the responsibility for restructuring to the managers while pursuing the project of domestic capitalism. In VSŽ, which dominated the steel sector and was responsible for nearly the entire steel production in Slovakia, domestic capitalism took on overtly clientelistic tones and became the symbol of the Mečiar regime.

Romania's Reform Pathway

In Romania, the initial low priority assigned to privatization, including in the steel sector, was a function of an ideologically cohesive communist successor coalition, which viewed privatization, especially of large industrial enterprises, with suspicion. Continued state ownership became a desirable alternative in the context of low levels of popular support for privatization, including to foreign owners.

Romania began the transition process with a set of legacies least conducive to democratic and economic reforms. In fact, many cite Romanian exceptionalism when discussing the country's post-1989 developments, especially the violent fall of the communist regime and the instability marring the outset of the transition process.[132] The popular uprising against Nicolae Ceaușescu, Romania's nationalist communist dictator, was exploited by his opponents in the Romanian Communist Party to gain power. The National Salvation Front (FSN), which emerged from the power vacuum following Ceaușescu's downfall, was led by Ion Iliescu, the former Romanian Communist Party Central Committee's secretary in charge of ideology, who fell from grace in 1970 after he opposed Ceaușescu's cultural policy.[133] Quite tellingly, at a time when communism was being denounced in other transition countries, Iliescu proclaimed, in his first postrevolution speech, that Ceaușescu betrayed the ideals of communism.[134]

Unlike the new governments in Poland and Czechoslovakia, which used the "time of extraordinary politics" to implement tough reforms, the Romanian leaders tried instead to appease the population, which had long suffered from privation under the Ceaușescu regime.[135] At the same time, the lack of organized anti-communist opposition provided little counterweight to the former communist apparatchiks, who acted with the support and to the benefit of the managerial and bureaucratic lobbies.[136]

The neoliberal faction was extremely weak in Romania, the nationalist and conservative factions being much more prominent among the opposition. Lest the general population start paying attention to small groups of disaffected students and intellectuals who protested against the FSN leadership, the transitional government increased consumption.[137] This was done by importing goods using an overvalued currency—financed through the depletion of foreign currency reserves—while displaced domestically produced goods accumulated as inventories.[138]

Facing elections in May 1990 against fledgling and splintered opposition parties that had adopted an anti-communist platform, the government did not even discuss radical reform measures. It pursued a course of gradualism instead, while conceding wage rises and maintaining price controls and an overvalued exchange rate. The strategy paid off: the FSN overwhelmingly won the May 1990 elections.[139]

Confronted with growing economic imbalance, in November 1990, the government, led by Petre Roman and representing a more reform-oriented wing of the FSN, carried out a partial liberalization of prices and devaluation of the lei. In early 1991, the government also introduced an IMF-supported stabilization plan, but the measures were unable to control inflation because of the unclear exchange rate policy and negative interest rates provided to select sectors.[140]

Romania's privatization law was also adopted in 1991 (Law No. 58/1991). The law's provisions reflected the apprehension with which the government viewed the privatization process as well as the influence of the more conservative faction of the FSN, which enjoyed the support of the bureaucratic and managerial interests. First, the effect of the law was limited. Based on the 1990 Law on State-Owned Enterprise Restructuring (Law No. 15/1990), the intended scope of privatization was limited to only 53% of total state-owned assets.[141] The 1990 law divided SOEs into two categories: those subject to privatization and the so-called *regies autonomes*, or "autonomous administrations," which were to remain state property. The latter were to be set up in strategic branches of the economy. By the end of 1992, a total of 950 *regies autonomes* had been set up, through both government decrees and decisions made at the local level.[142] Other than staying in state hands, the *regies autonomes* had the advantage of being shielded from future bankruptcy laws and thus enjoyed greater access to soft financing.[143]

Second, the complex institutional structure, with divided responsibilities, and relatively wide discretion concerning both restructuring and privatization opened the way for clientelism and capture by various bureaucratic and managerial factions. For example, the privatization law divided the responsibility for privatization among several entities: the National Agency for Privatization, the State Ownership Fund (Fondul Proprietăţii de Stat [FPS]), and five private ownership funds. The National Agency for Privatization was initially given the responsibility for the coordination, guidance, and control of the overall privatization process, until the establishment of the FPS and the private ownership funds. In 1992, however, the agency's responsibility was extended to include the supervision of small-scale privatization through its local branches, and it was later responsible for coordinating the mass privatization program. The private ownership funds were responsible for holding part of the assets of the enterprises included in the mass privatization program, and the FPS, established in 1992, was supposed to manage/restructure and privatize large SOEs.[144]

The FPS was to retain 70% of the enterprise capital, and the remaining 30% was to be divided among the five private ownership funds, which would serve as intermediaries in mass privatization.[145] Certificates of ownership were distributed to the adult population and could also be used to acquire enterprise property via MEBOs. Until the end of 1996, almost 29% of total privatizations were carried out via the latter method.[146] Mass privatization, however, was resisted by the management of the enterprises and by the red-brown coalition government under Nicolae Văcăroiu, which succeeded the Stolojan government that initiated the program.[147]

Third, the law greatly limited actual executive branch oversight and control over the privatization process. The first political battle within the FSN pertained to whether oversight would be delegated to the government or to parliament. With the more pro-reform faction dominating the government, the anti-reform faction in control of the parliament demanded that the FPS answer to the parliament. The anti-reform faction, benefiting from opposition support, prevailed.[148] Thus, the institutional structure that was to underpin privatization was complex and not conducive to rapid privatization. The process was left in the hands of the communist successor party coalitions, none of which embraced the privatization process, least of all, that of large industrial enterprises.

When the Democratic Convention government came to power in the aftermath of the November 1996 elections, it promised to break with the communist past and to accelerate reform. In 1997, the Ministry of Industry and Trade sent a note to the FPS and to the Ministry of Reform listing thirty key enterprises in Romanian industry that needed to be privatized rapidly. These included three steelworks: Sidex, Siderca Călărași, and COS Târgoviște.[149]

Privatization's prominent ranking on the government's official priority list was connected to the IMF and World Bank conditionality. Under the influence of the 1997 IMF standby agreement and in conjunction with a Financial and Enterprise Sector Adjustment Loan (FESAL) from the World Bank, the government's 1998 privatization plans were very ambitious; they entailed privatizing an unrealistic number, 2,745, of enterprises, of which 333 were to be big, 931 medium, and the remaining 1,481 small.[150] Prodded by the IFIs, the new government engaged in macroeconomic reform measures but made little headway with enterprise reform.[151] All in all, the center-right stint in power until the fall 2000 elections featured fractious, unstable coalitions that fell far short of societal expectations.

The steel sector exemplified resistance to privatization and the benefits accruing to the governing coalition from maintaining state ownership. The initial sectoral consolidation recommendations of foreign experts did not face ideological opposition by the governing party, but, as in the other countries examined in this chapter, the recommendations ran into strong opposition from the managerial strata. Consequently, the resulting documents nominally invoked coordinated restructuring in the sector. An official document, drafted by the Ministry of Industry and Trade in November 1993, based on the Sofres Conseil, Sofresid, and Cedres recommendations, paid lip service to "holding-type" strategic companies, Siderom and Romtub, ostensibly created in 1991. These were supposed to implement in the constituent enterprises the overarching restructuring strategy for the sector, to be devised and overseen by the Ministry of Industry and Trade.[152] In reality, however, these strategic companies lacked both power and financial resources. The idea of creating holding companies, initially one, then two such companies—one for flat products and the other for long products—was revisited in a proposal floated in 1995–1996. It, too, was met with outright managerial opposition.[153]

Managerial tactics were condoned and supported by the local politicians, who became the managers' close allies and business partners. According to interviews with union leaders as well as with a representative of the managerial strata, local politicians created a unified front with the managers in resisting any consolidation proposals. According to Petru Dandea, vice president of the Cartel Alfa trade union confederation, which represents the steel sector, it was practically impossible to have the managers of individual companies, each of whom was interested in preserving his own company, negotiate a holding. Rather, such a structure would require an administrative decision by the government, which, in turn, required political will. The problem was that each of the enterprise directors had local political support and backing. As a result, if the central government raised the issue of a holding, it would lead to a scandal and political rallying by supporters of the individual directors in an effort to protect the enterprises in question.[154]

This view was seconded by an interviewee from UniRomSider (Association of Steel Producers of Romania), who also referred to an alliance between the local politicians and the managers in opposing consolidation, when he said that there was

an adverse reaction from the local authorities in the counties. Senators and representatives from these counties blocked this [holding proposal] because this holding would be, somehow, an umbrella at the national level, with no power left to the local authorities; and there was a cry in parliament: "In my works, nobody will order [me around]!"[155]

It was not simply a question of politicians supporting enterprises in their local districts. Rather, politicians formed close alliances with the local managers and engaged in rent seeking at the expense of the SOEs. Rent seeking took the form of barter-based intermediary businesses that would transact with the SOEs by selling production inputs at marked-up prices and purchasing heavily discounted products that they would then resell at a marked-up price. These business intermediaries very frequently directly benefited political figures as well as their respective political parties. The parasitic relationship of the intermediaries with the SOEs received an evocative label of "tick companies," or *firme-căpușă*, and the phenomenon became a household name, frequently invoked by the interviewees. As one deputy defined it during a parliamentary debate, a "*firma căpușă* is nothing else than a[n] [enterprise] director's friend or an order given to the director."[156] Another deputy accusingly said to his communist successor Party of Social Democracy in Romania (Partidul Democrației Sociale din România [PDSR]) opponents: "How could you forget about the 'tick' companies, which brought the Romanian economy to bankruptcy and to which you are no strangers."[157] Shortly before the 2000 elections, the Romanian prime minister, Mugur Isărescu, referred to "political forces blocking the privatization process," and recognized that there are "heaps of tick companies benefiting from rents which they obtained or continue to obtain presently from state owned enterprises."[158]

The case of the steel sector shows that the status quo generated numerous benefits for the parties in power—no matter the political provenance—and became very difficult to change. Despite running on a reformist, anticommunist platform, the Democratic Convention government that came to power in November 1996 failed to deliver on its electoral promises of rapid reform. Rather, the politicians connected to the new governing coalition also became implicated in rampant rent seeking.[159]

Nonetheless, as laid out in the model of political economy of reform, the Democratic Convention government tried to optimize its reform plans by reconciling policy preferences for faster privatization with the benefits to be reaped from sustaining the status quo. The IFI-inspired, ambitious 1998 privatization plans included Sidex, COS Târgoviște, Siderurgica Hunedoara, Petrotub Roman, Industria Sârmei Câmpia Turzii, and CS Reșița.[160]

However, of these, the government left the privatization of the sector's behemoths for later and only made efforts to sell Petrotub Roman and CS Reşiţa, along with several other smaller—and less politically salient— producers in the sector. With two exceptions, both small pipe producers, these privatizations had disastrous effects, in large part due to the poor selection of investors, inadequate preparation of the privatization process, and unclear privatization contracts.

Conclusion

This chapter has shown that, faced with sectors in dire need of thorough restructuring, the governments of the four countries pursued incomplete reform that best suited their political needs, whether continued state ownership or privatization to domestic owners with questionable credentials. At the same time, these policy choices were part of the broader reform trajectory of the countries. In Poland, ideologically diverse coalitions that disagreed on the method and extent of privatization led, by inaction, to the steel sector remaining in state hands until foreign investors could be found at an unspecified future date. In Romania in the 1990s, state ownership of the sector reflected weak support for privatization by the successive coalitions, not to mention the powerful managerial lobby. By contrast, the governments of the Czech Republic and Slovakia supported privatization, albeit one that would lead to the rise of a domestic capitalist class. These initial policy choices created payoffs that made these policies self-sustaining, even in the face of changes in the governing coalitions.

While choosing these policies, the successive governments also rejected the idea of far-reaching intervention and redeployment of assets in the sector, and they devolved the responsibility for restructuring to the management of the enterprises in question. The next chapter examines enterprise-level restructuring as it discusses managerial attitudes to restructuring and the ways in which the state apparatus implemented the initial policies that were the subject of this chapter.

FOUR

Restructuring Behemoths

State Capacity and the Enterprise Level in Comparative Perspective

Each one [manager] was his own lord, on his own property.
  —Representative at the Polish Metallurgical
  Chamber of Industry and Commerce

"Better to be the first in Ostrava or Třinec than to be the third in Prague"—that's clear—an old Roman saying.
  —Civil servant at the Czech Ministry of Industry and Trade

They [managers] preferred to be the first violin in an orchestra playing really badly instead of being second violin in an orchestra, which is performing excellently.
  —Expert at the Organization of Romanian Steel Producers

The previous chapter discussed the factors that led the governments of the Czech Republic, Poland, Romania, and Slovakia to select (and maintain) an initial policy of either continued state ownership or privatization to domestic owners. It also showed that, in the absence of a coordinated sectoral approach, successive governments delegated the restructuring task to the managers of the individual enterprises. This chapter turns to policy implementation at the enterprise level; it focuses on the interactions of enterprise managers with the state apparatus.

Policy implementation requires the exercise of state capacity, and varying levels of state capacity lead to restructuring or privatization outcomes that produce different balance between efficiency and rent seeking (see figure 3). In other words, state capacity determines how initial policies are implemented, the payoffs they create for government and sectoral actors, and the sustainability of these policies. This chapter shows that restructuring outcomes in the region depended on state capacity rather than the ownership structure—whether state owned or private—by highlighting the contrasts between Poland and Romania, where state ownership continued, and between the Czech Republic and Slovakia, where privatization to domestic owners dominated. These contrasts underscore that privatization alone is no panacea.

Rather, at the enterprise level, restructuring outcomes turned on the ability of the state institutions to impose financial discipline on the enterprises, and to provide market-based institutional support for restructuring. Inability to rein in the unrealistic and inefficient ambitions of enterprise managers and to constrain rent seeking led to substantial financial ramifications for taxpayers in the countries with lower capacity. In the two states with higher state capacity, the Czech Republic and Poland, state aid per ton of steel produced in 1993–2004 was equal to $5.4 and $5.8, respectively. In Slovakia, which had medium capacity, state aid was $9.1 per ton of steel. With a price tag of $21.1 per ton of steel, restructuring cost taxpayers the most in Romania, the country with the lowest capacity. As chapter 6 will illustrate, the varying cost of restructuring, driven by the differences in state capacity, explains why initial policy choices persisted longer in the Czech Republic and Poland.

This chapter begins with a review of the theory of managerial prestige maximizing and then turns to the four country cases to flesh out the theory empirically. After exploring the vagaries of managerial behavior, the chapter examines how variations in the levels of state capacity affected restructuring. It does so by examining the degree of financial discipline imposed on enterprises (the type of budget constraint and credibility of the threat of bankruptcy), the types of institutional assistance in the restructuring process (debt workouts, market-based vehicles of support, and reorganization of assets), and the vetting of buyers during the privatization process.

Managerial Prestige Maximizing

The de facto delegation of the restructuring process to the managers, discussed in chapter 3, is precisely what the network perspective advocates.

After all, managerial resistance to reform blueprints drawn up by foreign consultancies is understandable, because the managers are the ones with the local knowledge. In other words, managers, knowing the strengths of their enterprises, will be best able to lead the restructuring process. As Stark and Bruszt note, "Where . . . [associative networks] are recognized, they can promote productive restructuring and provide a vital source of intelligence about real assets and liabilities, about mobilizable ties among organizations, and about the possible synergies of their locally coordinated projects."[1]

In practice, however, rather than use their knowledge to engage in a local coordination of projects, irrespective of the country, the managers pursued prestige-maximizing strategies aimed at preserving their power and expanding the production capacities and technology within their individual enterprises. Recall the four propositions presented in chapter 2 regarding managerial approach to restructuring:

P1: Managers resist "defensive" restructuring measures, such as layoffs and capacity closures, unless a crisis situation develops within the enterprise.

P2: Managers attempt to engage in technological restructuring; that is, they seek opportunities to upgrade and technologically enhance the capital stock of their enterprises, irrespective of market capacity.

P3: Managers eschew cooperation with the managers of other enterprises and resist sectoral consolidation measures.

P4: Managers demand protection by the state, in the form of bailouts and financial support.

Consequently, the managers resisted consolidation and coordination because it would have both diminished their individual power and prevented them from implementing their expansionary investment projects. As the subsequent sections demonstrate, this behavior occurred in all the countries examined, although in the case of Slovakia, due to the structural features of VSŽ's near monopoly, it took on the characteristics of empire building rather than refusal to cooperate. The empirical discussion starts with Romania and proceeds in ascending order of state capacity.

Managerial Prestige Maximizing in Romania

Interviewees representing a variety of perspectives, including those of managers, the trade unions, and the state authorities, pointed out that managers desired to maintain their production capacity, obtained government funds for investment, and refused to cooperate with each other to devise a common strategy for restructuring the sector. Managerial solutions to the challenge of restructuring did not entail sectorwide solutions; rather, they were myopic approaches focused on individual enterprises and aimed at prestige maximization. Consequently, the managers resisted any merger or holding proposals. According to a representative of the Ministry of Industry and Trade responsible for the steel sector, "The managers considered that it was an imposition, something imposed from the outside, imposed by the government, and they did not accept this type of association, considering that they would be stripped of their authority."[2]

The position of head manager of the enterprise was endowed with great prestige. As a journalist covering the steel sector at a major daily commented, "In Galaţi, the [head] manager of Sidex was even more famous than the president of Romania. . . . So if you are the [head] manager of a company, you are the king of the city."[3]

An interviewee from UniRomSider, the Association of Steel Producers of Romania, who represents the managerial strata, stated that managers

> preferred to be the first violin in an orchestra playing really badly instead of being second violin in an orchestra that is performing excellently because only one of them would be the general manager of the holding. So they preferred to be the captain of the sinking vessel instead of being the second one.[4]

The managerial desire to retain sole responsibility for the enterprise was echoed by the leader of the sectoral trade-union federation Metarom:

> It was a question of mentality . . . they [the managers] had a feudal mentality. . . . The feudal economy was based on self-sufficiency. And they were looking for the same thing: they wanted to develop their companies so that they would become independent from the others. . . . They were *masters, landlords.* So they did not want to collaborate with others, as the idea of a holding would imply.[5]

The managerial goal of maintaining production capacities—and lobbying for state support to do so—rather than of coordinating capacity closures to preserve the most efficient units under changed economic circumstances, was reflected in a PHARE-financed study of the Romanian steel sector in 2000. The study pointed out that the enterprises did not specialize and that there had been "an effort to produce in each factory the entire possible spectrum of steel products."[6] Moreover, in line with proposition 2, which posits that the managers are reluctant to engage in defensive restructuring, idle, obsolete assets were not destroyed, which led to very low utilization of enterprises' production capacity.[7] By 2003, when a significant proportion of Romanian industry had been privatized, capacity utilization of steel producers was still 62%, compared with the 81% EU-15 average.[8]

The managers also used local political connections to seek financial support for their enterprises. In the case of the Călăraşi-based enterprise Siderca, managerial pressure led to investments that were not supposed to be made. The director relied on a close relationship with deputies in the ruling party to obtain government funding to help finish a rolling mill for rails. Yet, instead of finishing the mill, he diverted money meant for the construction of a modern electric-furnace-based steel shop into the development of his own steel production capacity. As originally agreed, however, instead of carrying out its own production, Siderca was supposed to rely on steel from the Sidex Galaţi plant for further processing. Here, too, prestige maximization dominated, as the head manager never accepted that he would have to be dependent for his steel supply on the head manager of Sidex.[9] As a result of financial inability to finish either investment, Siderca experienced grave financial problems, and existing production ground to a halt at the end of the 1990s.[10]

Managerial Prestige Maximizing in Slovakia

Managerial prestige maximizing in Slovakia took the form of empire building rather than mutual competition. This resulted from VSŽ's near monopoly of Slovak steel production. VSŽ managers had an appetite not only for being the symbol of Slovak national capitalism but also for becoming a transnational enterprise with a diversified portfolio. Concerning steel production, in 1996 VSŽ purchased a 20% share in the Czech Třinec Steelworks, which it sold a year later. In January of 1998, despite an increasingly difficult financial situation, VSŽ purchased Diósgyör Steelworks, a financially troubled Hungarian steelmaker, which later went bankrupt.[11] Moreover, VSŽ entered into joint ventures with the Finnish Rautaruukki Oyj (Ranilla Košice, s.r.o.) in 1996 and with U.S. Steel in 1997 (VSŽ U.S. Steel).[12]

VSŽ's other purchases were far more conspicuous and ranged from banks and insurance companies to newspapers and sports clubs. In 1996, VSŽ held majority shares in eighty-six enterprises and minority (20% to

50%) shares in another thirty-five companies, including some investment funds and banks.[13] For example, VSŽ directly and indirectly held a 40% share of IRB, the third-largest Slovak bank. In 1998, VSŽ had shares in 102 enterprises in Slovakia and sixteen abroad.[14] Some of its more famous transactions included the purchase of the most prestigious Czech soccer club, AC Sparta Praha, as well as its Slovak equivalent, 1. FC Košice. VSŽ also owned different forms of media, including *Národná obroda*, a significant daily with nationwide circulation.[15]

Managerial Prestige Maximizing in Poland

The paired comparison of Poland and the Czech Republic provides the best test of the managerial prestige-maximizing theory. Unlike in Czechoslovakia, where the sector was bound by the comparatively rigid central-planning structures of the sectoral organization—Hutnictví Železa—until 1988, Polish steelworks had enjoyed a relatively high degree of autonomy since 1981.[16] Based on the 1981 Law on State Enterprises, Polish SOEs operated based on the "Three S" principle: self-sufficiency, self-governance, and self-financing. Consequently, in 1981, the Iron and Steel United (Zjednoczenie Hutnictwa Żelaza i Stali), the sectoral organization that coordinated the delivery of semifinished products between the steel mills for further processing, allocated raw materials, set prices, coordinated sales, and represented the mills before the relevant minister, was dismantled in favor of the toothless Iron and Steel Association (Zrzeszenie Hutnictwa Żelaza i Stali).[17] Mills became autonomous, and inter-mill cooperation greatly subsided.[18] Subsequently, when the sector entered the transition period, the enterprises had already been the locus of restructuring.

During the transition, the managers complained about the lack of sectoral consolidation and mutual cooperation, but they also resisted any capacity cuts that the process would entail.[19] By 1995, the chairman of the Metallurgical Chamber of Industry and Commerce, the steel producers' association, recognized that "restructuring has failed because . . . each director tries to introduce his own transformation programme."[20] In the words of other industry insiders, including representatives of the managerial circles, the "desire for continued independent existence" on the part of individual managers stood in the way of consolidation.[21] As one interviewee from the Metallurgical Chamber of Industry and Commerce quipped, "Each one [manager] was his own lord, on his own property."[22]

An integral part of the prestige-maximizing strategy was not only independent existence but also maintenance of production capacity and technological upgrading. According to Emil Wąsacz—the former minister of state treasury and Solidarity trade union leader and, later, director at Katowice Steelworks—in order to obtain permission for investment, the managers relied on false diagnostic studies in order to convince state officials to support their particularistic goals.[23] Indeed, these investments then rendered the suggested mergers less effective and also had detrimental consequences for other mills in the sector.

In the case of Katowice and Sendzimir Steelworks, the initial merger recommendations encountered strong resistance from the managers. Despite consultations with the management of both mills, neither signed on to a merger plan devised by the Arthur Andersen consultancy, and the discussions moved toward "greater cooperation" between the two mills, which never materialized.[24]

Managerial resistance to the merger also had significant consequences for the eventual privatization of these enterprises, for two reasons. First, potential investors interested in Sendzimir wanted a guarantee that Katowice would provide the necessary crude steel for further processing, which Sendzimir produced in inadequate amounts. Second, given the size of the Polish market, the investors wanted to be informed about Katowice's production and development plans, as these could impact their business strategy for Sendzimir.[25] With the responsibility for privatization negotiations devolved to the management under the Solidarity Electoral Action-Freedom Union coalition, the management kept in mind the interests of the company, rather than the sector as a whole.

The case in point was Austrian Voest-Alpine's attempt to invest in Sendzimir Steelworks in 1998. Although Voest-Alpine and the Dutch Hoogovens were interested in investing in both Sendzimir and Katowice Steelworks, British Steel was only interested in purchasing Katowice. Yet, the president of Katowice Steelworks, having authorization from the minister of state treasury to lead privatization negotiations, dismissed Voest-Alpine/Hoogovens's offer in favor of negotiations with British Steel, claiming that its offer was better for Katowice Steelworks: "The Austrian-Dutch proposal actually intended to consolidate the two enterprises but, from Katowice Steelworks' point of view, was too weak."[26] Despite experts' warnings that different investors would hurt privatization prospects, and despite the Austrian and Dutch bidders' warnings that they would pull out of Sendzimir privatization without the prospect of also buying Katowice, the Ministry of Treasury followed the managerial lead and pursued separate privatization routes for the two enterprises.[27] In the end, both privatizations failed; uncertainty concerning the government strategy toward Sendzimir was cited as one of the main reasons.[28]

Another important case of resistance to mergers was that of Częstochowa Steelworks and Zawiercie Steelworks, which set off a chain reaction in the sector. According to Wąsacz, the director of Częstochowa Steelworks used his political influence and trade union ambitions to build his own crude steel production facility, which already existed in Zawiercie Steelworks, about 40 kilometers away. Zawiercie Steelworks, now having excess crude steel production, installed a rolling mill for long products, which were, in turn, supposed to be produced by Ostrowiec Steelworks. Because Zawiercie beat Ostrowiec to restructuring, Ostrowiec was left in deep financial trouble.[29]

Indeed, from the beginning of transition, the head manager at Częstochowa Steelworks had been pushing through an ambitious investment program he had drawn up in the mid-1980s. While resisting any suggestions of a merger with Zawiercie, he spent $350 million on state-of-the art equipment and renovations in the 1990s, in both the primary steel and finished steel sections.[30] The investment program created high indebtedness, with the 1997 debts reaching $193.3 million (PLN 634 million). By 2002, the mill was caught in a debt trap, with a debt of $274 million (PLN 1.2 billion) surpassing its equity of $225 million (PLN 920 million). At that point, as the most modern steelworks in Poland, the enterprise stood at the brink of bankruptcy and was widely cited as an exemplar of "overinvestment."[31]

For its part, Zawiercie Steelworks—a semifinished goods producer with about a one-million-ton annual production and 7,000 workers— also resisted efforts to reduce its capacity and, from the very beginning of the transition process, insisted on creating a full production cycle.[32] In any case, the investments made in Częstochowa Steelworks, which made it self-reliant, necessitated an independent strategy by Zawiercie Steelworks. However, unlike Częstochowa's management, the management of Zawiercie was more frugal and devised a realistic restructuring program that acknowledged its own financial limitations.[33] The $13.5 million investment in machinery was financed by Impexmetal, the state-owned former international trade company, which in 1996 became Zawiercie's biggest shareholder by participating in a debt-equity swap organized by the Enterprise and Bank Restructuring Act. Zawiercie prospered relatively well under Impexmetal's ownership: by the end of 1998, it had paid back its outstanding debt and covered the remaining costs of the $50 million investments.[34] By 2003, it was considered one of the most successful steelworks in Poland.[35]

Ostrowiec Steelworks did not fare as well as Zawiercie, and its problems were largely related to the unrealistic expectations and investment projects. Like Częstochowa, Ostrowiec Steelworks was supposed to undergo capacity cuts and to close an antiquated steel shop, which nonetheless was kept in operation.[36] Ostrowiec was never able to recover from its financial difficulties during the early stage of transition, and, at the end of 1993, it entered into a bank conciliation procedure with its creditors. In 1994, it became nominally private, after being sold—through a debt-equity swap—to Stalexport S.A., a former foreign trade company controlled by the State Treasury.[37]

The new owner, wanting to turn Ostrowiec into a leading enterprise in its holding, engaged in profligate spending on the new property: between 1995 and 2001, Stalexport invested the hefty sum of $140.6 million (PLN 575 million), including bank loan guarantees.[38] By 2001, the enterprise had lost liquidity and was insufficiently using its production capacity—only 68% of the continuous casting production capacity in 1997, and 31% by 2001.[39] At that point, Stalexport decided to sell its shares in Ostrowiec and stop financing the steelworks.[40]

The case of Baildon Steelworks and its rolling mill investment provides yet another example of managerial prestige maximizing. The project was designed by domestic experts who were deeply influenced by the managerial vision. Based on the those plans, the state extended guarantees for bank loans that were used to finance the investment.[41] Whereas modernizing the plant had been proposed in the original sectoral study, the $54 million (PLN 220 million) investment envisaged by the managers not only went beyond the steelworks' financial means but also created an unsustainable production capacity.[42] In 1999 and 2000, production was equal to 14.8% and 13.5% of planned quantities, respectively; the continuous casting line was operating at 3% of its capacity, and the state-of-the-art rolling mill at 8.8%, in 2000.[43] Heavily indebted and unable to pay its social-security contributions, the mill declared bankruptcy in May 2001, after the state demanded that it either find external investors or file for bankruptcy.[44]

Thus, in the case of Poland, steelworks managers continued their individual restructuring programs, aimed at technical modernization of their plants. Yet, knowing the specificity of their industry and the attributes of the individual plants, they did not try to coordinate their strategies with those of the other managers. Rather, they attempted to compete with them, which had detrimental effect on the sector as a whole and on the individual plants in the medium to long run.

Managerial Prestige Maximizing in the Czech Republic

In contrast to the Polish steel sector, which was decentralized and fragmented at the beginning of the transition, the sector's circumstances in the Czech Republic were far more conducive to initial managerial cooperation. Unlike their Polish counterparts, whose enterprises had been self-managed and self-financed since 1981, the Czech managers belonged to the sectoral central-planning organization that coordinated production, Hutnictví Železa, until 1988. It was then dismantled as part of the perestroika-style reforms.[45] In 1990, the steelworks directors set up a new sectoral organization, bearing a strikingly similar name—Odvětvový Svaz Hutnictví Železa (Iron and Steel Federation). In 1992, the new organization created a joint-stock company, Hutnictví Železa, a.s. (Steel Federation, Inc.), and started to perform market-research activities; it also played a dual role as the employers' association in the sector.[46]

However, the managers' mutual interactions, facilitated by the sectoral organization, did not result in cooperation to foster restructuring. On the contrary, industry insiders from Steel Federation, Inc., highlighted the appetite for independence whetted by the first decentralization reforms of the late 1980s. As an interviewee from the organization's leadership summed it up, "The atmosphere was 'each goes his own way.'"[47] Similarly to managers in Poland and Romania, the Czech steel-sector managers preferred pursuing independent restructuring paths to recombining assets by deploying the network resources they had at their disposal. In fact, the initial recommendation to consolidate the Czech steel industry, devised by consultants at Roland Berger, was met with a sense of bafflement by Steel Federation. The organization's publication noted, "A rather opposite process is currently taking place in the Czechoslovak steel industry, which is the effect of an understandable reaction to the former, strict, central planning."[48]

However, what could be seen as a reaction to the former central planning quickly took on prestige-maximizing overtones. As a civil servant in the Ministry of Industry and Trade quipped, "'Better to be the first in Ostrava or Třinec than to be the third in Prague'—that's clear—an old Roman saying."[49] Indeed, the managers drew up investment wish lists and attempted to undertake similar investments.[50] According to a member of the top management at one of the North Moravia steel enterprises:

If only restructuring were carried out wisely at the beginning of the 1990s, instead of politicking. But each manager, management, looked at what could be gotten out of steelmaking. Who is better, Nová Huť or Třinec? . . . It is a pity that the managers did not agree among themselves. All are equally guilty. At that point, they could have reached an agreement, but after 1990 . . . one wanted this, another wanted this, and yet another also wanted this [referring to their desire to undertake similar investments].[51]

A case in point was the so-called minimill construction in the Ostrava region, for which Nová Huť and Třinecké Steelworks submitted competing proposals.[52] The scope of the investment necessitated state guarantees to obtain loans, and the government stressed that market capacity could only absorb the output of one minimill.[53] In the end, the government decided to assist Nová Huť by arranging a loan from the IFC. Disappointed by the government's decision, Třinecké Steelworks claimed that its survival depended on building its own minimill and insisted it could still find the means to obtain bank loans for a similar investment.[54]

Besides competing for new investment, the managers also resisted closing obsolete facilities and rearranging assets, as evidenced by the case of pig iron production in Vítkovice. After 1989, the local government demanded that Vítkovice's obsolete pig iron production facility be shut down because of its location in the city center and the resulting environmental liabilities and nuisance to inhabitants. The Ministry of Industry and Trade also pushed for Nová Huť's specialization in pig iron production. Nová Huť not only opposed a merger with Vítkovice but also resisted a joint venture, and preferred instead to supply Vítkovice with its own pig iron on a contractual basis.[55] For its part, Vítkovice was reluctant to shut down its facilities. Eventually, and only as a result of a government decision, on January 1, 1997, blast furnaces and thus pig iron production were separated out of Nová Huť and turned into a separate enterprise—Vysoké Pece Ostrava—with the intent that it would supply pig iron to both Nová Huť and Vítkovice.[56]

Not surprisingly, the big steelworks also resisted any overt merger ideas. In January 1994, Třinecké Steelworks and Nová Huť both opposed merger suggestions being floated around the ministries. Třinecké was particularly concerned about Nová Huť's lobbying power and claimed it would lose out on a merger deal.[57] After Nová Huť obtained the minimill loan promise, it unsuccessfully suggested cooperation, and even merger, with Třinecké, which was to continue and expand its production of long, rather than flat, products.[58]

As the financial pressure on the sector grew in 1998–99, so did the pressure for greater cooperation among the Ostrava region's enterprises. All along, the managers had realized that mutual cooperation was optimal given the circumstances: by taking advantage of synergy effects, they would be able to generate savings, and any further investments could be planned so as to prevent technological redundancies in the small Czech market. In 1999, the managers of the biggest three steel mills in the North Moravia region agreed to form a steel consortium, but the idea never materialized.[59]

Although the differences in the ownership structure at that point in time would have complicated consolidation, the interviewed managers who were close to the consortium project pointed squarely at managerial ambition as the reason for the project's failure:

> Everybody was too big of a manager, and for Vítkovice it was impossible to cooperate with Nova Huť, [for] Nova Huť it was impossible to cooperate with Třinec and [for] Třinec [it] was impossible to cooperate with Vítkovice. They signed some document, but the result was zero!!! Only some piece of paper . . .[60]

According to local press reports, even though all managers nominally supported the idea of consolidation, each had different priorities for the way it should be carried out.[61]

The Czech case shows that, as in Poland, the networks of personal relations existing among enterprise managers did not facilitate the development of a sectoral restructuring strategy or prevent prestige-maximizing behavior on the part of the managers. As a consultant from the Roland Berger Strategy Consultants, which helped the Czech government draft the eventual sectoral restructuring program for negotiations with the EU, commented:

> On informal basis, I think the cooperation [among the managers] was good, they knew each other, they were from the same region. But on the formal basis, when they went into negotiations and met at a table, in terms of business decisions, they were not very open, they were very individualistic and that was discouraging for us.[62]

The discussion of the Czech, Polish, and Romanian cases reveals that in all three countries, despite the differences in the type of communist regime, managerial behavior was very similar: the managers resisted consolidation, engaged in investment competition, and aimed at not just preserving but also expanding their existing capacities. This behavior transpired while the managers were publicly recognizing the advantages of greater cooperation during the restructuring process. Given managerial resistance to the rationalization of production, the problem was not that the managerial networks were endowed with *insufficient* power. If anything, the oversight of managerial activities was too lax after the governments de facto devolved the responsibility for restructuring to the company managements.

Given that the managers strived to maximize their own and their enterprise's prestige—understood in terms of size, not the balance sheet—rather than look at the interests of the sector as a whole, the restructuring outcomes depended on the level of state capacity to constrain managerial behavior.

State Capacity and Restructuring

State capacity affected restructuring through two mechanisms: the state's capacity to impose financial discipline on the enterprises and to provide market-based institutional support for restructuring. Specifically, *financial discipline* is understood as the ability to impose hard budget constraints and create a credible threat of bankruptcy. *Market-based institutional support for restructuring* is understood as the instruments of intervention available to state actors that can promote the restructuring of failing enterprises but that can also act as a lifeline to prop them up. These instruments include debt workouts, market-based vehicles of support, and reorganization of assets. Table 5 summarizes the flip side of the state capacity coin and notes financial disciplinary measures taken and examples of institutional assistance in the four countries. As the ensuing discussion illustrates, as state capacity increased, budget constraints became harder, but the forms of institutional assistance by the state also became more sophisticated. The discussion once again starts with Romania, where state capacity was the lowest.

Low State Capacity and Restructuring in Romania

In the case of Romania, poor state capacity manifested itself through a soft budget constraint and lack of a credible threat of bankruptcy, combined with poor oversight of managerial activities. At the same time, the state lacked market-based vehicles of intervention to support and restructure the enterprises.

Not only did the Romanian government devolve restructuring responsibility to the managers of the individual enterprises, but the Romanian

TABLE 5. Flip Side of the State Capacity Coin: Financial Disciplinary Measures and Institutional Assistance

Country	Overall Relative State Capacity	State Capacity I: Financial Disciplinary Measures			State Capacity II: Institutional Assistance	
		Budget Constraint	Credible Threat of Bankruptcy	Debt Workouts	Market-Based Vehicles of Support	Reorganization of Assets
Romania	**Low**	**Soft** Tolerance of large tax arrears and social security contributions Subsidized loans	**Low**	No formal debt workouts; informal mediation of negotiations with state-owned utilities providers	None noted	None noted
Slovakia	**Medium**	**Medium soft** Tolerance of tax arrears Rail transport discount	**Initially low, high** After cross-default declared at VSŽ by foreign lenders	No formal debt workouts	None noted	None noted
Poland	**High**	**Relatively hard** Low incidence of arrears in tax or social security contributions Very few government loan guarantees, but wide latitude to take out loans from state-owned banks	**Medium-high** Bankruptcy and liquidation of Baildon Steelworks Several threats of bankruptcy followed by debt workouts and restructuring campaigns at Sendzimir and Katowice Steelworks	Formal debt workouts at Sendzimir Steelworks, Katowice Steelworks, and Częstochowa Steelworks Debt-equity swaps at Zawiercie Steelworks and Ostrowiec Steelworks	Silesia Finance Society used to finance production at Katowice Steelworks and Częstochowa Steelworks	Reorganization of Warsaw Steelworks assets Reorganization of Częstochowa Steelworks assets Sectoral consolidation as Polish Steelworks
Czech Republic	**High**	**Relatively hard** No noted arrears of tax or social security contributions Few government loan guarantees, but wide latitude to take out loans from state-owned banks	**Medium-high** Bankruptcy of Poldi Steelworks Debt workout and restructuring at Vítkovice Steelworks	Formal debt workout at Vítkovice	Osinek used to finance production at Vítkovice; later purchased Vítkovice Steel	Separation of Vysoké Pece Ostrava from Nová Huť Separation of Vítkovice Steel from Vítkovice

state apparatus then failed to discipline the enterprises financially. An interviewee from UniRomSider summarized the activity of the Romanian state as follows: "No economic control almost at all, so no alert when losses overpass half of the turnover; no involvement in changing of management . . . because of bad results . . . Very low profile of the owner."[63]

The case in point is Sidex. Even though the State Ownership Fund (Fondul Proprietății de Stat [FPS]), responsible for the privatization process, announced that it wanted to privatize Sidex, the enterprise's management proclaimed, in 1998, that any discussions of privatization would only take place after "successful restructuring."[64] The management's vision of the role of future investors was unlikely to entice any:

> We are thinking about privatization only after we will become very attractive. In our opinion, it would be good if more investors would buy [a] smaller number of shares. We do not approve of a single investor buying an important packet of shares.[65]

The announced program was to entail employment restructuring and technological modernization at costs running at no less than $1 billion.[66] This massive restructuring effort was to be undertaken by a private company, Siderman, headed by Sidex's manager, to whom the management of Sidex was contracted out between 1994 and 1999. The ambitious restructuring plans were being unveiled, when only a year earlier Sidex was placed on the Ministry of Industry and Trade's list of enterprises in danger of bankruptcy and despite the FPS leadership being "totally dissatisfied" with Sidex's management.[67]

It was only in 1999 that, faced with Sidex's ever-deteriorating financial situation, the FPS decided to cancel the management contract with Siderman, citing lack of fulfillment of contractual obligations.[68] One of the criticisms of Siderman pertained to commissions paid to a joint venture with Balli International set up by Sidex—called Sidex International—for the distribution of Sidex products.[69] Balli International also acted as an intermediary by repaying a $100 million short-term syndicated loan Sidex had taken out from a consortium of foreign banks. In order to repay the loan, Sidex signed a disadvantageous contract to provide steel products to Sidex International, while Balli International sold Sidex's overpriced production inputs.[70]

However, Siderman's departure from Sidex did not do much to improve the situation within the enterprise, and there were as many as seven management reshuffles in 1999. The subsequent management continued to engage in rent seeking, and two managers who were in charge of Sidex in 2000 were later sentenced to ten-year jail terms for abuse of power and accepting bribes.[71]

It is difficult to estimate the scope of the financial losses due to the activities of *firme-căpuşă*, the asset-stripping "tick firms" like Balli International. Shortly before the privatization of Sidex in 2001, Romania's prime minister estimated the number of *firme-căpuşă* in Sidex alone at around 1,400.[72] The size of rents was significant. A former minister of privatization, Ovidiu Muşetescu, stated in 2001 that although the transport of iron ore from Brazil to the Romanian port of Constanţa cost about $3 per ton, the companies that transported iron ore from Constanţa to Galaţi would charge $18 per ton.[73]

Sidex illustrates an important consideration in the political economy of reform, discussed in chapter 3: the status quo generated numerous benefits for the parties in power, no matter the political provenance, and became very difficult to change. Thus, Siderman was run by a manager who became a Social Democracy of Romania (Partidul Democraţiei Sociale din Româ-nia [PDSR]) deputy during the 1996–2000 and 2000–2004 terms.[74] However, it was under the Democratic Convention government that came to power in November 1996 on a reformist, anti-communist platform that Balli International obtained monopoly over iron ore trade with Romania. At the time, Balli International bought the only Black Sea terminal in the country that took in metal shipments, Comvex, located in Constanţa, via a dubious intermediary. Illustrating the tightly woven political-economic nexus, the privatization minister, who resigned in 1998, had close personal ties to Balli International.[75]

The low capacity of the Romanian state failed to prevent financial abuse within the enterprises. It also failed to discipline them financially by harnessing the nascent market forces to foster restructuring through debt workouts. Rather, the Romanian state kept the enterprises afloat without accompanying restructuring measures. The Romanian treasury, facing serious fiscal constraints, did not disburse significant amounts in the form of direct subsidies. In fact, in the case of the seven integrated steel producers in Romania, direct grants totaled $15.6 million and were only given to three enterprises over the years 1993–2002. However, the state granted $49.6 million in long-term, interest-free loans to three enterprises, as well as $74.9 million in interest-free loans and $112.3 million in loan guarantees. According to interviews with the steel-sector privatization adviser to the minister of privatization and an analyst at the consultancy that helped prepare the government's 2004 restructuring strategy, these loans were not paid back.[76]

The most powerful means of assistance, and the epitome of both soft budget constraint and low state capacity, was the condoning of severe debt accumulation vis-à-vis the state, largely through arrears in taxes and in social security fund contributions. Although aid to the seven integrated plants from 1993 to 2000 totaled about $275 million, the subsequent penalty forgiveness for unpaid taxes was nearly $600 million, of which $517 million originated in Sidex.[77] Between 1993 and 2002, penalty forgiveness represented about 45% of total state aid. Moreover, according to the 2004 Restructuring Program for the Romanian Steel Industry, state aid expected to be granted between 2003 and 2010 entailed a further $524 million in debt forgiveness by the Romanian state (49% of anticipated state aid).[78]

In addition to tax arrears, the state also tolerated endemic arrears in payments to state-owned utilities, especially the electricity companies. Sidex, COS Târgoviște, and Siderurgica Hunedoara were among the most significant debtors.[79] The utility companies, increasingly under pressure from the IFIs to pay their own debts, threatened service cuts, and the government or state representatives intervened yet again to maintain continuity of production. According to an interviewed union leader from COS Târgoviște, the dialogue between the local utility companies and the steelworks was very informal and "was mediated by a state representative, the prefect, a member of parliament or a minister."[80] Although state authorities became involved in debt rescheduling, the new schedules were generally not respected, and the debts continued to accumulate. Often, in exchange for rescheduling favors, a company would give a financial token of appreciation to the intermediating politicians and their parties.[81]

This sectorwide pattern of state intervention that was not contingent on restructuring is best illustrated by the case of Sidex. In 1997, Sidex was at a break-even point, but it was one of the Romanian state's biggest debtors, having incurred enormous debt, especially to the social security fund.[82] In 1998, Sidex's debts to the state budget totaled over $136 million (ROL 1,200 billion), including more than $77 million (ROL 680 billion) to the social security fund and over $27 million (ROL 240 billion) to the state-owned electricity company, CONEL.[83]

By 1999, the company's debts to CONEL repeatedly threatened Sidex's electricity supply. As one particularly illustrative example of state intervention, the government issued an emergency ordinance in 1999 to underwrite a $34.7 million (ROL 400 billion) loan, to be used for the indebted Sidex's working capital and wages, which was to be granted by the state-owned Romanian Commercial Bank. In a move that underscored the Romanian state's weak capacity, the aid was not granted as part of a concrete restructuring plan. Despite the minister of finance's refusal to underwrite the loan, it was eventually underwritten by none other than the FPS president. The signatory of the loan shortly afterward compared Sidex to a black box, "We know what goes in, we know what comes out, but we don't know what happens inside. There is a certain disorganization of [the] financial and economic cycle, which is suspicious."[84] The same person later proclaimed that Sidex's commercial activities were controlled by *firme-căpuşă* run by former agents in the communist-era intelligence services.[85] Moreover, the government ignored and fudged the mounting debts and concomitant penalty payments, so as not to create conditions conducive to the declaration of bankruptcy.[86]

Thus, as far as restructuring is concerned, the Romanian state's poor capacity manifested itself in its failure to discipline the enterprises financially. There is no record of any targeted restructuring initiatives, such as debt workouts, aimed at improving the company balance sheets. Low state capacity also dogged the few existing privatization initiatives. As part of the World Bank's FESAL program, one hundred large enterprises were to be sold, with the international investment banks acting as intermediators.[87] In the steel sector, the program named the usual suspects: Sidex, COS Târgovişte, Siderurgica Hunedoara, Petrotub Roman, Industria Sârmei Câmpia Turzii, and CS Reşiţa.[88] However, during the process of selecting the investment banks, several privatizations of smaller, less politically salient producers in the steel sector took place through direct negotiations with the FPS. From the foregoing list, only Petrotub Roman and CS Reşiţa were put up for sale, along with four other enterprises. These—largely failed—privatizations, were unsuccessful as the FPS picked nonstrategic, unreliable investors, and the privatization process lacked transparency and clear contractual obligations. Moreover, most of the transactions lacked social dialogue with the enterprise unions as preparation for privatization was underway, a problem which will be discussed in the following chapter.

Table 6 summarizes the problems with the six privatizations that took place between 1998 and 2000.[89] Of the six transactions, two, both involving small pipe producers—Silcotub Zalău and Artrom Slătina—were successful. Four, however, ended in resounding failure and became a symbol of privatization gone wrong. Moreover, because of the subsequent cancellation of privatization contracts by the Romanian authorities, Romania gained a bad reputation for insecure property rights.

TABLE 6. First-Wave Privatization in the Romanian Steel Sector

Enterprise	Privatization Date and Buyer	Problems with Privatization	Result
Tepro Iaşi	1998 Železárny Veselí, Czech Republic	Lack of fulfillment of contractual investment obligations Allegations of asset stripping Investor deception concerning financial resources Poorly prepared contract: lack of clearly defined property rights, unclear asset pricing No initial consultation with the unions; violation of labor law and assassination of labor leader by the management	Cancellation of privatization contract by State Property Fund
Silcotub Zalău	1999 Tubman International, Gibraltar	Disagreement with the enterprise union concerning the scope of labor restructuring	Successful development
Petrotub Roman	1999 Tubman International, Gibraltar	No initial consultation with the unions regarding employment restructuring	Cancellation of privatization contract after veto by Competition Council (reason given: monopoly of the pipe market by the owner)
Artrom Slatina	1999 TMK, Russia	None	Successful development
Reşiţa Steelworks	2000 Noble Ventures, Inc., U.S.A.	Nontransparent privatization negotiations No consultation with the unions regarding employment restructuring Poor preparation of assets by the state Delayed rescheduling of enterprise debt by the state following privatization Financially weak investor: Nonpayment for the assets and for contracted investments Nonpayment of wages and utility bills	Cancellation of contract by State Property Fund; Romania won the case brought by Noble Ventures to the International Centre for Settlement of Investment Disputes in Washington, DC
Socomet Oţelu Roşu	1999 Marco and Stefano Gavazzi, Italy	Delayed rescheduling of enterprise debt by the state following privatization Financially weak investor: Nonfulfillment of contractual obligations Nonpayment of wage, utility, and production input bills Inability to meet rescheduled payment deadlines Allegations of asset stripping	Bankruptcy in 2004, assets auctioned off in 2005 to Ductil Steel Buzău

The first failed transaction was the 1998 privatization of Tepro, the Iași-based pipe producer. In July 1998, the FPS signed a $3 million privatization contract for a 51% stake in the company with Veselí Steelworks (Železárny Veselí), a Czech enterprise belonging to Zdenek Zemek's Z-Group. Privatization problems started with the enterprise workers' fears about their post-privatization employment prospects. Soon, the unions also accused the new owner of asset stripping and of not making the investments outlined in the privatization contract.

As time passed, the union accusations were supported by a group of local parliamentarians from different political parties and by a report by the government's Department of Control, which suggested that FPS examine whether the terms of privatization contract were being respected by the new owner. The report also noted FPS's incompetence and expressed concerns about the privatization contract itself—that is, about the lack of clearly defined property rights and questions regarding asset pricing.[90] Although the FPS initially denied the accusations, its local branch proclaimed in April 2000 that the new owners were "not acting in good faith."[91]

It increasingly looked like the new owner's goal was to shut Tepro down because it was a competitor for Veselí Steelworks.[92] The latter's fate seemed to be sealed when it emerged, in the summer of 2000, that contrary to the new owner's claims, Citibank did not vouch for the promissory notes submitted as guarantees for the investments to be undertaken at Tepro.[93] Moreover, the "investor" gave no signs of making the agreed-on investments.[94] After a dramatic turn of events at Tepro, when one of the managers was implicated in the assassination of the enterprise's union leader in September 2000, the Romanian authorities stepped up their efforts to cancel the privatization contract. The president of the FPS even admitted that Tepro privatization was a "failure."[95] The FPS turned to the Romanian Supreme Court, which decided in favor of FPS and declared the privatization contract invalid due to nonfulfillment of contractual obligations by Veselí Steelworks.[96]

The second significant privatization failure was that of Petrotub, yet another pipe producer, located in the town of Roman. Similarly to the case of Tepro, there was no agreement concerning post-privatization employment restructuring, which led to union protests and pressures to cancel the privatization contract following the announcement of layoffs. Although the privatization contract for the indebted enterprise was signed, the deal had to be approved by the Competition Council before the official takeover of the firm could take place.

The union protests quickly took on political overtones, with the local politicians, especially from the PDSR, supporting the protests.[97] The union claims were swiftly taken up by the Romanian Parliament's Committee for Industry and were also considered at the top levels of government, including by the prime minister, Radu Vasile.[98] The employees returned to work pending the outcome of the Competition Council deliberations, but with the lack of clarity as to Petrotub's future and no new orders forthcoming, the financial condition of the enterprise deteriorated.[99]

In the end, it was up to the Competition Council to determine whether Tubman International's takeover of Silcotub, Laminorul (a marginal pipe producer), and Petrotub would create a monopoly on the pipe market. In June 2000, the Competition Council approved Tubman International's takeover of Silcotub and Laminorul, but not of Petrotub, claiming that the ownership of all three plants would give the company a monopoly status with a 75% share of the pipe production market in Romania. The institutional mudslinging which followed reflected the perception of low state capacity.

On the one hand, the decision was quickly disputed by the FPS president who took issue with the Competition Council's particular definition of pipe market and accused the Council of acting under political pressure, as it had a "PDSR and PD [Democratic Party] component."[100] Moreover, he accused the Democratic Party of influence peddling and outright support for another prospective foreign buyer for Petrotub.[101] On the other hand, the FPS came under fire from PDSR and PD deputies for its poor privatization record and "lack of transparency and malversations."[102] As Petrotub reentered the state ownership portfolio, Romania was about to face two great privatization failures: Reşiţa Steelworks and Socomet Oţelu Roşu. Both cases involved leveraged buyouts by foreign investors, widely referred to in the interviews as "adventurers."

Reşiţa Steelworks, the oldest steel producer in Romania, was put up for sale in 1998, and it attracted the attention of a company based in Washington, DC, called Noble Ventures, set up by former Bethlehem Steel managers. The $85.2 million deal signed with Noble Ventures in June 2000 included $33 million in investments. The new investor also took on the enterprise's debts in the amount of about $52 million (ROL 1,100 billion), an amount four times greater than the value of its assets.[103] The US ambassador in Romania hailed the eventual signing of the privatization contract as "one of the greatest achievements made by Romania over the last ten years."[104]

The rhetoric of success notwithstanding, the deal was plagued by poor state capacity. First, the FPS did not inquire into the financial stability of the investor. After the fact, the FPS president claimed that he was suspicious about Noble Venture's financial position, but he was persuaded by the very strong recommendation of the US ambassador that the company was a solid partner. Second, the contract did not stipulate clear buyer-seller obligations. Third, the inclusion in the deal of assets that were not related to steel production made it apparent that no pre-privatization organizational restructuring had taken place. In fact, the value of the steelworks seemed to pale in comparison with the value of the other assets acquired in the transaction. These included a hydroelectric power station that supplied the local area with electricity and freshwater as well as prized real estate in Bucharest.[105]

The inclusion of these lucrative assets in the deal soon gave rise to speculation that the new owners' real intention was not to operate the steelworks, but to make profit on these additional assets. According to the UniRomSider representative, the reasoning was as follows: "I'll hire the power generator over there to the national system and I [will] recover the investment in two years. And the rest [of the assets] will have to be used by the filmmakers as a scene of apocalypse."[106]

The privatization turned problematic very quickly. On the one hand, the Romanian government did not reschedule the enterprise's debts in a timely manner. On the other, the new owners did not make the payments and the capital infusion they had committed to in the privatization contract. The investors' shortcomings were promptly pointed out by the enterprise union in January of 2001. The union's ire was fueled by the new owner's delay of wage payments and by interruptions of the steelworks' operation due to the nonpayment of utility bills.[107]

Time and again, the "investor" failed to continue production, which was being interrupted due to the inability to meet payment deadlines for rescheduled debts to the state-owned utilities.[108] The financial frailty of the new investor also manifested itself in problems paying wages. In an effort to pay salaries and make the required capital infusion into the enterprise, Noble Ventures even solicited state guarantees for a loan from a state-owned bank.[109]

In May 2001, the Romanian government rescheduled the enterprise's debts to the state, shortly before announcing that if Noble Ventures failed to make the company operational and to live up to its contractual obligations, then the government would cancel the privatization contract.[110] After renewed efforts to restart production proved futile, the Romanian state effectively renationalized Reşiţa in January 2003 by canceling the privatization contract.[111]

Noble Ventures subsequently sued the Romanian state for $353 million at the International Centre for Settlement of Investment Disputes in Washington, DC. In 2005, the tribunal unanimously ruled against Noble Ventures, dismissing the allegations that Romania violated the bilateral investment treaty between Romania and the United States, and noted infringements of the privatization contract by Noble Ventures.[112] It was a landmark victory, which Romania needed to rebut international criticism of the insecurity of Romanian property rights at an important moment before admission to the EU.

The case of Socomet Oţelu Roşu bears close resemblance to that of Reşiţa Steelworks. In April 1999, the FPS sold its 70% stake in this long products manufacturer to Italian brothers, Marco and Stefano Gavazzi. The shares were priced at $1.7 million and the new owners were supposed to invest $20 million in the plant as well as take over its debts, which the government promised to reschedule.[113] As with Reşiţa Steelworks, the government did not initially reschedule the debts in a timely manner, even though the investors were obviously short of cash.

From the very beginning, the Gavazzis did not fulfill their contractual obligations and stopped paying for utilities and production inputs, claiming that the Romanian government had failed to live up to its end of the deal by not rescheduling the debts. As a result, not only did the plant's debts continue to accumulate, but production ground to a halt after the utility companies cut off the electricity and gas supplies. With marked ambivalence about the impending halting of production, the Gavazzi brothers took off for Italy, leaving behind a desperate, unpaid workforce.[114]

The government intervened and tried to reschedule the debts, but it could not reach an agreement with the Gavazzis, who also seemed to be engaging in asset-stripping activities.[115] Although Socomet's debts were eventually rescheduled, the new owners were unable to meet the new payment deadlines.[116] With the plant's production grinding to a halt once again, the company entered into bankruptcy in October 2004, and its assets were auctioned off in 2005 to an Italian-owned Romanian enterprise, Ductil Steel Buzău.[117]

The relatively low capacity of the Romanian state was reflected in the absence of financial disciplining of the enterprises and a lack of institutional vehicles of intervention, and this carried a hefty price tag for the Romanian taxpayers. In addition to undermining the restructuring process, low state capacity also negatively affected the limited privatization efforts undertaken by a more reformist government through poor vetting of prospective buyers and unclear contractual obligations.

Medium State Capacity and Restructuring in Slovakia

In the Slovak case, state weakness manifested itself in two ways. First, the institutional checks on the clientelistic ties that tightly bound the managers of VSŽ and the HZDS government (discussed in chapter 3) and on the accompanying rent seeking were weak. Second, the Slovak state assisted the enterprise not only by making concessions, such as discounts on rail transportation, but also by tolerating tax arrears. The scope of this assistance was not as great as in Romania; hence, the budget constraint was harder. However, the Slovak state did not create a credible threat of bankruptcy in dealing with the enterprise's debts and, prior to the 1999 crisis, did not use any institutional vehicles to either force market adjustment or support the enterprise.

During the initial transition period, most notably from 1991 to 1993, VSŽ underwent a rapid process of spinning off subsidiaries, rightly criticized for its chaotic nature and for the purposeful creation of numerous asset-stripping opportunities that resulted in low profitability.[118] In a manner reminiscent of *firme-căpuşă*, members of the management used intermediary companies to funnel the financial resources out of the company in a process called "tunneling." By March 1993, VSŽ's president claimed that the enterprise was experiencing "significant financial difficulties" as "seven billion crowns [about $227.5 million], which VSŽ would need [were] missing," and it started to sell off some of its assets to raise funds.[119]

Ironically, as subsequent events would show, Alexander Rezeš, a member of the top managerial strata, gained popularity within the enterprise because of his criticism of the "wild atomization" and was elected by the employees to serve on VSŽ's supervisory board. By the end of 1993, the enterprise's shareholders had changed the top management to include Alexander Rezeš and Ján Smerek.[120] The Rezeš management outdid its predecessor in its predilection for tunneling. At the same time, the enterprise's ownership structure was becoming increasingly opaque because the managers began purchasing additional shares of VSŽ from other companies and then transferring them to other companies they controlled. Ultimately, companies holding 58% of VSŽ's securities were controlled by the managerial company ARDS.[121]

Given its large reserves, relatively modern technology, and production profile entailing flat products with high value added, VSŽ remained profitable following privatization. Quickly, however, it fell prey to elaborate tunneling schemes. These built on mechanisms initiated during the "atomization era" and included the use of intermediaries for purchasing overpriced inputs and selling underpriced final products; similar transactions were carried out with the VSŽ subsidiaries. For example, according to a former VSŽ vice president, at the end of 1996, a Lichtenstein-based enterprise called Trade Trans Import obtained monopoly rights for supplying VSŽ's production inputs, such as iron ore and coal. Shortly thereafter, prices per ton increased more than threefold (from $3 to $10). Given that VSŽ annually required 5.5 million tons of iron ore and 3 million tons of coal, the resulting profit margin was significant, even if one assumes a partially justified initial price increase.[122] Another example of managerial asset stripping was the case of Barkos, a firm registered in the Cayman Islands and closely tied to the Rezeš-Smerek-Drabik trio. The firm had exclusive right to sell VSŽ products in North America. In 1998, its debt to VSŽ reached $100 million.[123]

Not only were these asset-stripping activities not checked, the Slovak state also assisted VSŽ by tolerating tax arrears. This was in addition to the clientelistic exchange of favors between HZDS and VSŽ, such as discounted rail transport. The 1999 audit revealed tax arrears of about $115 million (SKK 5 billion), and the minister of finance proclaimed that the Slovak state was VSŽ's biggest creditor.[124]

As chapter 6 will examine in greater detail, the financial discipline came not from the Slovak state but from the world financial markets. In their expansionist zeal, the VSŽ managers not only relied on state-owned banks for loans but also used the services of the leading international banks. In 1995, VSŽ took out a $35 million loan from Merrill Lynch, and in 1996 it obtained a $125 million syndicated loan from ING Bank N.V.[125] The loan agreements included a cross-default clause, by which the lenders could accelerate their loans should the borrower default on any of them.

The financial situation of the firm deteriorated markedly in 1998. Following an extraordinary shareholders' meeting in February 1998, the well-known cast of characters consisting of Rezeš, Smerek, and Drabik took up positions on the supervisory board, while Rezeš's twenty-eight-year-old son, Július Rezeš, assumed the position of VSŽ's president.[126] Given that the average age of the managers on the executive board was less than thirty, this period was referred to in my interviews as the era of "kindermanagement." Early in 1998, it turned out that VSŽ's 1997 profit was $17.8 million (SKK 600 million), instead of the expected $41.7 million (SKK 1.4 billion). The lower-than-expected profit, combined with changes on the supervisory board, resulted in a drop in the share price from SKK 700 to SKK 300.[127] Yet, despite these financial difficulties, the enterprise continued pursuing its vision of becoming a transnational holding company, even changing its name to VSŽ Holding.

During the fall 1998 elections, VSŽ epitomized Mečiar's clientelistic politics, and its managerial financial abuses and empire-building proclivities became widely publicized. These claims were only validated by the fact that Rezeš the Elder ran HZDS's electoral campaign. Soon after the elections, in early November 1998, Július Rezeš announced that VSŽ was expected to incur losses for 1998, despite the $19.4 million (SKK 683 million) profit it had registered in the first three quarters of 1998.[128] Although he cited unfavorable steel prices and exchange rate developments as the culprits, observers overwhelmingly attributed the deterioration in the enterprise's situation to the asset-stripping activities and managerial incompetence.

The managerial bonanza was cut short soon after Rezeš's announcement; on November 9, 1998, VSŽ was unable to repay its $35 million loan from Merrill Lynch. Under the terms of their respective loans, the remaining international lenders declared a cross-default, which meant that these loans, amounting to about $450 million, were now payable on demand, instead of on the regular repayment dates in 2001 or 2002. VSŽ was on the brink of bankruptcy.[129]

The case of VSŽ illustrates the pernicious effects of clientelism and inadequate state control and regulation of economic processes accompanying privatization, all of which were subjected to scathing criticism during the 1998 electoral campaign. Responsible for 97% of Slovak crude steel production, VSŽ certainly dominated the economic landscape. However, Podbrezová Steelworks, the small pipe maker producing the remaining 3% of crude steel is the exception to the rule of the deleterious effects of domestic capitalism in the absence of high state capacity.

Like VSŽ, 97% of Podbrezová shares were offered for sale during voucher privatization. Due to the relatively low demand, about 35% of the shares ended up in the hands of investment funds, and the management purchased 54% of shares from the Slovak National Property Fund (FNM).[130] Following privatization, the enterprise encountered performance problems, in part caused by difficulties it was having obtaining scrap metal, the enterprise's key production input. Profitability fell from 2.29% net profit registered in 1992 to net losses of 12.89% and 8.38% in 1993 and 1994, respectively. Noting "management mistakes" and "personal issues," the shareholders then voted to change the company's president, but the management board membership remained stable, with five out of seven board members retaining their positions after 1992.[131] Given the stability of the management board, the change seems to have been more of a palace coup within the management.[132]

Rather than follow the asset-stripping route, early in 1995, the rearranged management adopted a revitalization plan drafted by a US consulting firm, and the company has been profitable since 1995. It even registered net profit of more than 2% during the difficult years 1998 and 1999. The management also engaged in restructuring, making annual investments of about SKK 300 million, with about 80% of this amount going into technological investments. As for employment restructuring, the management opted for the natural attrition route rather than mass layoffs, and total employment decreased from 5,570 workers in 1990 to about 3,800 in 2006, not counting the companies spun-off as part of the revitalization plan.[133] Moreover, in a region with an over 20% unemployment rate, the wages paid to the workers have been described as "good or very good" both by the company and the union organization.[134]

What sets Podbrezová's path apart from that of VSŽ? First, Podbrezová is a relatively small facility, which employed 5,570 workers in 1990. Thus, VSŽ was a much more lucrative target for rent seeking by the Mečiar government. Second, the enterprise relies on readily available scrap metal as a production input, making it easier to survive without strategic foreign investors. Third, humbler by the virtue of its production capacity and lacking domestic competitors, the company managers did not face the self-generated prestige-maximizing pressures that plagued VSŽ. Finally, Podbrezová's rare success has been attributed to the managers' desire to create a prosperous enterprise and their ability to resist the temptation to make a quick profit by tunneling the enterprise out. Lest we think that VSŽ, rather than Podbrezová, represents an outlier on the Slovak industrial landscape, it is noteworthy that the interviewed union leaders cited Podbrezová Steelworks as an illustrious exception in a region with a high unemployment rate, dotted with enterprises "tunneled out" by their managers.[135]

Relatively High State Capacity and Restructuring in Poland

In the Polish case, relatively high state capacity manifested itself in several ways. Even though Polish enterprises retained significant autonomy from the Ministry of Industry and Trade (later the Ministry of Economy), they were soon subject to a relatively hard budget constraint; there was a low incidence of tax and social-security-contribution arrears, as well as very few government loan guarantees. Moreover, even though the enterprises took out loans from state-owned banks, there was a credible threat of bankruptcy. At the same time, the state developed several forms of institutional assistance. First, through the Enterprise and Bank Restructuring Act of 1993, enterprises (and banks) were given an opportunity to restructure bad loans. Second, the courts were capable of handling complex debt workouts, accompanied by restructuring programs. Third, the state also deployed market-based vehicles to assist ailing enterprises. Finally, the Polish developments illustrated the lesson learned from the Romanian experience—namely, that the deployment of technocratic resources by the state was crucial to successful privatization via sales to foreign investors.

The pathbreaking privatization of Warsaw Steelworks (Huta Warszawa)—a relatively small manufacturer of special and alloyed-steel long-products located on the outskirts of Warsaw—illustrated many of these points.[136] It also foreshadowed that foreign investment was the way of the future for the Polish steel sector and for the region as a whole.

The presence of the hard budget constraint was felt relatively early in the Polish industrial sector, even though the enterprises were overwhelmingly state owned.[137] In 1991, Warsaw Steelworks' financial situation deteriorated so profoundly that it faced imminent bankruptcy. In fact, the enterprise was slated for closure in the Canadian restructuring proposal, to the great delight of the local government and the Warsaw population, which perceived it to be the epitome of the old regime's industrialization drive in the heart of the capital.[138] Thus, when the Italian Lucchini Group expressed an interest in acquiring a majority share of Warsaw Steelworks in the summer of 1991, the Polish government enthusiastically embraced the privatization initiative.[139]

Government officials, notably the minister of ownership transformation (minister of privatization) and the minister of industry and trade moved quickly and decisively in 1991 and 1992 to bring the privatization about. Taking advantage of the enterprise's difficult circumstances, the buyer insisted on changing the deal to a joint-venture: the working assets of the enterprise were subsequently transferred to a new entity, Huta L.W., in September 1992, and the debts and unusable capital were left in the hands of Capital-Accounting Agency, the SOE successor to the original Warsaw Steelworks.[140]

However, given the unprecedented privatization and its hurried nature, the ministry officials left it up to the management of Capital-Accounting Agency to clarify outstanding property rights questions. The management of the agency, which was in possession of the relevant documentation, acted in an obstructionist manner and delayed the resolution of the property rights issue in what was already a convoluted legal procedure.[141] Failure to step in decisively to clarify property rights, which eventually took place in 1995, caused a delay in restructuring of about two years, as the foreign banks refused to approve a loan for technological investment until the issue was resolved.[142]

Despite the shortcomings of this pathbreaking privatization, the case of Warsaw Steelworks highlights the contrast with the failed first-wave privatizations in Romania. In this transaction, state institutions displayed high capacity by stepping in and separating the working from nonworking assets, as well as by selecting a strategic investor with the wherewithal to restructure. The qualitative gap between the "adventurers" of the Gavazzi brothers' ilk and the—also Italian—Lucchini group was vast. Moreover, the subsequent ministers of industry and trade, and of ownership transformation, both defended their predecessors from the accusation that the privatization was a mistake or that it was poorly executed.[143] This, too, was a far cry from the institutional mudslinging and turmoil witnessed during the Romanian privatizations. Still, the ensuing difficulties and the criticism of the Warsaw Steelworks privatization served as a warning about the controversial nature of privatization.

Even as the state placed the task of restructuring the steel industry in the hands of the managers of the individual enterprises, it tightened the budget constraints. In fact, the managers repeatedly criticized the government for not giving the sector greater financial assistance. The state's investment guarantees were selective. According to Poland's Supreme Audit Office, from 1993 to 1997 the State Treasury guaranteed about $230 million (PLN 743.9 million) worth of loans (which, notably, were paid back on time).[144]

The overall state aid to the sector in the early years of transition was very limited, and the need for state aid only intensified after 1998, when the sector's financial troubles became prominent.[145] By mid-2000, when the sector found itself in a precarious financial position, tax and social-security-contribution arrears represented a relatively small fraction—7%—of a total debt burden of more than $1.6 billion (about PLN 7 billion).[146] It was equal to about $115 million, about the same amount that VSŽ owed the Slovak state in tax arrears in 1999.[147]

This relatively hard budget constraint was accompanied by a credible threat of bankruptcy. The embattled Warsaw Steelworks averted bankruptcy through early privatization to a foreign investor. In 2001, however, the indebted and overinvested Baildon Steelworks paid dearly for its managers' fanciful investment vision and profligate spending when it declared bankruptcy after state officials gave it an ultimatum: either find an investor or file for bankruptcy.[148]

At the same time as it hardened the budget constraint, the Polish state created institutional tools to assist embattled enterprises while still remaining within the confines of market discipline. First, to deal with the problem of bad debt faced by the banks and enterprises alike, it adopted the Enterprise and Bank Restructuring Act in 1993, which provided several means of restructuring indebted enterprises, including privatization through debt-equity swaps.[149] The effect of the act on the steel sector was relatively small, but two important steelworks, Zawiercie and Ostrowiec, were initially privatized through the program. Although the 2003 Supreme Audit Office's report on the steel sector considered the effects of the debt-equity swaps to be unsatisfactory because they did not lead to the development of these enterprises, the report admits that the swaps saved them from bankruptcy.[150]

The state-mediated and bank-led conciliation procedures designed by the Enterprise and Bank Restructuring Act bypassed the courts and thus sped up the debt workout process. After 1996, the enterprises had the recourse to the traditional court-led workouts. These were complex procedures, involving hundreds of creditors, and required substantial technocratic acumen to execute. Debt workouts canceled some parts of the debt and rescheduled others, and, importantly, they were also accompanied by restructuring programs that became an important constraint on the enterprises. Even though the effects of debt workouts and conciliation procedures proved temporary, they nonetheless brought about a degree of restructuring and a financial disciplining of the enterprises.

As examples, let us consider two workouts by Katowice and Sendzimir Steelworks. Over 1993–94, Katowice Steelworks was involved in a debt workout, as a result of which its debts were reduced by 30%.[151] By 1998, however, Katowice's financial performance had again deteriorated, and it took out a $100 million long-term loan from a consortium of Polish banks, which allowed the steelworks to pay outstanding short-term debts to its business partners.[152] The steelworks undertook restructuring, most significantly in the area of employment, and reduced its workforce by 7,000 people in 1999.[153] Moreover, over the years 1998–99, Katowice reduced its operating costs by more than $80 million (PLN 300 million).[154] However, these steps proved to be inadequate to solve the enterprise's long-term problems; by 2001, the enterprise once again found itself under the pressure from creditors and in danger of bankruptcy.

Sendzimir Steelworks underwent its first debt workout in 1994. As many as 1600 creditors agreed to reduce the enterprise's debt by 40% as well as to reschedule it. By the end of 1999, Sendzimir had repaid all the installments. However, at the beginning of 1999, after the management's talks with prospective foreign investors collapsed, the enterprise found itself in danger of bankruptcy, as it had problems paying off its last debt-workout installment and needed to repay a $16 million (PLN 60 million) revolving credit loan. At that point, the restructuring steps undertaken by its new management provided a short-term remedy. As in the case of Katowice, the most significant measure taken was a reduction of employment, this time by half the workforce (nearly 8,000 employees), to reduce employment costs that represented about 23% of total operating costs.[155] By mid-2001, Sendzimir was again in precarious financial shape, and in 2001–2002, it underwent another debt workout, with around 800 creditors. Although its creditors included such state-owned enterprises as Polish Railways and energy providers, the enterprise did not owe significant debt to the state budget itself.[156]

The perilous financial condition of the Polish steel industry in 2000 attested to insufficient and inefficient restructuring undertaken by managers who competed instead of cooperating in investment decisions. As a result, they had neither the money to finish the technological investments they had started nor the working capital to continue production. To make matters worse, the industry was now playing on a completely liberalized market of which it held a rapidly shrinking share. Between 1997 and 2001, the amount of imported steel used in domestic steel consumption doubled, rising from 21.7% in 1997 to 43.7% in 2001. The threat of bankruptcy loomed. In the early months of 2001, the sector's debt was approaching $2.36 billion (PLN 9.7 billion), of which 75% was short term.[157]

At that point, the state deployed more specialized tools to keep the enterprises operational. These took the form of a parastatal enterprise, the Silesia Finance Society (Towarzystwo Finansowe Silesia). Created at the end of 2000 to help the cash-strapped Katowice enterprise repay its loan interest to a Polish bank consortium, Silesia became a subsidiary of the state-owned Agency for Industrial Development. It was subsequently capitalized in the amount of PLN 250 million through the transfer of state-owned shares in prosperous enterprises.[158] According to its creator, the deputy minister of economy, Edward Nowak, Silesia was to function as a bypass, financing production inputs and selling Katowice's products, and assisting with the repayment of the interest on its loans.[159]

As chapter 6 will show, the use of these tools became more prominent as external pressures to change the status quo of state ownership mounted. Chapter 3 explained that the status quo was more convenient for the parties in power, for a variety of reasons, including pecuniary. Detailed fieldwork and examination of press articles failed to document rent seeking in Poland along the lines of Barkos in Slovakia or Balli International in Romania. Rather, rent seeking, noted the interviewees, seems to have taken a more veiled form than in those countries. Certainly, it had less onerous implications for the state budget, as evidenced by the far smaller amount of state aid granted to the sector, relative to the amount of steel produced.

Relatively High State Capacity and Restructuring in the Czech Republic

The Polish and Czech states shared several key similarities even though the initial policy choice in the two countries differed. First, the Czech enterprises were subject to a relatively hard budget constraint early in the process. Second, although the privatizations to domestic owners, discussed in the previous chapter, were clearly flawed, there was a credible threat of bankruptcy and expectations that the new owners would fulfill the conditions of their contracts. In terms of institutional assistance, the Czech state displayed a relatively high level of capacity through engagement in a complex debt workout, as well as through direct restructuring of enterprise assets. Like Poland, the Czech state developed targeted, market-based vehicles of institutional support for distressed enterprises.

Also as in Poland, the need for state aid became clear around 1998, two years after the Czech state could no longer issue aid without the assent of the European Commission.[160] However, enterprise liabilities were not to the state budget but to commercial partners. For example, in the 2001 Czech steel-sector restructuring program drawn up by Eurostrategy, a consultancy often used by the European Commission, the 2000 balance sheet of Nová Huť—the source of the biggest problems for the Czech government at the time—showed total liabilities of roughly $518 million (CZK 20 billion), comprising $363 million (CZK 14 billion) in bank loans and $155 million (CZK 6 billion) in trade creditor liabilities.[161]

The credibility of the bankruptcy threat became very clear in the case of the Poldi privatization, discussed in chapter 3. This earliest case of privatization in the sector admittedly revealed significant failures on the part of state institutions, such as inadequate vetting of the buyer and procedural mistakes in the approval of transfer of shares by the FNM. However, it also showed that the Czech state was quickly able to identify and attempt to rectify these mistakes.

State authorities stepped in as soon as the new owner failed to pay for his shares and when they received warnings of asset stripping. Attesting to the security of property rights, the court dismissed the government's argument that the initial privatization should be invalidated because the buyer had not met the payment deadlines and because the FNM had not properly approved the transfer of shares.[162] The Czech state then resorted to bankruptcy. The process, which ended in July 1997, was complicated by the owner's attempt to transfer assets to another entity.[163] Following the bankruptcy, the enterprise's assets were sold off to various companies in an effort to pay off the liabilities, equal to about $189 million (CZK 6 billion).[164] Some of the assets—Poldi Ocel and Poldi Construction Steel—were eventually acquired by the domestically owned long products specialist, Třinecké Steelworks, which were successful in restarting and developing production.[165]

If the Poldi case illustrates the credibility of bankruptcy, the case of Třinecké Steelworks illustrates the fulfillment of contractual obligations. It became a successful, restructured steelworks, even though, like myriad other companies that subsequently failed, it was acquired through a leveraged buyout. Moreover, although Třinecké's acquisition was tainted by a scandal involving campaign contributions, the actual purchase price—unlike its heavily discounted VSŽ counterpart—did not raise objections. Even more remarkable is that Moravia Steel, the owner of Třinecké, not only repaid its $99 million (CZK 2.6 billion, using the 1995 exchange rate) bank loan but did so two and a half years in advance.[166]

The enterprise underwent relatively successful restructuring, even though it backed out of some of the earlier investment plans and also obtained restructuring-related state support in the form of an approximately $15 million (CZK 514 million) subsidy for the reversal of environmental damage.[167] Třinecké's success is underscored by its profitability since 1997, even during the difficult year of 1998, when other enterprises, their management problems exacerbated by poor conditions on the world markets, folded.

Even though, as a successful large firm sold to domestic capitalists, it remains an outlier in the region, Třinecké Steelworks is not without foreign influence. In 2000, the owners sold an 11% share in the enterprise to a foreign investor, CMC Trading AG, the Swiss subsidiary of the US-based CMC Steel Group.[168] According to an interviewed Ministry of Industry and Trade civil servant:

> Moravia Steel did not have that cash, capital . . . it was necessary to look for some strategic partner—CMC. Třinec had problems with financing production and needed to borrow . . . banks had high interest rates . . . So it was clear that no factory could produce—it had to go bankrupt without the aid of some kind of foreign capital.[169]

Thus, Třinecké's seemingly domestic success also underscores the need for foreign capital in the region.

As the financial prospects of the sector worsened in 1998–99, the Czech state displayed a greater array of tools for intervening in enterprises that were on the verge of collapse. The most significant state intervention took place in Vítkovice. By the year 2000, Vítkovice stood on the brink of bankruptcy. Its total assets were worth less than $285 million (CZK 11 billion); its 1999 losses were equal to $251.7 million (CZK 8.7 billion), and 2000 losses equaled $233.2 million (CZK 9 billion). By 2000, creditors had produced 311 million dollars' (CZK 12 billion) worth of debt claims.[170] Although its dire financial situation resulted in part from the 1998 problems on the world steel markets, the enterprise's difficulties were also connected to unfavorable contracts that bore the marks of asset stripping.[171] For example, beginning in 1992, the company signed fictitious consulting-services contracts with a shell company located in Lichtenstein. All in all, the amount of money suspected of having been "tunneled out" of Vítkovice was estimated at $13 million (half a billion CZK).[172] Vítkovice's financial difficulties were exacerbated by a contract it signed with one of its major suppliers, Shiran. In exchange for lenient treatment of the enterprise's accumulating debt, the contract stipulated overpriced production inputs.[173] Under similar circumstances, Shiran also placed Nová Huť in a scissorlike bind, forcing it into signing a barter-type contract with overpriced production inputs and underpriced sales of final products.[174]

The Czech state became directly involved in Vítkovice's restructuring by reasserting control over the enterprise and by ensuring far-reaching financial and organizational restructuring. First, FNM canceled the privatization contract with the managerial company and called forth crisis management, led by Vacláv Novák, who had wide-ranging experience working abroad in crisis management. Second, given that Vítkovice lacked working capital, the Czech state used an FNM subsidiary called Osinek to deliver production inputs and to coordinate sales of final goods—very much along the lines of the Polish Silesia Finance Society.[175] To engage in this so-called tooling system, Osinek obtained a $46.6 million (CZK 1.8 billion) loan from Konsolidačna Banka.[176]

Third, the state also became involved in a debt workout aimed at a partial cancellation and rescheduling of Vítkovice's debts. Management filed for debt workout in June 2000, and the arduous process was finalized in September 2001. As part of the agreement, the creditors were guaranteed repayment of a minimum of 30% of their total debt claims within two years. A portion of the claims was bought up from more than 2,000 small creditors by Česká Finanční, a subsidiary of Konsolidačna Banka, which was capitalized by the state in the amount CZK 1 billion for that purpose.[177] The enterprise was to repay the remaining $62.2 million (CZK 2.4 billion) from privatization proceeds. Finally, in August 2001, in the crowning moment of the state-led restructuring of assets, the Vítkovice combinate was broken up into its steelmaking component, Division 200, also called Vítkovice Steel, and the heavy engineering division, both of which were subsequently put up for sale.[178]

Nová Huť's financial problems in 1998–99 were compounded by the initial complications related to the minimill investment, which experienced delay in full commissioning in 1999. All in all, Nová Huť's losses reached $160 million (CZK 4.7 billion) in 1999.[179] At the same time, as the financial problems of Nová Huť mounted, the Czech government came under pressure from IFC to guarantee repayment of its loan for the minimill project. Thus, it was clear that state intervention was also required in the case of the pseudo-private behemoth of the Czech steel sector. First, however, the state needed to wrestle control out of the hands of the managers who had been contracted out to manage the enterprise as part of the managerial privatization initiative. That effort, made in the face of mounting external pressures to come up with a comprehensive sectoral solution in the run-up to the closure of EU membership negotiations, is discussed in chapter 6.

Conclusion

Throughout the restructuring process, rather than use their knowledge of the sector and the existing ties among the enterprises, the managers engaged in prestige-maximizing strategies whereby they pursed myopic investment plans that had detrimental effects on the sector. Although all four countries made strides in the technological restructuring of their primary steelmaking facilities, the steelworks were unable to complete the restructuring process and invest in finished steel production. By the year 2000, the use of the obsolete open-hearth process had virtually been eliminated in all four countries.[180] In the same year, the use of continuous casting in total crude steel production was equal to 67% in Romania (up from 34% in 1989), 99% in Slovakia (up from 25% in 1989), 70.2% in Poland (up from 8% in 1989), and 87.1% in the Czech Republic (up from 2.3% in 1989).[181]

However, the effects of restructuring on the public purse in these countries differed. As table 3 shows, state aid was the lowest in the countries where state capacity was higher, namely, the Czech Republic and Poland, with $5.4 and $5.8 per ton of steel produced, respectively. Whereas the aid was about \$9.1 in Slovakia, it amounted to a whopping \$21.1 per ton of steel produced in Romania. These figures, in conjunction with the individual country-enterprise case studies, illustrate that where the state possessed greater capacity, it was able to bring about a market adjustment using market means, getting involved in debt workouts tied to restructuring. On the other hand, the state also engaged in targeted interventions, using state subsidiaries to avert the danger of bankruptcy, as was the case with the Silesia Finance Society in Poland and Osinek in the Czech Republic. This stood in contrast to the assistance in Romania, and to lesser extent in Slovakia, where the state tolerated tax arrears.

The contrast between the case studies involving the same type of ownership but different levels of state capacity also illustrates that state capacity, rather than ownership type, was crucial to the restructuring outcomes. In the Czech Republic, the clientelistic ties to the governing coalition were far more veiled than in Slovakia. First, there were more severe political ramifications following the discovery of illicit payments and favors, as the case of Třinecké Steelworks, compared to that of Slovakia's VSŽ, illustrates. Second, the new owners were expected to pay full price for the assets, as in the case of Poldi and Třinecké Steelworks. Moreover, when the new owners failed to abide by the terms of the contract, Poldi being the case in point, there were bankruptcy proceedings against the new owner.

In the case of state ownership, the contrast between Poland and Romania was stark in that even though there was politically connected rent seeking in Poland, it was not nearly to so great an extent as in Romania. In Romania, the financial consequences of the patrimonial capitalist pathway taken by the enterprises were shifted onto the shoulders of the Romanian state, which lacked the capacity to discipline the enterprises and became their main creditor. By contrast, rather than tolerate tax arrears, the Polish state became a partner in the debt-restructuring talks, while maintaining a credible threat of bankruptcy, as was seen in the case of Baildon Steelworks.

Although the primary steelmaking facilities were upgraded in all four countries, it had become apparent by the end of the 1990s that, with few exceptions, foreign investment was necessary to ensure the enterprises' and the sector's development and even survival. At the same time, the governments in power continued to pursue policies that favored domestic owners or state ownership and found themselves pressed by external economic and political pressures for radical policy change. However, their ability to resist these pressures was tied to the level of state capacity the governments had at their disposal. The story of that, ultimately failed, resistance to external pressures is presented in chapter 6. For now, we turn to the discussion of collective action in the sector—namely, to the story of labor. As chapter 5 illustrates, higher state capacity was associated with a more vibrant social dialogue at the sectoral level, which resulted in both a lower tendency to use labor instrumentally and in effective sectoral employment-restructuring programs.

FIVE

Acting Collectively

Social Dialogue, State Capacity, and Sectoral Restructuring

[The sale to U.S. Steel is] cheating the company, the workers, and
Slovakia.

—Metalurg union leader

[A]ll of the sudden, we woke up in a privatized firm.

—Reşiţa Steelworks union leader

[T]he unions were fully behind me, which is surprising because I
was so radical.

—Vacláv Novák, Vítkovice crisis manager

[W]e appeal to the Minister [of State Treasury] to take constant and
direct control over the privatization process of Polish Steelworks,
giving the enterprise an opportunity to develop.

—2003 letter from the Multi-Union Committee at the Polish
Steelworks to the Minister of State Treasury

Chapters 3 and 4 illustrated that the governments of the Czech Republic, Poland, Romania, and Slovakia eschewed sectoral policies and, instead,
devolved the responsibility for restructuring to the managers of the individual enterprises. The managers, far from cooperating with each other
to further the interests of the sector as a whole, pursued their individual
prestige-maximizing visions of restructuring. However, the course of
managerial restructuring, and its implications for the public purse, was

mediated by the level of state capacity. This chapter addresses the other important vested interest in restructuring, labor. Unlike the enterprise-level analysis of management, the discussion of labor is best conducted at the level of the sector, for two reasons. First, the financial feedback effect of the initial policy choice highlighted the need for sectoral-level solutions, and labor became the primary driver of the only sectoral programs adopted prior to the external push for radical policy change: employment restructuring programs. Second, the actions of labor are embedded in an institutional context of both enterprise- and sectoral-level social dialogue, and, as this chapter illustrates, encompassing sectoral-level social dialogue is conducive to restructuring.

This chapter examines union attitudes to restructuring in the institutional context of social dialogue. The theoretical discussion in chapter 2, concerning the role of social dialogue in restructuring, led to the following propositions, which are fleshed out empirically in this chapter:

> P5: Autonomous unions support (complete) reform when organized social dialogue exists at the sectoral level.
>
> P5a: Autonomous unions support (complete) reform when organized social dialogue exists at the sectoral level, even when the government is not spearheading reform.
>
> P6: Captured unions oppose (complete) reform, whether or not social dialogue is organized at the sectoral level.

The key claim is that social dialogue with autonomous unions, especially at the sectoral level, is conducive to restructuring but requires a state that has enough capacity to implement concluded agreements.

The propositions just enumerated question the validity of the executive insulation thesis, which advocates protecting executives from social pressures that could block reform.[1] As the propositions suggest, opposition to reform is characteristic of captured unions, and the reform-promoting solution is to engage in more, rather than less, social dialogue and to create encompassing sectoral structures. The evidence gleaned from process tracing suggests that the executive insulation thesis provides neither the right diagnosis nor the correct solution.

The chapter discusses the structure of social dialogue and its relationship to union attitudes to restructuring and privatization in each country. Captured unions in Romania and Slovakia opposed restructuring, while the autonomous, encompassing unions in Poland and the Czech Republic supported restructuring and privatization to foreign investors, even when the government was less than enthusiastic about the idea. The case studies illustrate the role of social dialogue in both industrial restructuring and privatization, and they also underscore the need for state capacity in implementing concluded agreements. As in the previous chapter, the discussion starts with Romania, where the level of state capacity was the lowest.

Social Dialogue and Restructuring in Romania

The Romanian case is distinctive for the initial government co-optation of the union in Sidex, the biggest enterprise in the sector, which undermined labor unity and substantially diluted the power of the sectoral organization. Despite early de jure institutionalization of social dialogue in the sector, it remained weak, and the state lacked the capacity and the will to implement the negotiated sectoral restructuring agreements. The weakness of state institutions exacerbated union distrust of privatization at the enterprise level because the unions doubted the quality of buyers. However, rather than acting as a deterrent, nearly all active union opposition to privatization transpired *after* the new owners failed to deliver on their privatization obligations. Before discussing the specifics of the Romanian restructuring and privatization process from the perspective of industrial relations, it is important to place the development of institutions of social dialogue in the context of the Romanian political transition.

The key labor organizations in the steel sector were the National Union Federation Metarom and the Galaţi Ferrous Metallurgy Workers' Trade Union Federation. The autonomous National Union Federation Metarom, which represented metal and heavy industry workers, as well as miners, as part of the Cartel Alfa Confederation, has remained avowedly apolitical.[2]

Early in the transition process, the illiberal National Salvation Front government aimed to weaken the nascent independent labor movement by forming alliances with powerful local unions and coaxing them to split from new, independent national and sectoral union structures. This was the case with the mine workers in the Jiu Valley, who were initially part of Cartel Alfa but subsequently formed their own organization.[3] In 1992, the enterprise-level union organization at Sidex, the most significant steel producer in Romania, withdrew from the Metarom federation. Like the Jiu Valley miners, it set up its own organization, the Galaţi Ferrous Metallurgy Workers' Trade Union Federation (Federaţia Sindicală a Siderurgiştilor din Galaţi), named for Sidex's location. According to the vice president of the Cartel Alfa Confederation, the leaders of the National Salvation Front

> were unhappy to see these important structures of civil society grow up and they were really unhappy that these structures would become very powerful and immediately decided to weaken these structures and used some trade union leaders in order to do that. The miners and Galați union are just two examples, but in early 1990s, these two federation structures were the most important federation structures in Romania.[4]

The Galați union gained federation status even though it only had local representativeness.[5] Throughout the transition, Sidex workers were particularly privileged in terms of wages and job security compared to other workers in the sector.[6] In fact, Sidex's 40,000-strong workforce was used as a shield by managerial and local interests opposed to restructuring and privatization. Metarom clearly dominated in the remaining enterprises.[7]

Although labor unity was undermined, Romanian social dialogue had officially been institutionalized at both the national and the sectoral levels early in the transition. The negotiations of the first branch-level collective agreement began in 1992, earlier than in any of the other country cases. However, the 1991 law guiding social dialogue was quite deficient, and the rules on social dialogue at the enterprise, sectoral, and national levels had to be clarified in 1996.[8] Throughout the transition, despite several short-duration general strikes and protests in the capital, the state representatives did not see the relationship with the federation as tense.[9]

Regardless of the formal structure of sectoral-level social dialogue, the enterprise level remained the core of union activity, and sectoral-level organization played a relatively peripheral role. According to Cartel Alfa's vice president:

> [T]he Romanian trade union structure is a very fragmented one. . . . In many situations, the superior structures, branch or national ones, have nothing to do with the activity developed inside the plant by plant trade union. Usually, we are not interfering with the activity, we are just participating in some activity, but just in case[s] [where] we had an official request by the plant trade union.[10]

Given the relatively weak sectoral-level institutions, undermined by an illiberal government early in the transition process, social dialogue was unable to become an effective instrument for introducing greater certainty into the reform process and elongating the time horizons for all the actors involved. The enterprise-level wage pressure persisted, despite agreements at the sectoral level. This attitude, however, changed over time, as the market mechanism became more palpable. According to a local union leader, the enterprise-level unions in Romania "opposed large workforce reductions and policies, which prohibited wage increases when the company had very bad results. Afterward, they realized that an equilibrium needs to be established between what can be given and what the firm's performance permits. . . . Otherwise, they don't have a way of getting the salaries."[11]

Union wage pressure is not a surprising outcome to proponents of the executive insulation thesis. However, the wage push did not come from the sectoral-level organization but, rather, from enterprises for which the local management supported the wage requests, as it continued to engage in dubious business activities within the enterprise. As one union leader put it, "They did not pay for the utilities but they were giving *absolutely* unjustified wage increases, without the company making profit. The balance sheets were fudged to show positive numbers to give a nice image to the company."[12]

By contrast, the sectoral-level organization Metarom realized that restructuring would involve job losses and focused its efforts on demanding compensation for workers leaving the sector, which included both severance packages and vocational retraining programs modeled after those in Western Europe. In June 1996, Metarom, along with the Galați-based union, signed the Social Assistance Agreement for the Restructuring of the Iron and Steel Industry (Convenția de Acompaniament Social al Restructurării Metalurgiei Românești), and the Hefaistos Employers' Association (Organizația Patronală Hefaistos), the precursor of UniRom-Sider, was the employers' organization representing the iron and steel producers.[13] Among other things, the agreement provided for the creation of an institutional support network, with regional branches (National Union for Reconversion in the Iron and Steel Industry–Uniunea Națională de Reconversie în Metalurgie [UNIRMET]), as well as a "national solidarity fund for the iron and steel industry," FONDMET, to support the vocational retraining program.

Underscoring the low capacity of the Romanian state, the agreement among the social partners remained on paper, despite Metarom's pressuring the government to develop concrete legal and financial implementation mechanisms of the social protection measures. Although Metarom's leader stressed that the government needed to accelerate the restructuring process because labor productivity in the Romanian steel sector was six times smaller than in the advanced capitalist countries, Metarom underscored that restructuring needed to entail job creation and training programs rather than simple severance payments for laid-off workers.[14] In other words, Metarom did "not want people to be paid to become unemployed."[15]

No measures implementing the Social Assistance Agreement for the Restructuring of the Iron and Steel Industry were put in place, despite a 1999 government decree and a 2001 law purporting to do so, because neither the government nor the employers paid the contributions they had promised.[16] At that time, however, privatization-related group layoffs, rather than job-training schemes, became the priority and created tension between the unions and the government. In what would have tremendous negative implications for union attitudes to privatization, group redundancies surrounding the first wave of privatizations were generally not negotiated with the unions.

Although Metarom recognized the need for restructuring, along with its enterprise-level affiliates, it took a more reserved approach to privatization. While, in principle, Metarom considered privatization to be "conducive to increased efficiency of the enterprises," the union opposed IFI-inspired rapid privatization carried out without the concomitant sectoral restructuring strategy demanded by the EU.[17] As Metarom's president put it, "We are not against privatization, but we seek that it be done in a real way, with true investors, and that the employees be given guarantees concerning workplaces or compensation."[18]

This reserved attitude can be attributed to the painful privatization experience of the Romanian steel sector in the late 1990s, which is summarized in table 7. As discussed in the previous chapter, the first privatization wave, from 1998 to 2000, attested to the low capacity of the Romanian state. It was also notable for its neglect of social dialogue during the privatization process and conspicuously silent concerning post-privatization labor restructuring. As far as the privatization-related labor relations were concerned, the privatization failures followed a similar pathway. First, the nontransparent privatization process and the resulting contract created a sense of uncertainty among workers. Second, mounting suspicions were usually confirmed by nonstrategic foreign investors, who lacked capital, announced enormous layoffs, trammeled worker rights, and in some cases, engaged in asset stripping. Third, rightly fearful about the workers' future but also instigated by local political figures, enterprise-level unions demanded the cancellation of the privatization contract.

This order of events contradicts the claim by proponents of the execu-

TABLE 7. Social Dialogue in the First and Second Privatization Waves in the Romanian Steel Sector, 1998–2003

Wave	Enterprise	Privatization Date and Buyer	Quality of Social Dialogue	Union Attitude to Privatization
I	Tepro Iaşi	1998, Železárny Veselí, Czech Republic	**Lacking:** unions not consulted; contract did not contain any provisions concerning worker protection; lack of schedule of employment restructuring or any concomitant severance payments; eventual management-inspired assassination of union leader	Union protests after privatization as a result of investor's layoffs in contravention of Romanian labor law, allegations (confirmed) of asset stripping by new owner, and lack of promised investments
	Silcotub Zalău	1999, Tubman International, Gibraltar	**Limited:** unions not initially consulted about immediate post-privatization layoffs; after tensions with the unions, negotiations regarding conditions of layoffs	Apprehensive, but no protest against privatization
	Petrotub Roman	1999, Tubman International, Gibraltar	**Lacking:** unions not involved in the privatization process; no accord concerning employment restructuring in the post-privatization period; eventual unlawful firing of union leader	Vehement protests over anticipated layoffs associated with privatization, questions about Tubman International's credibility as an investor
	Artrom Slatina	1999, TMK, Russia (via Staro Stahl-und Röhrenhandel GmbH, Austria)	**High:** unions consulted, "told what is ahead," involved in layoff and severance package negotiations	At first apprehensive, but privatization "embraced" by the unions as "the only hope to make the company function"
	Reşiţa Steelworks	2000, Noble Ventures, Inc., U.S.A.	**Limited:** union not consulted over privatization; after privatization, collective work contract signed with the union, stipulating no group layoffs until 2005	Very distrustful of the investor but no protests surrounding privatization; vehement protests started after a combination of delay in salary payments, problems with operating the steelworks, and nonpayment of contracted financial obligations by the new owner

	Sidex Galaţi	2001, LNM Holdings, Netherlands	**High:** union consulted during privatization; generous employment restructuring clauses part of privatization contract (departures via natural attrition for 5 years)	Resisted privatization for a long time (along with the management), but cooperative once negotiations started
	Special Steels Enterprise at Târgovişte (COST)	2002, Mechel Group, Russia (via Conares Trading, Switzerland)	**High:** union consulted during the privatization process; employment restructuring clauses part of privatization contract	Cooperative during privatization process
	Tepro Iaşi*	2003, LNM Holdings, Netherlands	**High:** union consulted during the privatization process; employment restructuring clauses part of privatization contract	Cooperative during privatization process
II	Siderurgica Hunedoara	2003, LNM Holdings, Netherlands	**High:** union consulted during the privatization process; employment restructuring clauses part of privatization contract	Union protests over the scope of layoffs related to privatization (resentment over provisions negotiated at Sidex); resolved through discussions with LNM over the schedule of severance payments
	Petrotub Roman*	2003, LNM Holdings, Netherlands	**High:** union consulted during the privatization process; employment restructuring clauses part of privatization contract	Initial union resistance against privatization due to concomitant employment restructuring (resentment over provisions negotiated at Sidex); union protests led to higher workplace retention
	Reşiţa Steelworks*	2003, TMK, Russia (via Sinara Handel, Germany)	**High:** union consulted during the privatization process	Cooperative during privatization process

* Second attempt at privatization after the first-wave failure.

tive insulation thesis that the executive needs to be isolated from the pressures of social actors who purportedly attempt to block reform. In the Romanian case, instances of privatization-related enterprise-level labor unrest overwhelmingly occurred *after* there had been signs that a specific privatization was experiencing problems, rather than prior to it. Moreover, in the two successful first-wave privatizations (Artrom Slatina and Silcotub Zalău), the engagement of the enterprise-level unions by the new management, at the time of privatization (Artrom) and soon thereafter (Silcotub), was much greater.

In the case of Tepro Iaşi, the unions were not informed about the privatization negotiations and the privatization contract mentioned neither the schedule of employment restructuring nor any severance payments such restructuring would entail. Early in 1999, the new management decided to halve the 2,840-strong workforce, and—faced with union resistance and in violation of Romanian labor law—fired the enterprise-level union leader and his assistant. Despite the union's protests and a strike, the new management, just as it had planned, laid off 1,300 workers and announced further layoffs.[19] The union leaders complained that the scope of layoffs violated the provisions of Romanian labor law and that they did not include severance payments.[20] They pointed to further violations of labor law, such as the fact that the board meetings took place in Prague rather than in Iaşi, making it impossible for the union leaders to attend.[21] They particularly blamed the FPS for not ensuring that the new owner would adequately protect the workers. To these allegations, they soon added accusations of asset stripping by the new owner as well as failure to make the investments to which it had committed in the privatization contract.

After the workers suspended their protest activities, the new owner announced new layoffs, this time entailing three-quarters of the remaining workforce—that is, 1,100 of 1,500 workers—claiming that they were intended to bring the enterprise up to modern efficiency levels. This time, the union promised harsher protest activities.[22] As the Romanian authorities began to recognize the new owner's asset-stripping activities, and as the FPS debated the different ways in which to cancel the privatization contract, the new owner chose "the mafia way to silence the opposition": Tepro managers were implicated in the assassination of the enterprise-level union leader, Virgil Săhleanu, in September 2000.[23] Soon thereafter, the Romanian authorities canceled the privatization contract.

As in the Tepro case, in the case of Petrotub there was no accord concerning employment restructuring in the post-privatization period, and the FPS also failed to involve the unions in the privatization process, a development that negatively affected the union's attitude to the deal. Consequently, about one thousand Petrotub workers blocked a national highway to protest against the signing of the contract without a concomitant protocol concluded with the union, which would have made a commitment to protect workplaces. Moreover, the union was skeptical as to the credibility of Tubman International, the new owner.[24]

Union protests were further fueled by the new owner's announcement that there would be layoffs, and that their extent would depend on the enterprise's efficiency following privatization. The new management said, "Presently, the efficiency of Petrotub is about 15% of international standards. In case we reach 30%, 1,500 workers will be laid off. Union protests only worsen the economic situation of the enterprise."[25] The union vehemently opposed the laying off of nearly 40% of the workforce, which had already seen 2,500 workers laid off in 1999. In addition to roadblocking, the enterprise-level union began a strike in opposition to privatization, which still had to be approved by the Competition Council.

As the enterprise deteriorated economically, and as the top government officials, including the prime minister, took an interest in the details of the privatization, the social atmosphere inside the enterprise following management's unlawful firing of the union leader became even tenser.[26] All in all, prolonged union protests resulting from the anticipated layoffs created strong pressure—supported by local political figures—to cancel the privatization contract. As discussed in chapter 4, in a contentious decision, the Competition Council eventually voted not to approve the privatization.

The events surrounding the privatization of Reşiţa Steelworks, an enterprise known for its militant workforce in the post-1989 period—the so-called Reşiţa phenomenon—also followed the pattern of labor unrest taking place only *after* a problematic privatization.[27] When the indebted Reşiţa Steelworks was put up for sale in 1998, the negotiations with the buyer, Noble Ventures, Inc., were shrouded in secrecy—a fact greatly resented by the enterprise union. As the UniRomSider representative said, there was a "total lack of transparency. It was a secret of the atomic bomb . . . nobody was allowed to know anything about the negotiations."[28] Indeed, the trade union leader claimed that the union was "not consulted, we were not even invited to the signing of the privatization contract. And all of the sudden, we woke up in a privatized firm."[29] From the beginning, the union was highly suspicious of the new owner, whom it perceived to lack the financial resources necessary to invest in Reşiţa.[30] Despite the union misgivings, however, there were no protests against privatization as such.

Soon after the privatization, Noble Ventures negotiated a collective work contract with the union, agreeing not to engage in group layoffs until 2005. Steelworks employees claimed in the interviews that they had been ambivalent about the privatization but hopeful: "[I]t's not that we didn't want the Americans here, no, we hoped that they would turn the enterprise around, but then they didn't pay us."[31]

The union quickly highlighted the new owner's deficiencies and became intransigent after the delays in wage payment and an interruption of the steelworks' operation due to the nonpayment of utility bills.[32] As far as the union was concerned, the privatization was a "total failure."[33] What ensued was a convoluted year-and-a-half-long social conflict inside the enterprise, which was stoked and rekindled by the repeated problems making wage payments. In addition to frequent protests, including two hunger strikes, there were two allegations of assault by the representatives of the new owners, who complained of "commando-type actions" on the part of union activists.[34] Even though the state authorities, as well as the Metarom and Cartel Alfa union organizations, stepped in at different points to attempt to diffuse the conflict, the enterprise-level union proved inflexible in its call for the cancellation of the privatization contract. At the same time, it also called for finding a new investor for the enterprise.

As in the case of Reşiţa, so, too, in Socomet: the sale of the enterprise to Gavazzi brothers resulted in problems with production, which eventually ground to a halt. The unpaid, abandoned workforce became increasingly hostile and, under the leadership of the enterprise-level union, engaged in repeated roadblocking, demonstrations, and even a hunger strike prior to the eventual cancellation of the privatization contract.[35]

If the failed privatizations of Tepro, Petrotub, Reşiţa, and Socomet were marked by a lack of social dialogue, the two cases of relative success, Silcotub and Artrom, illustrate its benefits. Silcotub, a pipe producer with a 200,000-ton production capacity, located in Zalău, was continuing a restructuring program when it was privatized. At the end of March 1999, Tubman International, a Gibraltar-based enterprise owned by Duferco, which specialized in steel production and distribution, purchased the enterprise.[36]

The enterprise union was apprehensive about the consequences of privatization, but it did not engage in protests to prevent it.[37] Based on the restructuring plan, employees expected that 257 workers would be laid off, at an enterprise having about 2,000 workers.[38] However, three months after privatization, the new owners announced that another 250 workers would be laid off. After the union leader "acted with all the legal means to forestall this, representatives of Tubman showed willingness to negotiate the number of workers to be laid off."[39] The union leader underscored that despite these developments, there was no doubt concerning the seriousness of Duferco's involvement in the enterprise.[40] This was clearly shown not only through Silcotub's increased profitability but also through investments undertaken by the enterprise.[41]

In the case of Silcotub, the new investors, to diffuse tensions, showed a willingness to engage in social dialogue following privatization . However, in the case of the Artrom privatization, social dialogue played an important role throughout the process. Artrom, a small pipe producer located in Slatina, had a 60,000-ton-a-year production capacity and 1,380 workers at the end of 1997, when the enterprise was first put up for sale. It did not have the ability to attract national attention and seek the government financial support that some of the bigger enterprises had received.[42] However, an interview with the enterprise-level union leader, as well as press accounts, indicate that social dialogue played a more prominent role than it had in the other cases of "first-wave" privatizations.

The union's initial attitude to privatization was reserved, as in the other cases, but the workers understood the gravity of their enterprise's economic situation. Whereas Artrom was barely profitable in 1997, it registered losses in 1998.[43] The enterprise was also burdened with a $15 million bank loan from a German bank. The union leader said, "At first we were afraid [of privatization] but the enterprise was functioning very poorly, so we agreed, we embraced the idea. Privatization was the only solution. It was the only hope to make the company function."[44]

The privatization process was completed in May 1999, with nearly 58% of the company's shares purchased by an Austrian investor, Staro Stahl-und Röhrenhandel.[45] The union felt that it had been involved in the privatization process:

> We were consulted. We were told what is ahead. We could not decide whether to sell or not, we could not decide about the privatization clauses, but we tried to keep as many people as possible and to ensure social insurance. First of all, we were interested in workplaces, and secondly, we wanted wages.[46]

In fact, shortly after privatization, shareholders adopted a restructuring plan that entailed closures of loss-making facilities. This, in turn, meant laying off 320 workers out of the remaining 1,040-strong workforce. These new layoffs were to involve severance payments. The union leader went on to say, "The foreign investor was afraid about how the workers would react. Little by little, people understood the new reality. Many were laid off but those who remained obtained salary worth 50% more than the previous level. That meant something."[47]

In fact, in the case of Artrom, the union leader also praised the improved quality of social dialogue inside the enterprise in the post-privatization period and, assessing the privatization, said, "A good thing was done, it was actually a miracle because a company, which was performing really badly, is now functioning very well. People earn well, they get bonuses."[48]

Unfortunately, the case of Artrom was an outlier in the first privatization wave, which was dogged by both poor state capacity and the absence of social dialogue in the run-up to the transactions. However, social protests transpired only *after* the privatizations had exhibited signs of failure, rather than in response to the privatization decisions as such. Where the opposition to privatization was strong ex ante was at the Sidex plant. The Sidex privatization opened the second wave of privatizations of the biggest— hence most problematic—enterprises. The external pressures that led to the decision to privatize these behemoths are the subject of chapter 6; the role labor and social dialogue played in these privatizations is discussed here.

Following the split from the Cartel Alfa Confederation, the Galaţi- based union enjoyed a privileged position, including wages that were substantially above the sector mean.[49] Throughout the transition period, union leaders turned a blind eye to the financial abuses taking place inside the enterprise. According to a Cartel Alfa vice president, the union leaders in Galaţi tried to preserve the status quo, because they, too, benefited finan- cially from it, courtesy of managerial largesse.[50] At the same time, however, the reputation of Galaţi as a "'social bomb' if the government did not ful- fill the trade union demands" served as a useful excuse for the managerial interests to maintain the status quo.[51] Any discussion of privatization was summarily delayed by the patently politically connected managers, who reaped rents from business exchanges with Sidex.

When, under external pressure, the government of Romania announced the privatization of Sidex in 2000, the Galaţi Ferrous Metallurgy Workers' Trade Union Federation acted in an obstructionist manner. Although one could argue that given the recent history of privatization failure, the union's caution was understandable, the union also made impossible demands. It may have been reasonable for the union to lobby the government to adopt a special decree for the privatization of Sidex. It was less so for it to demand that the new investor either deposit a guarantee at the Romanian National

Bank, equal to the value of the envisioned investments, or make an injection of working capital, in cash, equal to half the prospective investments. According to the union, the fulfillment of one of these conditions was seen as the only guarantee that the privatization would be proper and that it would not be followed by social unrest. Moreover, nonfulfillment of these wishes was seen as "a grave attack on the future of Sidex and the national economy."[52]

When, a day later, the FPS president said that the FPS board had approved the privatization announcement for Sidex and that the unions agreed with its contents, the union reaction was one of vehement opposition. The union leader stated that until all the unionists' conditions were accepted, the union would not agree to the privatization. He added that if the government and the FPS did not take union opinions into account, no FPS employee would be able to enter the enterprise.[53] Despite assurances from the government that the privatization process would take months, the enterprise-level union leaders resisted privatization, claiming that they were afraid the government would not be held accountable for a rushed privatization deal concluded shortly before the elections it was widely expected to lose.[54] In fact, in an attempt to postpone the privatization process until after the elections, the union actually blocked the FPS representative from entering the general shareholders' meeting at Sidex, claiming that the meeting would "take place neither today . . . nor tomorrow."[55]

However, when the new communist-successor-party-led government won the elections and continued the privatization process its predecessors had announced, the union relented. Statements, such as those by the minister of privatization, who, while pointing to the asset-stripping activities inside the firm, also proclaimed that the Sidex workers needed to understand that the "simple desire to cast steel" no longer sufficed in a globalized world, made the message of the new government clear.[56] At the same time, the new government went out of its way to ameliorate the social consequences of privatization, and the final privatization deal entailed employment restructuring based only on natural attrition for a period of five years.[57] As for the labor unions, with the political incentives for division gone as a consequence of privatization, the Galați Ferrous Metallurgy Workers' Trade Union Federation came back under the Cartel Alfa umbrella in 2003, and in 2006, it merged with Metarom.[58]

The second wave of privatization was generally much better prepared than the first one, in large part because, at the insistence of the IFIs, transactions were drawn up and prepared using professional consulting companies, buttressing the wanting capacity of the Romanian state.[59] However, at the initiative of the government, which had learned from the first-wave failures, they were also accompanied by much more vibrant social dialogue.[60] In 2002, the prime minister set up a tripartite interministerial committee intended to help coordinate the restructuring and privatization process. The committee included the Ministry of Industry and Trade; the Ministry of Finance; and the Ministry of Work, Social Solidarity, and Family, in addition to the Authority for Privatization and Administration of State Ownership (Autoritatea pentru Privatizarea şi Administrarea Participaţiilor Statului [APAPS], the 2001 successor to the FPS) on the government side, and Metarom, the Galaţi union, and UniRomSider on the social partner side. Nonetheless, according to Metarom, the committee's actual influence was limited.[61] Rather, it was up to APAPS to negotiate the social component of the eventual privatization package, including the exact conditions for redundancies.[62]

The sectoral union was not enthusiastic about the privatization process and bemoaned the lack of the kind of far-reaching pre-privatization restructuring that would make it possible to sell the enterprises at a higher price. However, it did not impede the process.[63] As for the enterprise-level unions, they were generally consulted, and post-privatization employment-related provisions had been specified in the privatization contract. The biggest contrast with the first wave was the case of Reşiţa, where both the APAPS privatization adviser and the union leader praised mutual cooperation throughout the process of selling the enterprise to the Russian group TMK. According to the Reşiţa union leader:

> [T]his time, we were consulted by APAPS, by Mr. Muşetescu [minister of privatization], by the Russian investor . . . the Russians were interested in what the trade union's opinion was and they said they would not come to a company, which had big social conflicts. . . . We had discussions with them for two months and struck a deal.[64]

The assessment of the union leader at the Special Steels Enterprise at Târgovişte (Combinatul de Oţeluri Speciale Târgovişte) was similar, "[W]e were consulted only as far as the social issues are concerned. These were reduced to the promise that for five years, there would be no layoffs. We negotiated this."[65] In the case of Tepro Iaşi, the privatization contract provided for retaining the 1,400 workforce for a period of two-and-a-half years.[66]

However, in the absence of pre-privatization sectoral employment restructuring, the enterprise-by-enterprise approach led to tensions over the expected size of the layoffs. In two of the five second-wave privatization cases, Siderurgica Hunedoara and Petrotub Roman, the negotiations were accompanied by worker unrest over the size of layoffs, and the unions at both enterprises were demanding the same employment restructuring conditions as Sidex.[67]

LNM Holdings—the prospective buyer of both enterprises—clearly stated that based on its own experience, it was "better if the government restructures [employment] before they privatize."[68] Therefore, LNM was willing to participate in some of the employment restructuring costs, but it stipulated that "before we closed the transaction they [the Romanian state] had to reduce employees, only then [would we] t[a]k[e] over the company . . . otherwise it was not attractive for us."[69] Indeed, following the conflict with the enterprise-level union over the scope of severance payments and the number of workers to be laid off, the union and the state found a mutually satisfactory solution, in which each employee would receive severance payments equal to between twenty and twenty-four months' wages.[70] The eventual privatization contract that followed several months later, however, provided for yet new layoffs that would reduce the total number of employees in the enterprise from 5,000 to 2,200, albeit with the same severance payment amount as in the previous round. Initial union protests over the layoffs were eventually resolved through discussions with LNM over the schedule of the severance payments.[71] Moreover, the government agreed to set up an industrial zone and to engage in some professional reconversion measures, with the help of EU funding, in order to create new workplaces.[72]

In the case of Petrotub, the privatization agreement stipulated the laying off of up to 1,700 workers, bringing total employment down to 3,100 workers from 4,800. The enterprise-level union also initially resisted the measures.[73] Union protests resulted in retaining one hundred more workplaces as well as no further layoffs for the next two and a half years. In general, the union at Petrotub considered the privatization to be an expression of the Romanian government's "lamentable lackey mentality" in accepting privatizations dictated "from outside."[74] That assessment, however, was far from universal among Romanian union leaders, some of whom—while bemoaning the weak Romanian state and the fact that "everything that was to be stolen was stolen already," called international influence "the last chance for Romania—the last chance!"[75]

We now turn to the Slovak case, where political incentives also brought about a split in the union movement and capture of the union in the major enterprise.

Social Dialogue and Restructuring in Slovakia

Tripartite social dialogue, created in Czechoslovakia at the national level in 1990, had accompanied Slovak industrial relations from the beginning of the transition process, and the centralized union structure adopted an ostensibly apolitical stance. Despite its formal corporatist structure, however, the union movement had proven unable to withstand government efforts to break it from within. VSŽ turned out to be the weak link as its union broke labor unity and established patrimonial relations with the illiberal Mečiar regime and his HZDS party.

VSŽ was the dominant member of the Slovak sectoral union organization, OZ KOVO, the strongest sectoral organization within the Slovak Confederation of Trade Unions (Konfederácia Odborových Zväzov [KOZ]). In addition to metallurgy, OZ KOVO represented workers in the engineering and electrical engineering sectors. As noted earlier, similarly to the management, the VSŽ union quickly established patrimonial relations with the Mečiar government.

After the September 1994 elections, the VSŽ-level union, under the leadership of Jaroslav Gruber, lobbied the FNM to sell the remaining 25% of VSŽ shares to Hutník, a company set up by the enterprise union organization. In February of 1995, FNM sold 10% of VSŽ shares to Hutník at the same heavily discounted price at which it "sold" the company's shares to the managerial company—Manager—that is, at 20% of their book value.[76] Moreover, in a surprising move for a union, together with the management, the union established a company, Ferrimex. A quarter of Ferrimex's shares were owned by Hutník, and the remainder by the managerial firm, ARDS. In July of 1995, FNM sold the remaining 15% of VSŽ shares to Ferrimex.[77] In essence, given the relatively large portion of shares at its disposal, the enterprise-level trade union became both a co-owner and a co-manager of the enterprise, a rather dubious mandate for a representative of worker interests.

Not surprisingly, the business activities of the VSŽ union did not receive the support of the other enterprise-level unions within OZ KOVO or the approval of the OZ KOVO leadership, which perceived this as a blatant co-optation by the Mečiar regime.[78] The co-optation was completed when, in November 1996, Gruber set up a rival enterprise-level union, Metalurg, and convinced an overwhelming majority of workers to leave OZ KOVO and join Metalurg instead. The official reasons given were dissatisfaction with the operation of OZ KOVO, including overly centralized structure, lack of democracy (apparently resulting from voting down the union leader's concerns), and bureaucratization within the organization. Yet, for the remaining unions within KOVO and KOZ, the move was an effort by the Mečiar regime to challenge union solidarity and weaken the union movement at a point when it was starting to become more critical of Mečiar's politics and policies and assertive in its demands for partnership.[79]

Seeing the government approach social dialogue in a high-handed manner, KOZ subsequently refused to sign the 1997 general agreement.[80] Moreover, in the 1998 elections, while not explicitly endorsing a single political party, KOZ, including OZ KOVO, supported the anti-Mečiar opposition and, initially, even the set of austerity measures it proposed.[81] These developments should come as a big surprise to the proponents of the executive insulation model. They illustrate that the autonomous sectoral union organization opposed the rent-seeking government and rogue union practices and supported continued economic reform.

By contrast, the captured Metalurg was clearly opposed to reform and implicated in the rent-seeking activities within VSŽ. According to an enterprise insider, Metalurg was "the gamekeeper-turned-poacher."[82] At the same time, the *short-term* interests of the workers were being protected, with no group layoffs and with higher wages in comparison with the other enterprises in the sector.[83] The workers, however, were growing disillusioned with the management and with the rampant theft they were witnessing.

The extent of Metalurg's complicity and collaboration with the VSŽ management and Mečiar's HZDS became clear after the declaration of cross-defaults by international lenders, which placed the enterprise on the brink of bankruptcy. Rather than support subsequent restructuring measures—all the more so because they did not entail redundancies— Metalurg did everything in its power to maintain the status quo.

By December 1998, the lender banks demanded a change in management. With the "kindermanagement" stepping down, an extraordinary shareholders meeting elected the foreign-bank-approved Gabriel Eichler as president.[84] From the moment of Eichler's entry into VSŽ, the struggle for control over the enterprise began. The task of the new president was to convince the banks to wait for the results of immediate restructuring and reprivatization to a strategic investor, rather than press for bankruptcy. He also started to prepare the enterprise for a sale to an investor who would ensure its long-term development.[85]

Eichler's restructuring as well as privatization efforts encountered strong resistance from the supporters of the old management, including the union leader. Representing Hutník, Jaroslav Gruber fought hard to retain control of the enterprise. Even though, in January 1999, Gruber proclaimed, "We will not solve this situation by ourselves, the entry of a foreign partner is necessary," his subsequent behavior served to obstruct this entry.[86] In the summer of 1999, reports emerged of a secret meeting in Spain of the Mečiar–Rezeš et al–Gruber triumvirate, intended to devise a strategy for retaining control of VSŽ. Gruber's explanations concerning the supposedly impromptu meeting seemed less than convincing:

> It's not that we were preparing a coup but that Mr. Mečiar wanted to gather together, one may say, dispersed former collaborators within VSŽ, that means Smerek-Rezeš-Gruber, where Mr. Drabik also happened to participate. That's it. In no way was there talk about a definite strategy for VSŽ.[87]

Yet, by August 1999, working in collaboration with the domestic managerial strata, including the pro-Rezeš former supervisory board members, Gruber, in an effort to fire Eichler, initiated a motion to call an extraordinary shareholders' meeting.[88] The outcome of this meeting was highly uncertain, as the shareholders' expected votes were split between the pro- and anti-Eichler factions, and the rest were undecided.[89]

The extraordinary shareholders' meeting confirmed the restructuring and privatization plan promoted by the Eichler management while repelling the pro-Rezeš lobby. In fact, Gruber was voted down from his supervisory board seat by a vote of nearly 79%.[90] Although Gruber threatened to bring a court case to invalidate the vote, when asked by reporters, he said he would not resort to organizing protest actions and strikes within the enterprise.[91] This attitude suggests that the Metalurg union, closely associated with the Rezeš management, did not have worker support. An enterprise insider claimed that the union had simply become irrelevant for the workers, whose main concern was the survival of the enterprise—that is, the retention of workplaces.[92]

Nonetheless, Gruber continued his opposition to the actions of the new management. After long negotiations between the new VSŽ management, the Slovak government, and U.S. Steel, in March 2000, VSŽ, U.S. Steel and the prime minister of Slovakia signed a memorandum of understanding outlining the terms of U.S. Steel's entry. These terms—including retention of full employment for the next ten years, and employment restructuring only via natural attrition—was decried by Gruber as "cheating the company, the workers, and Slovakia."[93] In the end, however, Hutník threw in the towel and ended up signing onto the privatization deal, voting, along with 99% of present shareholders, in favor of the entry of U.S. Steel.[94] Following the sale, Metalurg decided that Hutník's mission was complete and that it should end its business activities. After Gruber's departure from Metalurg, the relationship between the union and management improved.[95] One could also note a rapprochement between OZ KOVO and Metalurg, and an increasingly close cooperation at the sectoral level.[96] In fact, by the end of 2009, Metalurg had re-entered the OZ KOVO fold at both the enterprise and sectoral levels, after union leaders who had supported the original split retired.

Thus, in the Slovak case, as in the case of Romania, an illiberal government successfully co-opted a powerful enterprise-level union, which then acted to protect the status quo. Counter the outcome to be expected from the executive insulation model, the autonomous sectoral union organization, unlike the captured union, supported continued economic reform.

Social Dialogue and Restructuring in Poland

In the Polish case, autonomous union organizations supported the development of sectoral-level social dialogue. They also supported, and even took the initiative on, restructuring. In what reflected the unions' commitment to the welfare of their enterprises, they demanded clear pro-reform action at a time when the government was balking.

The evolution of Polish social dialogue took quite a long time. In the surprise of Polish transformation, the Solidarity trade union, which had led the struggle against communism, subsequently supported neoliberal reforms. Social dialogue at the national level was originally neglected based on the assumption that Solidarity's presence in government translated into a representation of worker interests. This logic was proven wrong by social unrest, in the summers of 1992 and 1993, that eventually led to the establishment of a tripartite commission at the national level.

Initiated in 1994, the tripartite dialogue at the national level was marked by political alliances of Solidarity and the OPZZ (Ogólnopolskie Porozumienie Związków Zawodowych, or All-Poland Trade Union Alliance, the successor to the communist-era trade union organization) with the post-Solidarity and communist successor parties, respectively. The union landscape was also characterized by numerous unions being represented within a single enterprise; the number ranged from two in Warsaw Steelworks to fourteen in Katowice Steelworks. The pronounced union politicization at the national level was far less significant at the sectoral and enterprise levels, where, in interviews, union leaders of various ideological orientations highlighted their relatively smooth cooperation.[97]

Early in the transition process, the identity of unions and management often blurred—a development reinforced by the initial ability of workers to elect the management of SOEs, such as Emil Wąsacz, a former Solidarity leader within Katowice Steelworks who was elected as the enterprise's director in 1991. In Sendzimir Steelworks, immediately following system change, Solidarity became an active advocate of restructuring and lobbied for employee ownership and the formation of enterprise subsidiaries.[98] In 1989, Solidarity created the Commission for Ownership Transformation at the enterprise because, in the words of the union leaders, "the management at the time had little desire to do anything in the direction of restructuring."[99] The commission's president later became the director of the Office of Restructuring at Sendzimir and thus entered the top managerial ranks. In the meantime, the new general director of Sendzimir, elected in 1992, enjoyed great union support.[100]

The unions' initial role as de facto co-managers of the enterprises is illustrated by the privatization of Warsaw Steelworks. There, the workers had demonstrated a commitment to their financially embattled enterprise, even at the cost of accepting temporary salary cuts.[101] Although the management was split in its support for privatization, the unions made an effort to attract the Italian investor to their company.[102] As a former Solidarity leader at the enterprise described it, as the unionists showed the Italian investors around, "We wanted to act as a host for the new investor."[103]

Although the unions were initially supportive of bringing in a foreign investor, they became disappointed in the aftermath of privatization. They resented the working conditions and the fact that the enterprise's productivity was rising far more quickly than the wages.[104] The unions also criticized investment delay, and in 1994, the two enterprise-level unions, affiliated with Solidarity and OPZZ, engaged in a two-month-long strike. According to both the management and union leaders—speaking from an almost ten-year perspective—the strike represented a power struggle that led to the establishment of a balance of power and delineated the respective roles of the management and the unions within the enterprise.[105]

In the remaining SOEs in the sector, in an environment of institutional uncertainty, the unions often supported managerial investment visions. According to Edward Nowak, former deputy minister of economy, there was a symbiotic relationship, an informal agreement, between the unions and the managers who felt that the unions could be used to exert political pressure.[106]

However, even when union and managerial interests coincided on technological—and even employment—restructuring, wages became a bone of contention at the beginning of the transition process.[107] According to a survey carried out in the sector, during the 1990s, there were strikes in 40% of the surveyed enterprises, all occurring prior to 1995.[108] These findings were seconded by an interviewed representative of the Association of Metallurgical Employers, who claimed that the peak of social unrest in the steel sector took place in the early 1990s—around 1992 and 1993—when the enterprise-level unions demanded salary increases and bonuses equivalent to those granted in other mills.[109]

In the absence of social dialogue at the sectoral level, the more radical proposals of the smaller unions within a single enterprise contributed to social volatility. Many of these were splinters from the Solidarity movement and did not share its more conciliatory stance toward economic reform. Even though their often spectacular protests were dismissed by the established unions as "attempts at marking their existence," they were nonetheless capable of radicalizing the overall worker demands within the enterprise. The most spectacular example was the eighteen-day June 1994 wage strike at Katowice Steelworks, directed against Wąsacz's leadership, organized by Sierpień 80 (August 80), which had only forty-nine members inside the enterprise. Nonetheless, it was able to rally the workforce behind it, while the dominant Solidarity union did not back its former leader.[110]

These wage pressures could be viewed as evidence of the nefarious influence of unions on the reform process, yielding support for the executive insulation model. However, in line with the critique of that model, the institutionalization of social dialogue at the sectoral level put an effective stop to wage strikes.

Social unrest at the enterprise level led to union pressure to organize social dialogue at the sectoral level.[111] In 1995, this resulted in the creation of the Tripartite Team for Social Conditions of Steel Industry Restructuring (Zespół Trójstronny ds. Społecznych Warunków Restrukturyzacji Hutnictwa), one of the very first tripartite commissions formed at the sectoral level. It entailed civil servants from four ministries, five union organizations, and the Association of Metallurgical Employers.[112] The Tripartite Team then negotiated a sectoral-level agreement, establishing minimum wage in the sector and regulating overtime wages, bonuses, and other financial benefits.[113] The signing of the Sectoral-Level Collective Agreement in 1996 resulted in much smoother relations at the enterprise level. In the words of the Association of Metallurgical Employers' representative, "[S]ince 1996 we [have] had no wage-based strike in the Polish steel industry."[114]

At the same time, the managers were reluctant to lay off workers, as any such measures remained contentious, and no sectoral-level compensation packages had been devised. The progressive decrease in the number of workers in the sector took place overwhelmingly via spin-offs and natural attrition. Speaking in a 1998 interview, the former Solidarity union activist and the director of the Sendzimir Steelworks' Office of Restructuring, claimed, "Throughout the restructuring process we did not fire a single worker. All got an opportunity to work."[115] The situation was similar at Katowice Steelworks. When the manager responsible for restructuring and development was asked, at the end of 1997, whether Katowice was considering the possibility of reducing employment instead of continuing enormous investments, he replied, "We do not even want to think about this. Our principle is development without layoffs."[116]

Nonetheless, as the EU pressure to comply with the EU *acquis* requirements in the steel domain increased, it became clear that more intensive restructuring steps were inevitable. According to a Solidarity sectoral-level leader, the unions realized that job losses would occur during the acceleration of the restructuring process. Consequently, the unions wanted to make sure that "no one would be left on the street"; hence, they collectively pushed for a sectoral solution for workers who would be laid off.[117] These efforts resulted in the negotiation of the Social Package for the Steel Industry (Hutniczy Pakiet Socjalny), signed in January 1999 by the Tripartite Team for Social Conditions of Steel Industry Restructuring. The package provided for the establishment of additional severance payments for specific high-risk groups of laid-off workers.

The agreement paved the way for the departure of 57,000 workers during 1999–2002, with 27,400 transferred to spun-off firms. The total cost of employment restructuring amounted to about $120 million, out of which the government directly covered about $28 million.[118] This package was followed by another one, which was to accompany the final restructuring process geared at meeting the EU viability criteria, the Activation Package for Steel Industry (Hutniczy Pakiet Aktywizujący), signed in 2003. In short, unions accepted the necessity of restructuring in the sector and aimed at minimizing its destructive social effects. They did not just react to government initiatives, but proposed policies of their own. In addition to the sectoral solutions, some of the union policies actually encouraged entrepreneurship and self-employment.[119]

Importantly, as the privatization of Warsaw Steelworks foreshadowed, the unions were not an obstacle in the way of privatization to foreign investors. Rather, as the financial pressures on the government grew, and as the EU pressure intensified, successive governments vacillated as to how to respond: whether to engage in sectoral consolidation, coordinated privatization, or both. The details of these policy choices will be discussed in the next chapter, but suffice it to say that by December 2002, the government had merged Katowice Steelworks and Sendzimir Steelworks, which, along with the much smaller Florian Steelworks and Cedler Steelworks formed Polish Steelworks (Polskie Huty Stali). The unions not only did not protest against the postmerger privatization, but actually criticized the government for stalling.

Privatization talks resumed only in the spring of 2003 and the sale took place in October 2003, just in time to administer the state aid negotiated with the EU. As the government engaged in nothing short of brinksmanship in its negotiations with potential investors, the unions grew restless. They were dismayed by the government's slow progress regarding privatization and concerned by the uncertainty of their enterprise's future. The Solidarity sectoral-level union leader said at the time, "The feeling right now is that we just want to know who the investor will be." The unions could then proceed to negotiate the privatization-related social packet.[120]

The unions in Polish Steelworks went even further, forming a multi-union committee and issuing a letter to the minister of state treasury, responsible for the privatization process, in which they expressed "great concern" about the tempo of the "decisions related to privatization, for which we have been waiting for a number of months."[121] The committee bemoaned the "lack of clear decisions of the Polish government regarding the future of Polish Steelworks and the attempts carried out by certain circles in the last weeks, aimed at stopping the decisions pertaining to the future of the firm."[122] Furthermore, the unions appealed to the minister to "take constant and direct control over the privatization process of Polish Steelworks, giving the enterprise an opportunity to develop, which would make it possible not only to retain existing workplaces, but also create an opportunity for their increase."[123]

In the Polish case, although the government dithered, the unions clearly favored an acceleration of the selection of the foreign investor. As the following section illustrates, a similar development transpired in the Czech Republic.

Social Dialogue and Restructuring in the Czech Republic

Several key developments characterized social dialogue in the Czech steel sector. First, the unions were autonomous and nonpoliticized, with no noted co-optation by the government.[124] Second, they were organized at the sectoral level and participated in social dialogue with the government. Rather than block reform in the sector, the unions served as partners in the restructuring process, devising proactive measures to ease the social effects of restructuring. However, the Czech unions went further, and in advocating privatization to strategic foreign investors in the late 1990s, they seemed to be more reform-oriented than the government.

The national-level tripartite dialogue started in 1990 in Czechoslovakia, and it helped to reduce tensions in the industry by assigning clear social-partner roles to unions and managers. Although the assessment of the capacity of social dialogue to impact economic reform in the Czech Republic differs, some seeing it as a source of transformative power and others perceiving it as a rubber stamp of neoliberal reform, there exists relative consensus concerning social dialogue's ability to maintain social peace in the face of reform.[125]

The national-level social dialogue set the stage for sectoral-level dialogue between the employers' association and the union organization Odborový Svaz KOVO (OS KOVO), which eventually began in 1995, when the first sectoral-level collective agreement was signed. Moreover, the enterprise-level unions, supported by the sectoral- and even national-level union organization, have played a watchdog role vis-à-vis management and overwhelmingly defended the restructuring process. In a statement that summarizes the unions' attitudes well, an interviewed OS KOVO leader said, "We do not have the goal to manage the company. . . . That is the issue for the owner and for the management. But we will judge the steps of the management . . . and if need be, suggest some other ones."[126]

In voicing their dissent, the unions rarely resorted to direct strike action and tended to proclaim strike emergencies or engaged in short-duration (e.g., hour-long) work stoppages instead. The interviewed trade union leaders, employer association representatives, and the Ministry of Industry officials involved in social dialogue at the sectoral level confirmed that the relationship with the trade unions remained harmonious throughout the transition process.[127]

Reflecting on the gradual changes in the sector, the former minister of industry pointed to union strength in the industrialized North Moravia region.[128] Yet, while admirable in principle, social sensitivity masked the opaque buy-out deals with the management of the biggest enterprises in the sector—Nová Huť and Vítkovice.[129] Ultimately, these deals failed and left the enterprises in dire financial straits. The unions made it clear that they understood the gravity of the situation at both the enterprise and sectoral levels.

At the enterprise level, the 1996 Poldi debacle foreshadowed union attitudes. When the new owner, Vladimír Stehlík, fought fiercely to retain control of the failed enterprise, he called government actions attempts at "eradicating a Czech businessman," and organized overtly political rallies in the center of Prague.[130] These drew several thousand protesters from the enterprise and the town of Kladno. In stark contrast to the developments in Slovakia, the enterprise-level union not only did not rush to defend Stehlík but also pushed for his departure.[131] It stood firmly against worker participation in these demonstrations, seeing them as efforts to manipulate the workers.[132] Also in contrast to the Slovak case, the enterprise-level union cooperated closely with both the sectoral-level federation and with the national confederation, the Czech-Moravian Chamber of Trade Unions.[133] In the meantime, the enterprise-level union became involved in a "crisis team," with the participation of government actors, to consider possible solutions to Poldi's grave situation.[134]

When financial crisis began to impact the sector, the unions, under the OS KOVO umbrella, became more involved in the restructuring process. Like their Polish counterparts, they pushed for the establishment of a sectoral-level tripartite commission, the so-called Consultative Team (Konzultativní Tým) to establish financial support for workers who were bound to be laid off as part of employment restructuring in the sector.

Organized in 1999 and composed of union leaders, employers, and the Ministry of Industry and Trade civil servants, by 2000, the Team created the Associated Social Program (Doprovodný Sociální Program) intended to accompany the restructuring process until the end of 2003, then extended until the end of 2006—that is, the end of the official restructuring period negotiated with the EU.[135] The program provided support for laid-off workers meeting specific conditions, mostly pertaining to early retirement schemes not made explicit in the enterprise-level collective agreement.

From 2000 until 2003, the state paid over \$30 million to support 5,400 workers who used the program.[136]

While the unions realized the critical circumstances of the sector, the successive Czech governments, both on the right and on the left, attempted to retain the Czech steel sector in the hands of domestic owners, rather than search for foreign investors. Nonetheless, the previously mentioned management buyout schemes ended in resounding failure. In an effort to save the majority state-owned Vítkovice from bankruptcy in the aftermath of this failed initiative, the Czech government eventually brought in a crisis manager, Václav Novák. Novák and his team subsequently faced a complicated relationship with state officials for whom the pace of change in the enterprise was too fast.

If the advocates of the executive insulation model were correct, one would have expected the unions to block reform. However, Novák underscored that his greatest ally in restructuring was the company union, supported by the sectoral organization, "[T]he unions were fully behind me, which is surprising because I was so radical."[137] According to Novák, for the unionists, "[T]he company was their life and they would do a lot for the company." "More than anyone else," they realized the danger of bankruptcy. Social dialogue became the cornerstone of the radical restructuring strategy:

> I knew that it was going to be important and that's why I tried to have them on my side. That's why I went to their meeting every two weeks, explaining the progress. I had a progress report which was much more detailed for them than for the supervisory board.[138]

The situation was similar in Nová Huť. With the plant facing liquidity problems, the Czech government put forth a series of controversial initiatives, attempting to consolidate the sector and to create one giant enterprise to be restructured and later sold to a foreign investor. The initiative met with opposition from numerous quarters, including other enterprises and unions. In what may seem a surprising move, the Nová Huť union explicitly demanded the sale of the enterprise to a foreign investor capable of ensuring the steelworks' future. Nová Huť steelworkers even protested in Prague and declared a strike emergency to support the entry of such an investor.[139] As will be discussed in the following chapter, under heavy external pressure, the government decided to scrap its consolidation program and opted for quick privatization deal with LNM Holdings, before the upcoming 2002 elections. However, it is important to keep in mind that this was not the first time the union had raised such demands. In 2000, the union organized a warning strike as it demanded that the management step down to open the way for privatization to a strategic investor.[140]

Summing up, one can say that the Czech unions were a supportive force for enterprise reform. As economic circumstances called for more radical restructuring, unions cooperated in a proactive way as they insisted on social support for laid-off workers. They went even further, however, spurring on a reluctant government to privatize to foreign investors.

Conclusion

The history of social dialogue in all four country cases provides strong evidence against the executive insulation model. The lack of significant wage-related social unrest was associated with the development of tripartite social dialogue at the national level in Czechoslovakia early in the transition period. The Polish case, however, goes beyond correlation and clearly illustrates a causal relationship between the institutionalization of social dialogue and social peace. Both union and employer-association representatives in Poland attributed the absence of wage strikes to the sectoral-level agreements concerning wages. Thus, the Czech and Polish cases strongly support proposition 5 and the insights of critics of the executive insulation model.

However, the evidence in the Czech and Polish cases goes further than what is stated by the critics of the executive insulation model and indicates that the autonomous organizations supported restructuring and privatization *even in the absence* of a squarely pro-reform government. In both countries, the sectoral labor organizations took the initiative to organize financial and job-training support to alleviate the negative social effects of the anticipated employment restructuring/layoffs in the sector. Second, in the key privatizations in both countries, the enterprise-level unions actually demanded that the government speed up the privatization process, so as to secure the future of the workers of the enterprises in question. Thus, the Czech and Polish cases provide support for proposition 5a, which does not predicate union attitudes on the pro-reform inclination of the government.

The Slovak and Romanian cases provide a stark contrast to the Czech and Polish ones, though there are notable differences between them. In Slovakia and Romania, the fate of social dialogue was tied to a lower quality of democracy. In Slovakia, the illiberal Mečiar regime managed to split the labor movement by capturing the union at the biggest steelworks in the country. The captured union's opposition to privatization clearly supports proposition 6. However, in what provides evidence against the executive insulation model and supports proposition 5a, the severely weakened sectoral-level organization remained supportive of the economic reform process. It saw reform as the key to providing long-term developmental prospects for the economy. Therefore, the sectoral organization not only supported the privatization of VSŽ to U.S. Steel but also endorsed the pro-reform forces in the crucial 1998 elections.

The Romanian case is more nuanced. As in Slovakia, only earlier, the illiberal government split the union movement to weaken the nascent civil society. The captured union was certainly not supportive of reform, but the politically connected Sidex managers preempted any radical moves on the union's part, effectively sabotaging any privatization announcements. Only after the anti-reform coalition within the enterprise saw opposing privatization as a losing fight, did the union acquiesce to it. Thus, the Romanian case clearly offers support for proposition 6.

The evidence against the executive insulation thesis is also strong in Romania. The autonomous Metarom Federation advocated, for a long time in vain, employment restructuring support measures similar to those of its Czech and Polish counterparts. Rather than block reform, the organization tried to be a partner in the restructuring effort, even if its sway over the individual enterprise unions was relatively weak. Moreover, the fact that the lack of privatization-related social dialogue at the enterprise level in the first privatization wave only created union uncertainty and fueled, mostly justified, suspicions clearly contradicts the executive insulation thesis. By contrast, far wider consultation with social partners during the second wave of privatization created a much smoother and more transparent privatization process.

The low capacity of the Romanian state also explains why the support for proposition 5a in Romania may be weaker than in the other cases. Although Metarom wanted the state to be a partner in restructuring, taught by painful experience, it did not see privatization to foreign investors as the best solution because it simply did not trust the state to select *strategic* investors. Moreover, the low capacity of the Romanian state explains the nonimplementation of sectoral-level employment restructuring measures similar to those in Poland and the Czech Republic.

Thus, the case studies illustrate the correlation between quality of democracy and state capacity. Social dialogue requires substantial state capacity. At the same time, the evidence in this chapter suggests that there is no trade-off between economic efficiency and democratic values as far as social dialogue is concerned. Rather, democratic practice reinforces economic efficiency. This is because other aspects of governance encountered in countries displaying a high quality of democracy, such as well-developed vertical and horizontal accountability, constrain the government, making it difficult to create patrimonial ties with individual unions. Union autonomy, on the other hand, combined with functioning social dialogue at the sectoral level, creates encompassing organizations that enable the unions to look at the long-term interests of the sector's workers. Thus, rather than simply being engaged by pro-reform governments as social partners in the restructuring process, the autonomous unions pushed stalling governments to implement long-term reform in the sector, one which would preserve jobs and wages in the future.

From the union perspective, however, involvement as a partner and a potential facilitator of reform carries the risk of marginalization. The answer, of course, is not collaboration in an effort to stop reform, à la Metalurg or the Galați union, which represents one way of making the union irrelevant. The resulting irrelevance was clear in the case of Metalurg. In the case of the Galați union, even though it reconciled and merged with Metarom in the aftermath of privatization, its leadership in the enterprise became challenged by the rise of a rival union. That, in turn, led to internal conflict among enterprise-level unions, which hardly inspires the confidence of workers.[141] The cautionary tale applies also to the loyal partners of restructuring. According to a survey carried out by the Warsaw School of Economics in the run-up to privatization, 56% of responding steelworkers noted that Solidarity (49% for OPZZ) actively participated in the preparation and realization of the restructuring program at their enterprise. At the same time, 42% of the responding workers felt that nobody represented their interests well.[142] This means that unions need to be clear about their role as the representatives of the interests of workers, rather than stewards of the enterprise as such. In what may be a fortuitous—from the unions' long-term perspective—consequence of transnational capitalism, the new owners drew the distinction between labor and management very clearly, thereby forcing the unions to focus on worker representation.[143]

SIX

Rise of Transnational Capitalism

State Capacity and External Pressures

Privatization of Sidex will be a great blow to the mafia.
  —Radu Sîrbu, president of State Ownership Fund in Romania

This is your problem. If you want to solve it, you better get cracking.
  —European Commission negotiator to the Czech government on steel restructuring

[EU entry was] a pistol held to our head.
  —Civil servant at the Polish Ministry of Treasury

The previous two chapters discussed ways in which state capacity mediated restructuring and privatization outcomes at the enterprise and sectoral levels. The implementation of the initial policy choice, whether continued state ownership or privatization to domestic investors, produced a financial feedback effect that put additional pressure on the governments of the four countries. In the case of Poland and the Czech Republic, the subsequent adjustment took place via sectoral employment restructuring measures, as well as through targeted involvement in the individual enterprises via debt workouts, market-based vehicles of support, and reorganization of assets. In Slovakia, the state did not engage in much in the way of restructuring, while in Romania, the response of the new Democratic Convention government was limited to several—mostly failed—sales of smaller steelworks to foreign owners. Even though the restructuring process had very different effects on the public purse of the four countries, as evidenced through tax arrears and

142

social security contributions, all four sectors found themselves in precarious financial condition. As a result of inadequate adjustment, which prevented these enterprises from reaching the gold standard of restructuring—that is, the operation on fully liberalized markets without state aid—the governments of Romania, Slovakia, Czech Republic, and Poland came under strong external pressures to resolve the sector's problems.

With the enterprise level and the sectoral level considered in the two preceding chapters, this chapter examines the third level of analysis: the interaction between states and external actors. The external pressures for policy change were in many cases mutually reinforcing, but the goals of external actors often differed, as did their sway over the governments. This chapter demonstrates that the specific pressures that were dominant in the ultimate decision to sell a key enterprise to foreign investors were related to the state capacity of the country in which the enterprise was located. Low state capacity, exemplified by Romania, was associated with levels of rent seeking that made it difficult to maintain macroeconomic stability, which triggered the pressure of the IFIs. Medium state capacity, as illustrated by Slovakia, enabled the government to keep the IFI pressures at bay, but it was not able to retain the trust of international lenders and withhold the pressures of the international financial markets. The relatively high level of state capacity in the Czech Republic and Poland allowed these countries to withstand external pressures the longest. Thus, they were able to both withstand the pressures of the IFIs and to negotiate with actors on the financial markets. The reluctant sales of the Czech and Polish enterprises to foreign investors resulted, ultimately, from the requirement that these countries cease all forms of assistance to the sector as a condition of accession to the EU.

This chapter illustrates the extrication from incomplete reform in the four countries and the process of convergence on transnational capitalism and ownership by foreign owners. Thus, it discusses the interaction of the governments in the four country cases with external pressures, and it examines ways in which the specific states exerted the effort to keep these pressures at bay. As in the other chapters, the discussion begins with the case of Romania, where the level of state capacity was the lowest and addresses the cases in the order of increasing capacity.

Privatization in Romania and the Driving Influence of the IFIs

The low capacity of the Romanian state had several implications for economic reform. First, it resulted in the macroeconomic instability of the country, and second, it undermined both the reforms attempted by the government and those inspired by the IFIs. Not surprisingly, in the steel sector, this led to the enterprises' inability to adjust to the market conditions and to failed privatization attempts. As this section illustrates, in the end, it took targeted, micro-level policies and the assistance of the IFIs to sell the enterprises to foreign investors.

The low capacity of the Romanian state, as previous chapters showed, manifested itself in a fragmented and contested institutional framework that led to repeated enterprise bailouts that were not conditional on enterprise restructuring. Consequently, the macroeconomic imbalance, external deficits, and inability to raise funds in the private sector, led to Romania's dependence on the IFIs for financing. Between 1991 and 2004, Romania signed a total of seven standby agreements with the IMF. However, the first five agreements were never completed.[1] The World Bank was also actively involved throughout the transition process in Romania. By 2004, the World Bank was Romania's largest creditor.[2] However, the implementation of the IFI programs was also dogged by low state capacity. Romania's Restructuring Agency provides an example of how low capacity adversely affected reform efforts funded by the IFIs.

Microeconomic adjustment and, thus, enterprise restructuring became the central element of discussions with the IFIs, so the government created the Restructuring Agency to coordinate the process. However, the new agency's efforts were hindered by two factors. First, the FPS[3] refused to give up its control of the restructuring process. Second, the arrangement for the institutional oversight of the Restructuring Agency reflected the competition between two entities vying to control it: the Reform Council, which was tasked with coordinating the entire reform process, and the Ministry of Industry. In the end, the Restructuring Agency was subordinated to the Reform Council but located within the Ministry of Industry. Critically, however, to placate the FPS's interests, the Restructuring Agency was limited to drawing up recommendations.[4]

The problems with this arrangement became apparent during the implementation of the World Bank–inspired Enterprise Isolation Program between 1994 and 1997.[5] The program, coordinated by the Restructuring Agency, the Reform Council, and the Ministry of Industry, had attempted to engage in restructuring prior to privatization. However, the World Bank switched its approach to squarely promoting privatization when the enterprises, rather than become more effective as a result of fund infusion, began demanding more funds.[6]

Chapters 4 and 5 presented the feedback effect of the initial policy choice and the very limited efforts by the Romanian government to rectify the resulting problems of indebted enterprises. The Democratic Convention government's privatization attempts did not stretch far enough to encompass the loss-making behemoths. As for the smaller enterprises, their privatization attempts were largely unsuccessful. These privatization failures resulted from insufficient state capacity, which was apparent in the inability to find financially credible partners for the enterprises and to draw up clearly specified contracts, and inadequate pre-privatization restructuring, be it organizational or employment. In this respect, the steel industry was quite representative of the Romanian manufacturing sector as a whole and the consequences for the Romanian economy were dire.

At the end of 1998, the leading credit agencies reduced Romania's credit rating. Fitch IBCA downgraded Romania from a BB– to a B rating, placing Romania on the same level as Moldova and Turkmenistan—only Russia, which was undergoing the 1998 financial crisis, ranked lower.[7] Moreover, the Fitch report reaffirmed Romania's reliance on the international community, stating that there was "little prospect" that Romania would be able to raise from private creditors the $5 billion it needed to cover external debt payments and the current account deficit. Without the IMF and World Bank assistance, Romania risked defaulting on its debt.[8] However, given Romania's poor record in implementing reforms, the IFIs were reluctant to commit the new funds that the Romanian government badly needed because its reserves had fallen to a precariously low level.[9]

Thus, during the second feedback loop, the IFIs intervened by taking a micro-level approach to accelerate the privatization process.[10] No longer content solely with numerical benchmarks that were part of the FESAL program, initiated in 1996, the World Bank drew up lists of specific enterprises to be privatized or liquidated should they fail to find an investor as part of the Public Sector Adjustment Loan (PSAL) program.[11] The PSAL I program, arranged with the World Bank in March 1999, selected sixty-four enterprises that were the most problematic for the Romanian state budget and Romanian macroeconomic stability. In the steel sector, the list included Sidex, Siderurgica Hunedoara, COS Târgovişte, and Industria Sârmei Câmpia Turzii—the most important steel industry enterprises. As the FPS president put it, "[V]ery few important enterprises from our portfolio have remained outside of PSAL. . . . All or the majority of everything that generates loss and financial blockage in the economy has been included in the PSAL."[12]

In what amounted to the outsourcing of state capacity to the IFIs, the World Bank also supervised the selection of privatization advisers for the particular deals, including setting the criteria by which the advisers were chosen. Even the FPS president stated that international investment banks would "do a better job than the FPS."[13] Moreover, the World Bank also extended technical assistance loans "to ensure that the country had the right kind of technical expertise to implement the reforms."[14]

The IFI presence in Romania was far more pronounced than it was in the other three countries, and government statements were framed in reference to commitments made to the international lenders. For example, in the case of the failed Reşiţa Steelworks privatization, when APAPS, which had succeeded the FPS, considered bringing a court case to cancel the contract with Noble Ventures, the announcement stated that "dramatic situation of Reşiţa will not stop the development of a privatization program according to the commitments made to the International Financial Institutions."[15] Thus, at the turn of the century, as Romania was attempting to establish macroeconomic stability and build the structures of a market economy, the IFI restrictions were clearly driving Romania's policy choices. Starting in 2000, after Romania began EU accession negotiations with the European Commission, the position of the IFIs was further strengthened.

As in the other countries, EU membership was the overarching developmental goal of Romania's successive governments, and Romania needed to obtain the status of a functioning market economy to conclude its membership negotiations with the EU. Although the Commission had the final say over the granting of this status, it was de facto relying on the IFIs' stamp of approval. By mid-2001, the World Bank's assessment of Romania's qualifications was still highly reserved, as the institution pointed to the daunting tasks ahead and clearly expressed its dissatisfaction with the progress it had observed: "After almost a decade of uneven reform efforts, the new Government that took office in December 2000 *has begun laying plans* for reforms needed for a successful economic transition, and eventual accession to the European Union (EU). The key challenge ahead is turning these plans into sustained actions."[16]

Nonetheless, the priorities of the IFIs and the EU were different, which became clear in the case of the steel sector. Whereas the IFIs pushed for speedy privatization or liquidation of unviable enterprises, the EU, which cared less about the property form, was far more attentive to state aid and concomitant capacity closures. Production capacity closures, as part of a restructuring program, were not as significant a problem in Romania as they were in Poland and the Czech Republic. According to a European Commission desk officer for Romania who dealt with the steel sector, reduction in capacity was "not a huge issue because so much of the capacity had been reduced already."[17] The UniRomSider expert concurred, saying that the EU

> put before us the task to limit our production from, for example, 17.5 million tons to 9 million tons of crude steel. We never went past 6.4 million. So the threshold was far enough from our real performance. So we didn't feel any pressure from them. Except to rule somehow the privatization . . .[18]

The main challenge the Commission set out for the Romanian government was to create a comprehensive sectoral restructuring plan that would allow the Romanian government to retroactively grant enterprises state aid and also provide them with a clear vision for attaining viability by the year 2008. Officially, state aid could not be granted before the EU approved the sectoral restructuring program.[19] Thus, the formulation of the restructuring program should have preceded the case-by-case privatizations advocated by the IFIs, which required state aid in the form of debt write-offs.

Although the Romanian government was well aware of the need to draw up a sectoral program, its development was both slow and contentious. As we have seen, proposals to create such a program, including the formation of some combination of holding companies, had been presented prior to 1996 but were not officially adopted as laws. After the 1996 elections, new proposals appeared, most notably in 1997; but a clash ensued between the statist proposal developed by the Ministry of Industry and Trade and the IFI-advocated plan for rapid privatization, which the FPS paid lip service to supporting. The flagrant confrontations between these two bodies highlighted the fragmented nature of the Romanian state, a symptom of its weakness.[20] In the meantime, the status quo persisted.

A replay of the 1997 duel between the Ministry of Industry and the FPS took place in 2000. Seeing that the government, pressed by the IFIs, was planning quick privatizations, the EU insisted on the prompt submission of a sectoral restructuring program. In July 2000, after the FPS had submitted a bill providing for the privatization of Sidex Galați and the other steelworks included in the PSAL program, a PHARE-financed study, prepared by Usinor Consultants, suggested adopting an étatist model of restructuring—that is, halting the privatization process and restructuring the enterprises before selling them.[21] This strategy was seen as being more profitable in the long run than immediately selling enterprises in financial distress would be. The funds needed for restructuring were to come not only from the state budget but also from the EU's PHARE funds.

Seeing the incompatibility of the PSAL commitments with their (and Usinor's) proposals, the Ministry of Industry, supported by Metarom and the employers' organizations, insisted on removing the steelworks from the PSAL framework, thus flatly countering the FPS proposal.[22] The statist approach advocated by the Ministry of Industry was quickly renounced by the IFIs and by the FPS.[23] The European Commission negotiators also distanced themselves from the recommendations, insisting nonetheless that the government had to develop a sectoral plan. Under strong IFI pressure, the government rejected the sectoral restructuring alternatives proposed by the PHARE-financed study and chose to go the route of seeking individual sales.[24] For its part, the EU recognized the advantage of solving the problem of Sidex and the other loss-makers and implicitly agreed to deal with the state aid issue retroactively.

The landmark case was the privatization of Sidex Galați to the LNM Group (the future ArcelorMittal) in 2001. Although initial talk of privatizing Sidex had started around 1996, no concrete proposals had emerged, and the enterprise remained in state hands.[25] As chapter 4 discussed, over time, Sidex came to symbolize the ills of the Romanian economy: *firme-căpușă* and the overlap between the political and economic spheres, a lack of financial discipline, and the resulting arrears in tax and social security contributions. Even though Sidex was on virtually every IFI-inspired "hit list," both the government and the FPS were resistant to the idea of taking concrete steps to privatize the giant enterprise. This is not to say that privatization was not discussed; on the contrary, in 1998 the government considered Sidex privatization a "priority." It even teased the IFIs by choosing a privatization consultant, creating an illusion that privatization was imminent.[26]

By 2000, however, the situation had become dire, as Sidex's accumulated debts now exceeded $1 billion, and losses generated in 2000 reached a million dollars a day.[27] IFI pressure to act according to the provisions of the PSAL program and to conclude the privatization process intensified. In fact, Sidex privatization became one of the conditions the government put in its letter of intent to extend the fifth standby agreement with the IMF in 2000.[28] The government recognized that the fulfillment of the terms of a new standby agreement with the IMF and of the new PSAL II program with the World Bank (a continuation of PSAL I objectives) was necessary to gaining market-economy status and being recognized as such by the EU. Eventually, keeping their commitments to the IFIs, state representatives highlighted the extent to which their hands were tied. The chairman of the FPS proclaimed that "Sidex cannot wait any longer without the money the Government can no longer provide."[29] He stressed, moreover, that the government would no longer be able to provide bailouts like the 1999 ROL 400 billion guaranteed loan:

> The moment the Governments repeats the decision to make capital injection into Sidex to prolong its existence, it will encroach upon all agreements with the international financial institutions, with the World Bank and the IMF in the first place, which will put the Government into a grave impasse. For that reason, delay with privatizing Sidex will mean the killing of the enterprise . . . only privatization can provide the saving capital infusion.[30]

Nonetheless, the government did not take up the privatization process before the elections. It took two significant steps in that direction, however, which largely constrained the future government. First, in June 2000, it signed a contract with Fieldstone Private Capital Group and Deloitte & Touche consortium for privatization consulting. More significantly, about two weeks before the election, it published an official privatization announcement, soliciting bids for the enterprise.[31]

As expected, the communist successor Party of Social Democracy of Romania emerged as the winner of the 2000 elections, and it was soon joined by the splinter Romanian Social Democratic Party to form the Social Democratic Party (Partidul Social Democrat [PSD]).[32] One of PSD's goals was to shed its heavy historical baggage and association with the slow pace of reforms during the 1990–1996 period. Given that the announcement with its request for bids was official and could only be canceled by a separate legal act, withdrawing it would have sent a negative signal to the international community. It would have amounted to canceling the privatization process, just when the government was attempting to establish its reputation as a reformer. In fact, PSD insiders interpreted the publication of the announcement right before the elections as an effort to not just tie PSD's hands, but also as a somewhat mean-spirited parting gift by the former government.[33]

The PSD government certainly took up the challenge and continued the privatization talks but not before changing Sidex's management to include its close allies. This process meant placing a PSD deputy from Galaţi on Sidex's management board as well as a phoenix-like rebirth of Siderman—the managerial company whose performance had been detrimental to Sidex in the late 1990s—whose representative was also placed on the board. In the meantime, the new privatization minister went out of his way to deny that there was a "PSDization" process underway at Sidex.[34]

Initially, several foreign investors were interested in Sidex. LNM Holdings' interest was the strongest, but there was also interest from Usinor, in a consortium with Turkish Erdemir and German Salzgitter. A consortium of domestic capitalists, composed of former foreign-trade enterprises involved in the lucrative transport of Sidex's products, submitted its bid as well. The APAPS decided to extend the submission deadline to give this consortium a chance to prepare the bid. This move could perhaps be explained by one of the companies that formed the consortium belonging to the president of a small junior coalition party, the Romanian Humanist Party.[35] The stakes seemed too great, however, for the closely watched PSD to risk its reputation by selling a strategic enterprise to a financially weak consortium of political insiders.[36]

Eventually, only LNM put forth a binding offer to privatize the enterprise. The Usinor consortium, by contrast, was interested only in buying a 25% stake in the company and taking over its management, with the option to purchase a subsequent 26% over the next two years.[37] Sidex privatization reached the top levels of political lobbying, with Tony Blair sending a controversial letter on LNM's behalf to his Romanian counterpart, Adrian Năstase.[38] Although it is doubtful that the letter made the difference in the final decision to sell Sidex to LNM, the APAPS quickly concluded the deal just prior to a visit to Bucharest by the French prime minister, Lionel Jospin, during which he was expected to lobby on behalf of Usinor.[39]

On July 24, 2001, the government signed a privatization contract with LNM Holdings, the fourth-largest world steel producer at the time. The contract entailed LNM's purchase of shares for $52 million, a $351 million investment in the first decade, and $100 million in working capital. A debt-equity swap was to clear half of the debt, giving LNM 90% of equity; the rest of the debt (out of $1.2 billion total) would be written off. As discussed in the previous chapter, the deal also provided for employment restructuring based only on natural attrition for a period of five years.[40]

The privatization of Sidex was duly noted by external actors, including the EU. One European Commission civil servant perceived the sale as a "huge achievement" by the government, which "has been clear as to the goals and desire to bring about change."[41] Nonetheless, well after the privatization of Sidex, the Romanian government had yet to develop the restructuring program necessary for closing the competition-policy chapter with the EU. The European Commission rejected the first such program, put forth by the Romanian government in 2002. As the preparation of an acceptable sectoral program continued, privatizations of the other major steelworks placed in the PSAL program, COS Târgovişte, Industria Sârmei Câmpia Turzii, and Siderurgica Hunedoara, were underway.

Like Sidex, all three were a source of significant losses for the Romanian budget and utility companies. The first enterprise to be privatized was COS Târgovişte, sold in August 2002, followed by Industria Sârmei Câmpia Turzii in March 2003. Both of these heavily indebted firms were sold to Conares Trading, a Swiss-based subsidiary of the Russian Mechel Group, the largest specialty steel producer in Russia and the Commonwealth of Independent States. In both cases, financial restructuring of the enterprises' debt preceded the transaction, and the majority of debts were written off.[42] Both contracts also stipulated, in clearly defined clauses, maintaining employment levels in the post-privatization period.

LNM Holdings has certainly been the most aggressive of the foreign investors; its shopping spree across the Romanian steel sector effectively brought about the sector's partial consolidation. After purchasing Sidex, LNM Holdings took over Tepro Iaşi, which had been put up for sale again after the acrimonious conflict and cancellation of the privatization contract with Veselí Steelworks. Prior to the second privatization, the enterprise was placed under special administration, which took some employment and financial restructuring steps, including forgiving a large portion of utility company debts, in preparation for privatization.[43] In May 2003, APAPS sold Tepro to LNM Holdings in a contract worth $15.6 million, with over nine million dollars' worth of investments. At the same time, LNM took over the enterprise's $9 million commercial debt, as well as $15 million in rescheduled debts to the state. Importantly, from the perspective of the workers and enterprise-level union, the privatization contract committed to retaining the 1,400 workforce for the period of two-and-a-half years.[44] Thus, the second privatization attempt seemed to be rectifying the mistakes made during the first try concerning the choice of investor, clarity of contractual obligations, and social dialogue.

LNM Holdings next acquired Siderurgica Hunedoara and Petrotub Roman. Both enterprises had been put under special administration, given their precarious positions and the need to prepare the upcoming privatization process.[45] Siderurgica Hunedoara, a long products producer famous for manufacturing railroad tracks, was part of the PSAL program, but it was probably the most challenging enterprise in that it suffered from a collection of obsolete assets and overemployment. Faced with disastrous economic performance, which included a 2003 debt load of $260 million and losses equal to a half million dollars a day, APAPS, with the assistance of the World Bank, had undertaken some restructuring steps to prepare the enterprise for privatization. Knowing that Siderurgica was hardly a coveted asset, APAPS put it in a package with Petrotub, which LNM Holdings was greatly interested in.

LNM was adamant that it would only acquire Siderurgica if it underwent employment and organizational restructuring prior to privatization. Speaking in the context of pre-privatization organizational restructuring, an interviewed senior manager at what is now ArcelorMittal Galați said:

> In Sidex, we have taken over the whole company, we also have some assets, which are not [operating] but some of them we might operate in the future . . . we could see that it made sense. . . . But in Hunedoara, most of the assets were not operational and they were not going to be operated ever. Many have been dismantled. . . . So it did not make sense to take over [the company]. The government itself formed different companies. . . . Without it, privatization was not possible. If you give any buyer the whole site, with many non-operative assets, which are never going to operate, and there are environmental liabilities, nobody is going to [buy it].[46]

As far as employment restructuring was concerned, the same manager noted that "before we closed the transaction they [the Romanian state] had to reduce employees, only then we took over the company . . . otherwise it was not attractive for us."[47] Indeed, in 2003, prior to the signing of the privatization contract, 2,500 people in Siderurgica were laid off, much to the dismay of enterprise-level union that saw the strikingly different treatment of the Siderurgica and Sidex workers. The eventual privatization contract for Siderurgica and Petrotub, signed in October 2003, was worth $126.4 million, of which $43.4 million was allocated for Siderurgica's portion of the transaction, and the rest for Petrotub.[48]

Reșița Steelworks was not on the list of PSAL I or PSAL II enterprises. However, given that it was a very large enterprise—defined as one with over 1,000 workers—its sale, along with three other enterprises in the same size category, was a precondition for the conclusion of a new standby agreement being negotiated with the IMF.[49] The enterprise, this time with the support of its union, was sold to the Russian TMK group. The privatization was much better prepared this time around—the Romanian government specified the investments to be made while canceling about 38 million dollars' worth of tax arrears and engaging in some organizational and social-assistance-supported employment restructuring prior to privatization.[50]

With the steel-sector privatization virtually complete, the Romanian government finalized the sectoral restructuring program, which would normally be the basis for negotiating state aid provisions with the European Commission. The final program, submitted to the Commission in December 2004, served as the basis for the closure of the membership negotiations concerning competition policy.[51] The resulting program pertained to a "framework of a completed privatization process," and was based on the business plans of the individual enterprises benefiting from state aid, their details negotiated with the European Commission.[52] These largely pertained to state aid amounts and concomitant production-capacity cuts (mostly retroactive), projected investments, and prescribed employment reductions.[53]

Thus, in Romania, the privatization of the steel sector was overwhelmingly driven by IFI conditionality. The EU, although important as a general developmental influence in Romania, was preempted by the IFIs as a decisive force in the restructuring of the steel sector because of the dire fiscal predicament of the Romanian state, unable to introduce a hard budget constraint or institutionally assist the enterprises. In fact, the significant restructuring that took place, notably the organizational restructuring of Siderurgica Hunedoara, happened under the aegis of the World Bank technocrats who lent their advice. These developments stand in stark contrast to those in the Czech Republic and Poland, where the state intervened in a variety of ways to bring about market adjustment. The Romanian case also differs from the developments in Slovakia, where top-down IFI pressures were not the dominant force in the sales to foreign investors—the bottom-up forces of the international financial markets were.

Privatization in Slovakia and the Driving Influence of the International Financial Markets

Despite the reformist platform of the post-Mečiar coalition, the post-Mečiar years were marked by the mistrust of foreign lenders vis-à-vis the capacity of the Slovak state. Even though the government had averted a decisive IFI intervention in the aftermath of the lavish government spending under Mečiar, the incomplete reform status quo could not withstand the pressures of the financial markets. Although EU membership was the overarching goal of the Slovak government after 1998, it was not the driving factor in the decision to sell Slovakia's biggest steelmaker to foreign investors.

Throughout the Mečiar years, the Slovak economy remained quite robust. Under the influence of the Slovak National Bank, the Mečiar government retained the macroeconomic policy of its predecessors and maintained a low inflation rate. In part, the latter was facilitated by procrastinating on unpopular measures, such as the liberalization of the prices of utilities. Impressive economic growth, by regional standards, was at the same time ensured by expansionary fiscal policy, including public works projects and government-encouraged investment in nominally private enterprises.[54] Much of the latter was financed through foreign loans.

The resulting fiscal deficit endangered the financial stability of the country, with gross foreign debt reaching over 60% of GDP in 1998 (compared to about 40% in Poland and just over 20% in the Czech Republic) and a current account deficit equal to 11% of GDP (by comparison, it was about 4% in Poland and 2% in the Czech Republic).[55] The proportion of nonperforming loans held by banks rose to 44% of all loans in 1998 and was equal to 28% of GDP.[56] Slovakia's financial vulnerability was exacerbated by the structure of foreign debt, of which 40% was short-term debt. The majority of this debt was held by the private sector, and 60% of the loans were underwritten by the state.[57]

Thus, when the new anti-Mečiar coalition came to power after winning the September 1998 elections, Slovakia was in an extremely fragile financial condition. In December 1998, it lost its investment-grade-status rating; despite the new government in power, Fitch IBCA downgraded Slovakia's rating to the speculative grade of BB+ and claimed that Slovakia would face a tough challenge in 1999 in raising the $3 billion it needed to finance the current account and budget deficits.[58] At the same time, unlike in Romania, IFI influence was limited in Slovakia. The role of the IMF remained advisory. The World Bank stepped up its involvement after 1998, most significantly with a €200 million Enterprise and Financial Sector Adjustment Loan in 2001; but rather than play the role of the government's agenda setter, the bank's role was to reinforce it. Secondly, the loan was largely focused on reforming the banking system and its bad loans.[59]

In contrast to Romania, in the case of VSŽ the dominant force for sales to foreign investors came from bottom-up financial market pressures. After VSŽ, having found itself in deep financial trouble, was unable to repay the $35 million loan to Merrill Lynch, the other foreign banks invoked cross-default clauses in their loan agreements. This meant that the loans were now payable "on demand," not on the originally agreed on date. With the declaration of cross-default and a debt burden of $450 million to foreign banks, the enterprise could have been declared bankrupt at any time.

As discussed in the previous chapter, by December 1998, the lender banks were demanding a change of management, and they pressed for the candidacy of Gabriel Eichler, who had wide-ranging experience in international banking and corporate restructuring, as the new president and CEO of VSŽ. Because bankruptcy was in nobody's interests (including the lender banks'), Eichler's task was to turn the enterprise around so as to ensure loan repayment in short order. Given the total enterprise debt burden, the foreign banks saw the entry of foreign investors as the only realistic option.

The new government found itself in a difficult position. VSŽ's bankruptcy would mean economic, social, and political catastrophe in Košice, Slovakia's second-largest city, and in eastern Slovakia more generally, where about 100,000 families depended on the steelworks. The entire Slovak economy would be affected because VSŽ was responsible for 14% of exports and its turnover was equal to 8% of GDP.[60] However, given the Slovak government's precarious financial position, it had limited financial options for assisting the enterprise.

At the same time, of the four governments, the post-1998 Slovak government had the least interest in maintaining the status quo. Quite the contrary, VSŽ epitomized Mečiar's clientelism and was the main financier of Mečiar's activities.[61] It made sense for the new Slovak government to try to gain control of the enterprise and, given its pro-reform electoral agenda and fiscal constraint, embrace reprivatization to a foreign investor. However, the attempt to take over VSŽ revealed both the considerable weakness and fragmentation of state institutions and a characteristic ambivalence to change. Rather surprisingly, given the government's efforts to portray itself as a reformer, Eichler encountered resistance not only from enterprise insiders but also from state representatives as well. According to Eichler, at the outset of his tenure at VSŽ, most of the shareholders did not want him to be there. As for the state, the support for his leadership depended on the individuals who were representing the state and on their relationship with the former management. Given that these representatives were actually making deals with the latter, the only credible threat was by the banks, who said that if Eichler were removed, they would have no choice but to declare bankruptcy.[62]

In fact, it was not until September 1999 that the state representative fully backed Eichler's efforts. The prime minister, Mikuláš Dzurinda, became Eichler's ally during his second year of crisis management and was actively supportive of privatization to a foreign investor. U.S. Steel was seen as a choice candidate and, controversially, was granted exclusivity in the negotiations, despite the efforts of Lakshmi Mittal's LNM Group to take part in the privatization. U.S. Steel's main advantage was the belief that if a conservative US company were to invest in Slovakia, it would send a positive signal to other foreign investors. Moreover, given Slovakia's NATO membership aspirations, putting the US flag near the Ukrainian border was seen as particularly helpful.[63]

The government, however, was divided. As the prime minister's support for Eichler grew, the other ministers openly opposed Dzurinda; one even attempted to take Eichler's job, claiming that his own metallurgical background made him more qualified to run the company. In short, Eichler was very clear that because VSŽ had been turned around, some members of the government saw opportunities to engage in rent seeking.[64] Eichler's assessment of the Slovak state was rather grim: there was no state as such but rather a collection of individuals driven by all kinds of motives.[65]

The capacity of the Slovak state left a lot to be desired during the takeover process, which was fraught with problems. First, even though the trade in VSŽ shares on the Bratislava Stock Exchange was suspended in November 1998, Alexander Rezeš obtained a release to sell a significant portion of his stock in VSŽ, which he owned via another company. A technical error subsequently enabled him to sell these shares to three foreign companies, which were supposed to hold them until the situation improved.[66] These shares were eventually returned and ended up on the account of a company controlled by the state.[67] Another example was the loss of the state's majority standing in Priemyselná Banka (Industrial Bank), which held 9.3% of VSŽ shares, by letting the management increase the bank's capitalization. Even though the bank eventually voted with the state and even transferred its VSŽ shareholder voting rights to the Ministry of Privatization, this was preceded by a period of uncertainty.[68] Finally, as the government was attempting to buy up enough VSŽ shares to give it control of the enterprise, an investment group called the Penta Group acquired at least 8% of VSŽ's shares for speculative purposes.[69]

In the end, however, state authorities, and foremost, the minister of finance, cooperated with the crisis manager in reaching an agreement with the foreign banks that would ensure continued financing and support for the sale of VSŽ to U.S. Steel, which was being negotiated throughout the restructuring process. In addition to the fragmented state authorities, still marred by clientelistic relations, there was also overt pressure on the proreformist elements. For example, as the deal with U.S. Steel was being negotiated, and in the run-up to a special shareholders' meeting, which aimed at unseating Eichler, the president of Slovakia proclaimed that VSŽ's problems could be solved without the entry of foreign capital.[70] In the end, however, the pro-reform forces prevailed.

As the government-backed Eichler's restructuring plan and negotiations with U.S. Steel progressed, in March 2000, VSŽ, U.S. Steel and the prime minister of Slovakia signed a memorandum of understanding outlining the terms of U.S. Steel's entry, much to the chagrin of the pro-Rezeš forces inside the enterprise. After the state exerted considerable effort to shore up votes in support of U.S. Steel, the general shareholders' meeting on October 12, 2000, was a resounding success, with 99% of the shareholders present voting in favor of the sale.[71]

The final contract stipulated that U.S. Steel pay $60 million up front, assume debts of $325 million, pay $15 million in tax arrears, and pay shareholders between $25 million and $75 million by 2003, as well as invest a minimum of $700 million over the following five to ten years. Critically for the region, the new enterprise, U.S. Steel Košice, would retain full employment for ten years, basing any employment restructuring only on natural attrition. In exchange, the buyer would profit from full tax holidays for five years, followed by a 50% reduction in taxes for the following five years. Originally unlimited, the total value of tax holidays was later capped at $500 million as a result of the EU Accession Treaty negotiations concerning state aid.[72]

The state aid issue begs the question of the role the EU played in the VSŽ privatization. In short, the EU consideration in the government's decision to reprivatize VSŽ was marginal. Although integration with the EU never ceased to be HZDS's official goal, the fulfillment of the conditions for membership became secondary when it conflicted with the short-term pecuniary or political interests of the government. The HZDS coalition-led efforts to establish clientelistic relations went hand in hand with the transgression of democratic norms. Their violation became clearer as the phenomenon spread to other domains, such as minority rights and the abuse of the security service for political ends. Following numerous warnings, at the December 1997 Luxembourg Summit, Slovakia was excluded from the first wave of applicant states on account of its poor democratic credentials and consequent noncompliance with the Copenhagen criteria for EU membership.

Thus, the 1998 election turned on the question of Slovakia's growing international alienation and, above all, its exclusion from the first wave of EU accession countries. In the aftermath of the election, Mikuláš Dzurinda formed a government supported by a four-party coalition composed of the Slovak Democratic Coalition on the right, the communist successor Party of Democratic Left, the center-left Party of Civic Understanding, and the Hungarian Coalition Party. In 2000, the Christian-Democratic Movement joined the coalition. The government made a special effort to speed up the reform process, concentrating on measures aimed at securing accession to the EU in the "first wave."[73] Its efforts paid off, and the Commission favorably assessed Slovakia's compliance with the Copenhagen political criteria in its 1999 Regular Report.

The relations between the new government and the EU, however, did not seem to play a role in the international banks' calculus. Even after they had secured a working relationship with a CEO they could trust, the banks did not turn to the government as a partner in the process, and they did not try to share their risk, presumably because they knew that the government was not able to play that role. As for the EU, it did not seem concerned about the privatization deal. Rather, the EU consideration became prominent several years *after* the reprivatization deal took place, during the membership negotiations concerning the competition chapter and state aid. In 2004, a disagreement ensued between the European Commission, on the one hand, and the Slovak government, along with the U.S. Steel Košice management, on the other, over the interpretation of the agreed-upon production limits appearing in the Act of Accession. The wording of competition policy provision of Annex XIV of Slovakia's Act of Accession, which placed caps on U.S. Steel's production and sales, gave rise to "good faith misunderstanding" about whether the provisions stipulating an annual 3% increase in production caps were valid as of January 1, 2002, or as of the accession date. In the end, bringing visible relief to the Slovak government, U.S. Steel agreed to have the total amount of state aid reduced by $70 million.[74] The controversy revealed just how closely the EU followed all state aid disbursed to producers in accession states.

Thus, unlike Romania, the Slovak state's capacity was sufficient to avert direct IFI pressure to rectify Slovakia's problems. However, before the EU concern over state aid had a chance to grow, the Slovak state, with an important political and economic stake in the enterprise, came under pressure of the international financial markets and became directly involved in efforts to transform the enterprise into a subsidiary of a major steel producer. The discussion now turns to two cases where state capacity enabled the governments to withstand both IFI and financial market pressures.

Privatization in Poland and the Driving Influence of the EU

As chapter 4 showed, the relatively high capacity of the Polish state meant that the enterprises faced a fairly hard budget constraint but that the Polish state also possessed the tools to step in and assist the enterprises. Consequently, the IFIs were kept at bay and their role was reduced to that of supporters of the reform agendas of the successive Polish governments. After playing a supportive role in the initial macroeconomic stabilization, the IMF ceased to be involved after the conclusion of the 1994 standby agreement, after which Poland paid off its debt to the institution.[75]

The role played by the World Bank was more substantial over the years, but by the end of the 1990s, its role was to buttress the institutional reforms undertaken by the Polish government and lend support for the development of specific areas. According to the World Bank,

> By 1999, as that year's CAS [Country Assistance Strategy] Progress Report reflected, the Bank's program had evolved to one sharply focused on the structural reform agenda for EU accession and real convergence. Since 1999, this has included *advisory assistance* on the overall reform agenda, public expenditure composition and transparency, incentives for agriculture reform, anti-corruption and roads financing; and *lending assistance* for coal sector restructuring, railway reforms, rural development, infrastructure (roads, ports), environmental sustainability and energy efficiency [emphasis in the original].[76]

Although IFI pressures did not play a role in the Polish government's approach to the steel sector, the steel sector—and the government—found itself under strong market and EU pressures. The EU pressure to produce a comprehensive sectoral restructuring program had intensified after repeated demands by the Polish government for a relaxation of trade-liberalization schedule adopted in the Europe Agreement.[77]

The successive proposals submitted by the Polish government were not accepted by the European Commission. The 1996 proposal was rejected for not meeting the EU standards of economic analysis and for trying to preserve too much—13.5 million tons—capacity.[78] The officially adopted 1998 Restructuring Program for Polish Iron and Steel Industry was also rejected, for two reasons. First, it tried to preserve too much capacity (13.2 million tons). Second, it did not pay enough attention to the strategy for privatizing Katowice Steelworks and Sendzimir Steelworks, which the government had planned to privatize by the end of 2001.[79]

Successive governments proved remarkably resistant to the EU pressure. In response to the criticism of the 1998 program, the minister of economy said, "We want to engage in quick privatization—quick, but not mediocre. . . . As far as privatization of both mills is concerned, the decision regarding whether they will be merged or not and about when this will take place, will be made by the Minister of the State Treasury."[80] Nonetheless, as explained in chapter 3, the government maintained its policy of delegating restructuring and privatization decisions to the management, with negative consequences.

With the failure of the Katowice Steelworks privatization effort in 2000, the Polish state extended lifelines of institutional support to the struggling enterprises. The contrast with the Slovak and Romanian cases is great. By mid-2000, the Slovak government was finalizing discussions with U.S. Steel, and the Romanian government had already selected a privatization adviser. In Poland, by contrast, rather than intensify efforts to find external investors, the government turned its attention to finding ways to assist the mills through quasi-market means.

The main vehicle of state intervention in the sector was the parastatal Silesia Finance Society (Towarzystwo Finansowe Silesia). As discussed in chapter 4, this subsidiary of the state-owned joint-stock company called the Agency for Industrial Development was created at the end of 2000 to help the cash-strapped Katowice Steelworks repay its loan interest to a Polish bank consortium and prevent default. Silesia was subsequently capitalized in the amount of $57.5 million (PLN 250 million) through the transfer of state-owned shares in prosperous enterprises.[81] According to its creator, the deputy minister of economy Edward Nowak, Silesia was to function like a "bypass" that would both supply production inputs and market the final products, circumventing the need for working capital.[82] Silesia's role then expanded into other companies. In addition to using parastatal companies to keep the market pressures at bay, the Polish government also engaged in sectoral consolidation as a stalling tactic in the face of EU pressure to create an effective solution for the steel sector.

After the failure of the 2000 privatization of Katowice Steelworks, the government took a long-overdue coordinated approach to privatization and restructuring. A renewed privatization effort was to be coupled with sectoral consolidation, an idea that had been promoted by experts for more than a decade. Reviving the idea of a merger, the new privatization offer involved both the Katowice and Sendzimir Steelworks, along with Florian Steelworks (producing flat products) and Cedler Steelworks (producing long products). The last two were small but valuable companies, with a high value-added production profile. The government had for years been resisting foreign interest in these enterprises because their sale would have meant "getting rid of the pearls" that could later be used as a bargaining chip when privatizing the larger enterprises.[83]

As a backup, in case the privatization talks failed, the deputy minister of economy developed a sectoral consolidation plan intended to merge the four enterprises, and possibly others. The program, called the "Update of the [1998] Restructuring Program for Polish Iron and Steel Industry" and adopted as the Law on Restructuring of Iron and Steel Industry in August 2001, became the basis for final negotiations with the EU.[84] In contrast to its predecessor, the new program stipulated that the 2005 maximum production capacity be lowered to 11.6 million tons.[85]

The privatization offer attracted significant interest, including from LNM and U.S. Steel. However, advanced four-way privatization talks were suspended following the change of government after the September 2001 elections. On coming to power, the new Democratic Left Alliance–Labor Union–Polish People's Party (Sojusz Lewicy Demokratycznej–Unia Pracy–Polskie Stronnictwo Ludowe [SLD-UP-PSL]) coalition decided to postpone privatization as it prioritized the sectoral consolidation outlined by its predecessors as a contingency plan.[86] The consolidated Polish Steelworks (Polskie Huty Stali [PHS]), comprising the four steel producers, with a 70% share of Polish steel production, first took the form of a holding company, registered in May 2002, and became a full-fledged merger in December 2002.

The government stalled, despite being faced with the reality of closing the competition chapter. The "2002 Regular Report on Poland's Progress towards Accession," the last report prepared by the Commission prior to the closure of the membership negotiations, stated that "[i]n order to complete preparations for membership, Poland's efforts now need to focus on seeing through the restructuring process, notably with respect to steel and other traditional industries."[87] The problem was not a lack of potential foreign investors, because U.S. Steel, LNM Holdings, and Arcelor had all expressed interest in purchasing PHS. In the summer of 2002, for example, a spokeswoman for LNM said, "We are waiting for the decisions of the Polish government concerning the beginning of the negotiations."[88] In early fall, Lakshmi Mittal, the president of LNM Holdings (now ArcelorMittal), said: "We have not yet made any official proposals because the government has not yet adopted the PHS privatization program. We are counting on it happening soon."[89]

However, the government's attention was elsewhere. For example, in the summer of 2002, SLD's Silesian "baron," Andrzej Szarawarski, in charge of steel sector's consolidation and privatization under the SLD-led government, began organizing PHS's own distribution network to bypass the intermediaries.[90] The project was unnecessary because the new owners would presumably rely on their own distribution networks, but it could be explained by rent seeking: Silesia purchased a financially troubled, private, domestic steel distribution company at a highly overvalued price.[91]

The government's procrastination also plagued negotiations with the EU over the sectoral program. Prior to the Copenhagen summit of December 2002, where the EU accession negotiations were to be finalized, the enterprise managers, the government, and the Commission engaged in an intensive shuttling of the continuously revamped and adjusted sectoral program drafts between Brussels and Warsaw. Here, too, the government encountered managerial obstructionism as individual managers attempted to obtain maximum aid in exchange for minimum capacity cuts. The Commission, dissatisfied with individual steel mills' proposals for restructuring, threatened not to close the competition chapter negotiations if they were not corrected.

In the end, the chapter was closed based on a draft version of the sectoral plan. The final program, called "The Restructuring and Development of Polish Iron and Steel Industry till 2006," was adopted by the Polish Council of Ministers in January 2003, but its revisions were not finalized until the end of March. According to the final version of the program, in addition to PHS, to which was allocated 87% of the state aid granted over the years 1997–2003, seven other steel mills were to obtain public aid, mostly in various forms of sometimes retroactive debt relief, presumably in conjunction with the privatization of PHS.[92] Over the years 1997–2006, the beneficiary companies' total production capacity in finished products was to be reduced by more than 1,351,000 tons, including a 1,026,000-ton reduction to take place at PHS.

When negotiations with the EU were finalized, the privatization of PHS began anew. Although the government knew that under its agreement with the EU, it could only grant the negotiated state aid monies to the sector until the end of 2003, it postponed the privatization decision as long as it could, to the dismay of the foreign investors and the enterprise-level unions. The new privatization announcement was finally made early in 2003, and the government changed the date for selecting the final negotiations partner no fewer than five times. In the meantime, to forestall the threat of PHS bankruptcy, the government used the Agency for Industrial Development to float State Treasury–guaranteed bonds, which it then used to buy up PHS's bank debts in the amount of $118 million (PLN 459.5 million).

Shortly before the decision was announced, some commentators had feared that, given the improved situation on the markets, the government would once again forego privatization. According to a journalist at a major daily who closely followed the steel sector, there were opinions among the PHS leadership that if privatization were to fail, PHS would "'somehow manage'—there is some bizarre faith that the state will restructure."[93] Another journalist claimed that "ever more frequently, there are unofficial opinions that PHS will manage without an investor."[94] That, however, would have required capitalization of the enterprise by the State Treasury, in the amount of $175 million (PLN 680 million), which would have breached the terms of the agreement reached with the EU.[95] In the words of a senior civil servant at the Ministry of State Treasury, EU entry represented a "pistol held to our head."[96]

Eventually, under great investor and public pressure, the minister of state treasury announced, in mid-July 2003, that he had granted exclusivity in the privatization negotiations to the LNM Group and had also asked U.S. Steel to prolong its offer, as a backup in case the negotiations with LNM failed.[97] The privatization contract, signed in October 2003, totaled almost $2 billion for 60% of PHS shares (PLN 7.7 billion). The financial arrangements were broken down as follows: LNM was to purchase a 60% stake in the enterprise for $1.5 million (PLN 6 million) and to settle enterprise debts in the amount of $160 million (PLN 621 million) to principal SOE creditors, and in the amount of $95 million (PLN 370 million) to the Agency for Industrial Development; it was supposed to increase enterprise capital by more than $205 million (PLN 800 million); provide $77 million (PLN 300 million) in working capital loan, along with $617 million (PLN 2.4 billion) in guaranteed investments; and settle 874 million dollars' worth (PLN 3.4 billion) of debt to other creditors.[98]

Instances of using state capacity to perpetuate the status quo also occurred at the local level. A particularly pernicious case in point is Ostrowiec Steelworks. The enterprise, as discussed in chapter 4, was initially "privatized" in 1994 through a debt-equity swap to the state-owned Stalexport; but by 2001, Stalexport had decided to sell its shares in the financially embattled enterprise.[99] In addition to bad investment decisions, rent seeking by the local business-political elite nexus via subcontracted firms that overcharged Ostrowiec for their services, contributed to the steelworks' financial woes.[100] When, at the end of 2001, Stalexport's new management decided to sell its shares in Ostrowiec, it had the option of selling the enterprise to foreign investors, several of whom had expressed interest.[101]

At the beginning of 2002, however, the local government, through the Agency for Local Development, offered to purchase Stalexport's shares in

Ostrowiec Steelworks. National-level SLD top brass not only supported the proposal but also—successfully—pressured Stalexport to "sell" the shares to the Agency for Local Development.[102] Three months later, the steelworks filed for bankruptcy; the local court declared the enterprise bankrupt in one business day, a record. About a week before the bankruptcy, the local soccer club, under the leadership of Ostrowiec's SLD president, registered a new company, Zakłady Ostrowieckie Huta Ostrowiec (ZOHO), and teamed up with a local businessman involved in the steel sector.[103] The company then leased Ostrowiec's assets from the bankruptcy manager and engaged in asset-stripping activities, for which its management was later charged.[104] Although ZOHO subsequently attempted to acquire the steelworks through a leveraged buyout in a privatization tender announced by the bankruptcy manager, the president of the city of Ostrowiec had a change of heart and not only distanced himself from ZOHO but also ended up supporting the entry of Celsa, the Spanish foreign investor. Obviously, in the face of EU restrictions on state aid, Celsa was much more likely to make investments and preserve workplaces—and hence, the president's job—than the local government.[105]

The cases of the PHS and Ostrowiec privatizations demonstrate that, faced with an opportunity to sell to foreign investors, successive governments exerted substantial state capacity to maintain the status quo of incomplete reform. The enterprises were kept afloat by various institutional lifelines, which shielded them from the bottom-up market pressures. These were not sufficient, however, to shield them from the EU's regulatory rules. At the same time, however, institutional maneuvers characteristic of relatively high state capacity also gave the government extra time to negotiate with the investors, even after EU accession. The case of Częstochowa Steelworks is a case in point.

As discussed in chapter 4, Częstochowa Steelworks was an exemplar of "overinvestment" due to the investment ambitions of its managerial strata. By 2002, the mill, which had not benefited from state aid, stood on the brink of bankruptcy, and its debt of $274 million (PLN 1.2 billion) surpassed its equity of $225 million (PLN 920 million).[106] It was not included as a beneficiary company in the 2003 sectoral restructuring program agreed on with the European Commission because the mill was slated for bankruptcy. However, after much debate about how to handle the mill's dire predicament, the government eventually adopted a restructuring program that entailed a debt workout coupled with asset revamping and a search for a foreign investor. Because the assets of the enterprise would inevitably

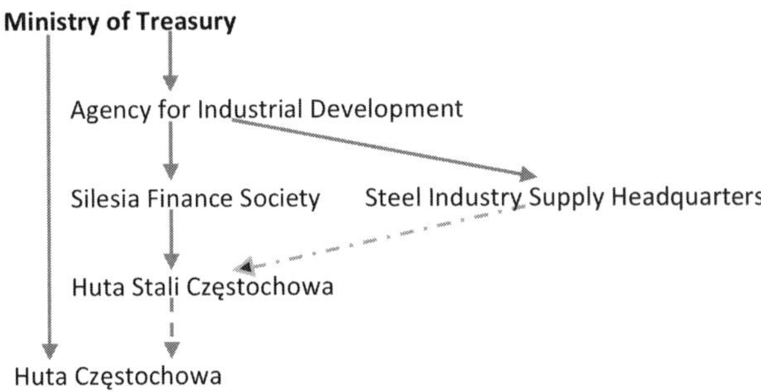

Relationship:

⟶ : Ownership

⇢ : Lease

⇢ : Credit

Fig 6. The Visible Hand of the State: Intervention Tools at Częstochowa Steelworks

be damaged if the production process were stopped, the state exerted an institutional tour de force, illustrated by figure 6, to maintain production.

The enterprise's assets were leased by a newly created entity with a remarkably similar name, Częstochowa Steel Foundry (Huta Stali Częstochowa), and its employees put on the latter's payroll. As figure 6 indicates, Huta Stali Częstochowa was owned by the Silesia Finance Society. However, being purely an operating firm, the company had virtually no capital and therefore experienced difficulties obtaining credit. To alleviate the problem with credit, the Agency for Industrial Development, which also happened to be Silesia's owner, lent a helping hand using another one of its subsidiaries, Centrala Zaopatrzenia Hutnictwa (Steel Industry Supply Headquarters). Centrala obtained bank credit and then transferred it to Huta Stali Częstochowa to continue production.[107] This set up allowed the enterprise to function until privatization in 2005. Although the privatization process was fraught with controversy, the comparison of bids strengthened the government's hand in negotiating the terms of the final deal.

Privatization entailed a choice between LNM Holdings, which was already negotiating the PHS deal, and the Ukrainian-owned Industrial Union of Donbas. In February 2004, the minister of state treasury controversially gave exclusivity to LNM Holdings, even though the offer from Donbas was better financially and the privatization committee recommended Donbas rather than LNM. After Donbas protested, the negotiations were called off and the privatization process frozen. In a subsequent court case, the court decided in favor of Donbas, determining that the rejection of Donbas's offer violated the terms of the tender.[108]

The Ministry of State Treasury reactivated the privatization process in December 2004, with Mittal Steel—the successor to LNM Holdings, also controlled by Lakshmi Mittal, and the predecessor of ArcelorMittal—once again obtaining exclusivity. The ministry used Mittal Steel's inability to finalize the social packet with the unions as the official reason for eventually turning to the Ukrainian partner, who promised better social protection. A more convincing reason seems to be that the Polish government did not want to alienate its new post–Orange Revolution Ukrainian counterpart, which was strongly lobbying for the Ukrainian enterprise and urged the Polish government to show support in more than symbolic way. Union support for the deal was an additional much-welcome bonus for the government. The deal was finalized in July 2005, with the Ukrainian partners agreeing to pay $385.8 million (PLN 1.25 billion) for the mill, invest $135.8 (PLN 440 million) over the following seven years, and return EUR 4 million in public aid.[109] The workers obtained a ten-year employment guarantee, and a 40% higher privatization bonus and a higher pay raise than that offered by Mittal Steel.[110]

Interestingly, in the meantime, the European Commission launched a probe into a possible violation of state aid rules by the Polish government. After a year-long investigation, the Commission concluded that the "write-off of public claims was compliant with normal market behavior and therefore does not involve state aid."[111] The Commission did, however, find that the government had in the past assisted the steelworks with EUR 4 million, which needed to be returned. In commenting on the decision, the competition commissioner stated: "This decision illustrates that the Commission takes its responsibilities to monitor state aid in the steel sector very seriously."[112]

Thus, the Polish case stands in stark contrast to the Romanian and Slovak ones examined previously. Because of its institutional prowess, which the Polish state deployed when politically convenient, the Polish steel producers only privatized under the constraints of the EU competition rules. Whereas in Slovakia, the privatization question was settled long before accession, and in Romania, too, the submitted restructuring program pertained to a "framework of a completed privatization process," the privatization journey in Poland was only beginning at the time when the Polish government was finalizing the accession negotiations.[113]

Privatization in the Czech Republic and the
Driving Influence of the EU

As in the case of Poland, so, too, in the Czech Republic EU accession was the propelling force behind convergence on transnational capitalism in the steel sector. Although the International Finance Corporation (IFC), the financial arm of the World Bank, also exerted pressure on the Czech government, this group of international lenders treated the government as a partner in negotiations, and their bargaining power increased precisely because of the strictures of the EU competition policy chapter.

In terms of the IFI pressure, consistent with the idea of the "Czech way," the role of the IFIs in the Czech Republic was marginal, even in the aftermath of the 1997 banking crisis.[114] During its history of relations with the IMF, the Czech Republic concluded one stand-by agreement in 1993, but it did not utilize the funds. Out of the four countries examined in this book, the Czech Republic's use of World Bank funding was the lowest, and by 2005, the country announced its plans to "graduate" from the World Bank lending programs and to become a donor nation.[115]

As chapter 4 showed, the Czech government used state capacity to intervene in the sector when it found itself in a difficult financial situation in 1998–1999. However, as it became apparent that the steelworks' problems could not be solved without decisive state involvement, the EU started exerting pressure to present a sectoral restructuring program that would ensure viability while coupling state aid with capacity cuts. The first such program, developed in 1999, was rejected by the European Commission because it inadequately examined the market conditions, and maintained too much production capacity and employment.[116] The EU subsequently provided PHARE funds to finance another sectoral study, carried out by the London-based Eurostrategy Consultants, a consultancy close to the European Commission, which later also advised the Commission in its steel sector negotiations with Poland.

In the meantime, in 2000, the Czech state became actively involved in Vítkovice's rescue and restructuring. As detailed in chapter 4, like its Polish counterpart, it engaged an operating firm, Osinek, to deliver production inputs and to coordinate sales of produced goods. The state also became involved in a far-reaching restructuring program and debt workout. Before the Vítkovice combinate was broken up into its steelmaking component, Vítkovice Steel—also called Division 200—and the heavy engineering division, the crisis management attempted to find a buyer for the steel section.[117] Several interested buyers had emerged already in October 2000, but the FNM mandated that the management not proceed with the negotiations.[118] In mid-2001, facing obstructionism from the minister of industry and trade, and decreasing support from the FNM, Novák and his managerial team tendered their resignation.[119]

In addition to Vítkovice, the Czech government also faced the challenge of Nová Hut's financial predicament. Nová Hut's financial problems in 1998–99 were compounded by the initial complications related to the minimill investment, which experienced delay in full commissioning in 1999. All in all, Nová Hut's losses reached $160 million (CZK 4.7 billion) in 1999.[120] As the financial problems of Nová Hut' mounted, the Czech government came under pressure from the IFC to guarantee the repayment of IFC's 1997 loan for the minimill investment.

In 1997, the IFC made a $250 million loan for the minimill project. As discussed in chapter 3, the loan was made to an only nominally private enterprise, as an 18.25% stake in the enterprise, just enough to reduce the state's share to 49%—thus making it nominally private and technically eligible for the loan—was placed with Credit Suisse First Boston, an international commercial bank, for sale. However, per agreement with the bank, should it be unable to sell the shares on the market, the bank would sell the remaining shares back to the FNM. As a result, a steel-sector analyst stated in an interview, "there is really no point in mentioning it as privatization."[121] Indeed, by 2002 the bank transferred the shares back to the state.[122] In the meantime, the IFC had started exerting pressure on the Czech government.

At first, in the early months of 2000, the IFC agreed with the minister of finance, Pavel Mertlík, to accelerate the privatization of the enterprise. To open the way for privatization, the IFC demanded a change of management and, ironically, firmer state control over Nová Hut'. According to Mertlík, the two sides agreed to "very quickly cooperate in seeking a solution, meaning in obtaining a strategic partner."[123] The minister expected the process to take about a year. He also mentioned that the IFC was not clear about the possibility of debt capitalization, but was "open to discuss it with us further. It will depend on how quickly the government will proceed with Nová Hut' privatization."[124]

Faced with Nová Hut's liquidity problems, the government stepped in to assist the enterprise. Its first step was to reassert control over the enterprise by changing the management and by attempting to cancel the privatization contract with the managerial team, Petrcíle, which blocked any ownership change until the end of 2001.[125] The cancellation of the contract eventually proved possible before the end of 2000, but the option of sale by the state of the 15% stake promised to the management awaited a court decision until 2005.[126] As for the new management, its task was to prepare the enterprise for privatization to a strategic investor.[127] This goal was also supported by the enterprise-level union. Finally, the Czech state assisted Nová Huť through a bank loan to support production. Specifically, it guaranteed a loan in the amount of $19.4 million (CZK 750 million), which was later declared to be illegal aid by the Czech Competition Office.[128]

After Mertlík left the government in April 2001, the Czech government placed the plans to seek a foreign investor on the back burner. In fact, the resignation of the economically relatively liberal minister and vice-minister of finance meant a victory of the statist-leaning minister of industry and trade, Mirosláv Grégr.[129] In the meantime, the IFC not only refused to consider a Nová Huť debt workout à la Vítkovice, but it also tried to shift its business risk onto the shoulders of the Czech government.[130] In an interview, the former chairman of the FNM, who negotiated with the IFC, related his discussions with its representatives and said that the IFC representatives had argued that the state's substantial stake in the company acted like an invisible guarantee. His retort was that because, according to its own rules, the IFC cannot finance state institutions, it cannot use this argument.[131] In fact, in exchange for state guarantees, which the Czech government refused, the IFC considered the rescheduling of debt repayment.[132] Therefore, the relationship between the Czech government and the IFC could be characterized as one of partnership, involving bargaining and negotiations over how to minimize the losses of either party, rather than a fully asymmetric one, where, as in the Slovak case, the government found itself with its back against the wall of a bankruptcy threat.

The complicated relationship between the IFC and the Czech government became even more complicated after the publication of the Eurostrategy sectoral restructuring proposal, which was presented in May 2001. The proposal seemed to be a compromise between the neostatist sectoral consolidation and market-based, private-sector-led restructuring. It recommended sectoral consolidation under the leadership of the private Třinecké Steelworks, with the ultimate goal of subsequently finding a strategic investor. Individual sales to foreign investors was the second-best option because the consultants believed that it would take Nová Huť at least a year and a half to complete the sale process which, given the enterprise's financial difficulties, was seen as too risky.[133]

The proposed option of Třinecké-led consolidation immediately became a lightning rod for criticism of what was seen to be a state-financed gift to Třinecké. According to the Eurostrategy plan, all shares of Nová

Huť, Vítkovice Steel, and Vysoké Pece Ostrava (the pig iron producer separated out of Nová Huť) were supposed to be acquired by the state and the enterprises' debts largely reduced (except for the IFC debt which Třinecké would have had to repay), following which they were to be "sold" to Třinecké for a symbolic crown each. Třinecké was then to lead the restructuring process, having de facto been given the assets of the entire Czech steel industry, along with about $894 million (CZK 34 billion) in state aid, not counting the social and environmental expenditures.[134] The resulting consolidated Czech Steel Company (Český Ocelářský Podnik) would eventually seek a strategic foreign investor.

The idea of consolidation under the leadership of Třinecké raised much opposition within the domestic steel community. Skeptics doubted Třinecké's competence and financial ability to lead the highly complex restructuring process, especially since its own debts were not yet repaid. Both the managers and the trade unions were mistrustful of Třinecké which, with the stigma of privatization scandal still attached, was now supposed to be "getting the other steelworks for free."[135] They attributed the Eurostrategy proposals to Třinecké's and Moravia Steel's lobbying efforts, both in Brussels and in Prague.[136] Finally, there was also the opposition of the managers of Nová Huť and Vítkovice Steel against becoming dominated by Třinecké's management, which would obviously be in charge of the Czech Steel Company.[137]

Not surprisingly, the IFC also opposed the program and there emerged an informal managerial-union-IFC coalition against the Eurostrategy proposal. Despite Eurostrategy's misgivings that a suitable partner would quickly be found, the IFC pushed for a sale to a foreign investor as the preferred option. Surprisingly, its most ardent ally in the process was the enterprise-level trade union.[138] Although the EU was well disposed toward the Eurostrategy suggestion the IFC opposed, it was open to any other proposal that would credibly ensure market viability of the struggling Czech steel industry. With the Eurostrategy proposal an obvious nonstarter, the government faced the need to present a credible program to the Commission that would simultaneously appease the IFC.

The government and especially the empowered minister of industry and trade, Grégr, were squarely in favor of sectoral consolidation.[139] The government claimed that there were no quality foreign investors interested in the mills, but it never made a legitimate effort to advertise its intention to sell the enterprises to strategic investors.[140] In fact, by mid-May 2001, the government even called off the privatization consultant selection process.[141]

Rather, the Ministry of Industry and Trade proposed a new program which was to combine the entry of foreign investors with partial sectoral consolidation, this time leaving Třinecké out of the picture. Under the new plan, Nová Huť and Vítkovice Steel, and Vysoké Pece Ostrava were supposed to be consolidated within the Czech Steel Company, but the brunt of the estimated CZK 78.8 billion in restructuring costs was to be borne by the foreign investor and the enterprises themselves. This plan was adopted as an official government program by a government resolution at the end of August 2001 and was promptly submitted to the EU for approval.[142]

The EU rejected the plan, having criticized its vague projections of market conditions, but primarily because of concerns that it was unrealistic. First, the state aid to be provided was seen as inadequate to attain viability. Secondly, it was doubtful whether a foreign investor would enter into enterprises that were so heavily indebted. Thirdly, it seemed that the indebted companies would not have had the means of generating the funds necessary to finish the restructuring process. In other words, the fear was that the government would not be able to find an investor willing to bear the financial burden, leaving the state to foot the bill for restructuring.[143]

The EU's refusal to accept the program cast a shadow on membership negotiations. According to the negotiations schedule, the competition chapter was supposed to be closed by the end of 2001, a prospect that seemed impossible after EU's rejection of the sectoral program draft in October.[144] This gave rise to speculations that the steel issue might delay not only the closure of membership negotiations but even membership itself.[145] After all, the European Commission's "2001 Regular Report on the Czech Republic's Progress towards Accession" rather dramatically stated that

> [t]he **restructuring of the steel industry** has not yet begun. . . . The implementation by the government of a viable restructuring programme and individual business plans for the steel sector, which is fully in line with the state aid *acquis*, should now be a high priority. The Czech steel industry urgently needs comprehensive restructuring to become internationally competitive.[146] (emphasis in the original)

On top of piercing EU criticism came the IFC demands for loan repayment. Resiliently against the IFC pressure, the government, in response to EU's rejection of its latest plan went on a statist offensive, this time with the FNM in the lead. In January 2002, the FNM, with the agreement of the Ministry of Industry and Trade, presented a plan which proposed sectoral consolidation using Osinek, the Vítkovice Steel operator and FNM subsidiary, as the leader of the consolidation process. Osinek was to be capitalized in the amount of $244 million (CZK 8 billion) and was to manage the merger of Nová Huť, Vítkovice Steel, and Vysoké Pece into the Czech Steel Company.[147] During the restructuring process, Nová Huť was to engage in a debt workout, and the Czech Steel Company, more generally, was to undergo financial, employment, and organizational restructuring. The state, in an effort to mollify the Commission, would put CZK 15 billion into financial restructuring alone. Subsequently, within two to three years, the FNM would seek a foreign investor for the restructured Czech Steel Company.[148]

Before the plan could officially be adopted, let alone presented to the EU, it was taken off the government's agenda in response to heavy IFC and union pressure.[149] As former member of the top management at the Revitalization Agency said, the "IFC wanted to have 100% return on its loan."[150] Thus, the idea of a debt workout was certainly not to IFC's liking. As for the union leadership, it claimed that its opposition to the plan was driven by the desire to protect workplaces, as consolidation would have meant that only selected assets would have been included in the Czech Steel Company.[151] The Nová Huť steelworkers even protested in Prague and declared a strike emergency to support the entry of a foreign investor.[152] Perhaps more surprisingly, the Nová Huť union was mistrustful of the state authorities, which, it claimed, would not be able to solve the enterprise's problems.[153]

The Czech government's principal justification for the proposal was that there was no serious foreign investor interested in entering the privatization process immediately, a claim that surprised the prospective privatization consultant, Roland Berger Strategy Consultants. Although the government eventually decided to retain the consultancy's services back in November 2001, it had not signed a contract with the company. In the words of the director of the Czech operations:

> We have not yet signed a contract with FNM, what is more, there is no sign that anybody would want us to try to find a foreign investor. The world steel industry is now in a complex situation, but I am surprised that we are not trying to find an investor, that with this, we are also holding back the privatization of Nová Huť.[154]

This opinion was also reflected in an interview with a Prague-based analyst with Roland Berger.[155]

Therefore, similarly to the Polish case, when faced with EU pressures, the Czech government attempted to use the relatively high capacity of the Czech state to maintain the status quo of retaining, even temporarily, the Czech steel industry in Czech hands. In part, the decision to do so may have been ideological, but the close relationship between the top strata of the Czech Social Democratic Party (ČSSD) and a company belonging to the Shiran Group, which certainly benefited from the steelworks' financial predicament, also raised the question of pecuniary incentives.[156]

The use of state capacity to maintain the status quo can also be seen in the approach to the Vítkovice Steel debt settlement issue. According to the workout agreement reached in 2001, the company needed to repay its bigger lenders $63 million (CZK 2.4 billion), with the money to be generated from the proceeds from the sale of the enterprise's assets. To fulfill this commitment, the government put the enterprise up for sale using a public tender. However, the government organized the tender at the time when it considered the Czech Steel Company proposal as the preferred option. Given that they preferred to purchase the entirety of the Czech steel sector and that Osinek, which was de facto operating Vítkovice Steel, was less than forthcoming with information on the company, the interested parties, including U.S. Steel, Třinecké Steelworks/CMC, LNM Holdings, and Shiran, walked away from the bid.[157] In the end, Osinek became the only company to make a bid to purchase Vítkovice from the FNM, in the amount of $86.8 million (CZK 3.3 billion). This was just enough to cover the CZK 2.4 billion in outstanding rescheduled debts to Vítkovice lenders as well as $18.4 million (CZK 700 million) to the state-owned Czech Consolidation Agency.[158]

In the meantime, in a stunning move, in February 2002, the government gave LNM Holdings exclusivity in the Nová Hut privatization talks until the end of May, without ever holding a tender.[159] The fact that other companies, such as U.S. Steel (alone or in a consortium with Třinecké Steelworks and its American investor, CMC), or Salzgitter also expressed interest in purchasing Nová Huť further indicates that the claim about the lack of investor interest in the industry served as an excuse for keeping the status quo.[160] In fact, in a 2002 interview, the LNM spokeswoman revealed a long-standing interest by LNM, further undermining the government's claims about the lack of interested foreign investors, "We are the only long-term important investor, which is negotiating with the Czech government since 1997. . . . Then, of course, the situation wasn't conducive to privatization. Since that time, we have been in regular contact with the interested sides and have repeatedly been expressing our interest."[161]

The quick decision to grant exclusivity to LNM reflected a shift within the government against Grégr's consolidation idea. The decision was spurred by a combination of EU pressures, IFC loan security concerns, all exacerbated by the prospect of upcoming elections.[162] Exclusivity was widely criticized by the enterprise union, industry analysts, and local politicians.[163]

As the elections neared, the pressure intensified. The EU's impatience with the situation was reflected by the Commission negotiator's comment to the Czech government, "This is your problem. If you want to solve it, you better get cracking, and there are several ways to solve it, but time is running very short. . . . If you don't make up your mind over the next few weeks, it will be too late."[164] A quickly negotiated deal meant that the roadblock to the completion of the competition chapter and state aid negotiations with the EU would be removed while satisfying the IFC, which began applying concerted pressure on the government, including sending a "final warning" to the Czech cabinet and even threatening Nová Huť with bankruptcy in early May 2002.[165] Thus, according to an interviewed former deputy minister of finance, selling the company to LNM provided an exit strategy.[166]

On May 29, 2002, the government accepted LNM Holdings' bid, thereby opening the way for the closure of the negotiations with the Commission regarding the competition chapter, which was finalized on October 24, 2002.[167] According to the government's resolution, the privatization package entailed $9 million for a 67.25% stake, different forms of financial injection into the enterprise in the amount of $64.8 million, investment of up to $3.05 million (CZK 100 million) in the development of the North Moravia region, and minimum capital investments in the amount of $242.8 million to be made between 2003 and 2012. Employment in the enterprise was to be maintained at the minimum of 8,860 employees until December 31, 2005. Very importantly, LNM Holdings was to reach an agreement and present a memorandum of understanding between the enterprise and the IFC, along with the consortium of Czech banks, concerning the repayment of the respective loans. In a controversial and criticized exchange, the Czech government decided to sell the credit claims in the amount of CZK 3.8 billion, held by the Czech Consolidation Agency, to LNM for CZK 380 million or $10.9 million. According to the interviewed Consolidation Agency official, the "creditors were put into a corner and Mittal bought the debt at 10% of its value."[168] In addition, the Consolidation Agency assumed the responsibility for issued bonds in the amount of CZK 1 billion. By contrast, the IFC obtained a repayment of its loan in full. Despite providing a solution to the Nová Huť problem, the deal was subsequently criticized for being overly generous to the purchasing party. Without holding a tender and comparing bids, however, the government left itself open to this kind of criticism.[169]

With the biggest problem in the sector solved through the sale of Nová Huť, the government's attention turned to Vítkovice Steel. In a testament to the Czech state's capacity, following the restructuring measures undertaken by the emergency management under Novák and continued by his successors, the enterprise became profitable while under the tutelage and eventual ownership of Osinek. However, Vítkovice still required a financial injection and therefore had to rely on Osinek for financing production. After Osinek's acquisition of Vítkovice, the FNM gave exclusivity to LNM Holdings for the purchase of Vítkovice, expecting to obtain a bid of at least CZK 3.3 billion, in other words, Osinek's purchasing price. However, LNM Holdings refused to pay more than $88.2 million (CZK 1.25 billion) and even threatened that it might reconsider the Nová Huť deal. State officials remained unmoved and decided to postpone the Vítkovice deal in an effort to obtain a better price in the future, keeping the enterprise, for the time being, in the hands of Osinek.[170]

Learning from past mistakes, the new ČSSD government held a public tender in 2005.[171] This time, the bids were much higher, with sums of up to $375.6 million (CZK 9 billion) being offered for Vítkovice Steel. In July 2005, the government decided to sell the enterprise to the Russian Evraz Group S.A. for $294.2 million (CZK 7.05 billion), with additional $104.3 million (CZK 2.5 billion) in promised investments. Thus, in the Czech case, similarly to the Polish privatization of Częstochowa Steelworks, institutional maneuvers characteristic of relatively high state capacity gave the government extra time to negotiate with the investors, even after EU accession.

Overall, the Czech case shared with the Polish case the attachment to the status quo of incomplete reform, which persisted in the face of market pressures for a decisive solution to the steel sector woes. Similarly, the timing of the sale to foreign investors was determined by the restrictions of the EU's regulatory rules and demands to present a strategy that would ensure the sector's viability on international markets without further state aid.

Conclusion

This chapter has shown that state capacity was not only the central factor responsible for the restructuring outcomes, but ultimately, for the manner of extrication from incomplete reform. Incomplete reform provided certain benefits to power holders throughout the transition process, but the governments' ability to maintain it differed. In the case of Romania, the massive rent seeking was limited not just to the steel sector but occurred throughout the economy. This triggered targeted IFI action to force the hand of the government to take decisive measures to remove the source of rents—that is, sell the enterprises to strategic investors. Faced with a dire macroeconomic climate, EU regulatory concerns at that point were secondary; however, the prospect of EU membership made the government more likely to abide by the IFI tutelage.

In the case of Slovakia, a new reformist government faced the threat of bankruptcy from international lenders in a critical enterprise, albeit one controlled by its political adversaries. In their pressure on the enterprise, the international lenders displayed little faith in state assistance, and state authorities indeed displayed significant institutional shortcomings when both attempting to take over the enterprise and deal with the crisis. In the Slovak case, although the EU membership considerations certainly shaped the overall reform trajectory of the post-Mečiar government, the EU was not the driving force behind the privatization to strategic investors. The Slovak case is best contrasted with the Czech Nová Huť debacle. There, unlike in Slovakia, the international lender, the IFC, wanted to shift its business risk onto the Czech government by having it underwrite the loan. When the Czech government did not take the IFC up on the offer, the IFC then used the opportunity afforded by the EU accession pressures for a sectoral program to intensify its own pressure to sell to a foreign investor. In the meantime, the Czech government assisted the enterprise and also used elaborate institutional tools to both restructure and operate another major steelmaker, Vítkovice.

The Polish case shares remarkable similarities with the Czech case as far as the sophisticated use of parastatal companies to assist problematic enterprises is concerned. Thus, Silesia Finance Society and the Agency for Industrial Development, like Osinek and the Consolidation Agency, enabled the government to give a lifeline to enterprises that were experiencing the threat of bankruptcy. However, by dampening the financial pressures on the enterprises, they also enabled the incomplete reform to persist longer, until it was checked by the regulatory rules of the EU.

This chapter showed that the four countries shared the same outcome of transnational capitalism, and that the EU played different roles in bringing this outcome about. Whether this role was direct or only indirect was related to the level of capacity of a given state. The following chapter places the journey to transnational capitalism in a broader regional context.

SEVEN

Conclusion

European Capitalisms in Comparative Perspective

We are a Luxembourg-based company with European cultural values . . . [Alexei Mordashov is] a true European.
—Arcelor CEO

I see plenty of scope for growth in developing countries and plenty of opportunities for consolidation in developed countries.
—Lakshmi Mittal, ArcelorMittal CEO[1]

In a more open world, emerging economies are spawning their own giants . . . investment now flows increasingly from south to north and south to south, as emerging economies invest both in the rich world and in less developed countries.
—The Economist, January 10, 2008[2]

The type of capitalism emerging in postcommunist Europe has become the subject of considerable scholarly debate in the field of political economy of postcommunism. Although labels and categories differ, depending on the aspect of the capitalist system being considered, it is clear that a transnational capitalist system, dominated by foreign capital, emerged in the region. The rise of transnational capitalism, however, was marked by variation between and within countries, both with regard to the timing of foreign investment and sectoral distribution.

This book examined the political economy of the emergence of transnational capitalism in the extremely sensitive steel sector. The sector's communist legacy made it a particularly politically, socially, and economically challenging sector to restructure. At the same time, the transition states faced external pressures, including those from a concerned EU. The latter had itself recently undergone a costly and painful process of steel-sector restructuring and was confronting strong pressures from new market entrants from the emerging economies. Thus, the steel sector provided a prism through which to examine the complex interactions between governments and the domestic and external pressures, and to see how different levels of state capacity either enabled or constrained the governments in power.

The chapters thus far have process traced the reform from the initial policy choice to its implementation at the enterprise and sectoral levels. After examining the restructuring process at both levels, the analysis moved to the intersection of national and international levels. I have shown that the four countries followed different reform pathways, distinguishable by the predominance of differing external pressures, to arrive at the common outcome of transnational capitalism and that the level of state capacity determined the pathway each country followed.

The purposes of this concluding chapter are sixfold. First, I reiterate the key findings of the book. Second, I address alternative explanations that could shed light on the developments I examined. Third, I present other cases in the region and discuss how far my argument travels. Fourth, I compare the developments in the four country cases to the domestic capitalist pathway followed by Russia and, to lesser extent, Ukraine. Fifth, I turn my attention to developments in Western Europe, which also faced severe pressure to restructure and rationalize its steel industry, and I assess how the consolidation processes, and the resulting capitalist system, differ from the ones in East Central Europe. Finally, I address the question of the relevance of state capacity once privatization to foreign investors has taken place.

Summary of Key Findings

The three levels of analysis in this project—enterprise, sectoral, and national/international—have enabled me to address different literatures and debates concerning the political economy of postcommunist transition and of economic reform more broadly. The focus on a single sector, in turn, enabled me to evaluate the relevance of the variables these literatures identified as significant during the reform process and to examine their interaction. My analysis revolved around a key intervening variable that has been neglected in the literature on postcommunist transition: state capacity.

As I showed in chapter 3, the four countries made different initial policy choices, which were rooted in their political contexts and reflected a combination of the ideological preferences of those governments and pragmatic political considerations. Thus, while Poland and Romania maintained state ownership, the Czech Republic and Slovakia pursued privatization through sales to domestic owners. In either case, though, the governments rejected the idea of far-reaching intervention and redeployment of assets, and devolved the responsibility for restructuring to the management.

It was state capacity, rather than ownership type—whether state or private—that mattered for the quality of restructuring. As shown in chapter 4, which focused on the enterprise level, restructuring outcomes turned on the ability of the state apparatus to constrain rent seeking and to rein in the unrealistic ambitions of enterprise managers. In addition to the ability to impose financial discipline on the enterprises, restructuring also depended on market-based state institutional support for restructuring.

Chapter 4 also showed that even though the managers had a thorough knowledge of the industry, they were unwilling to cooperate. Contra advocates of making industrial networks the agents of restructuring, I showed that rather than cooperate to further the interests of the sector as a whole, the managers pursued enterprise-specific, prestige-maximizing visions of restructuring that had pernicious effects on the sector.

At the level of the sector, examined in chapter 5, I focused on the relationship between labor and the state. I showed that social dialogue with autonomous unions, especially when it is conducted at the sectoral level, is conducive to restructuring but that it requires a state capable of implementing the concluded agreements. By contrast, opposition to reform is characteristic of captured unions, and the reform-promoting solution is to engage in more—rather than less—social dialogue and to create encompassing sectoral structures. Thus, chapter 5 showed that the advocates of insulating the executives from social pressures in an effort to promote reform provided neither the right diagnosis nor the correct solution.

The national/international nexus was the subject of chapter 6, in which I examined the ways in which the state mediated external pressures for full reform, and analyzed each country's extrication from incomplete reform. I showed that in Romania, where state capacity was low, the inability to maintain macroeconomic stability led to micro-level IFI pressure to sell to foreign investors. Slovakia, a state with a medium level of capacity, was able to withstand IFI involvement because its macroeconomic imbalance was not critical, but it lacked the tools to assuage the pressures from the international financial markets and from bank lenders who demanded repayment of their loans.

I also demonstrated that the relatively high state capacity of Poland and the Czech Republic served as a double-edged sword. Based on the steel sector, I showed that although institutional strength and sophistication can be employed to discipline firms into engaging in market-oriented behavior, it also enables political actors to shield enterprises from market pressures and promotes the pursuit of political and personal preferences that are potentially inefficient. Thus, relatively high state capacity is a useful tool in the hands of a government, allowing it to pick and choose which reforms to engage in so as to stave off external pressures and cater to the domestic audience or its own personal interests. As long as enough economic sectors fare well, particular areas of inefficiency can be absorbed by the entire economy without disastrous effects that would warrant intervention by multilateral lenders. For this reason, countries with relatively high capacity can maintain incomplete reform until it is checked by external regulatory pressures. In the country cases in this study, such pressures emanated from the EU. Thus, the book highlighted the EU's role in propelling transnational capitalism in the region, as well as examined the complex interaction between the different external pressures acting on the EU accession states.

Alternative Explanations

The evidence presented in this book supports the argument that state capacity plays a central role in mediating various external pressures that lead to convergence on a transnational capitalism. Having presented the nuances of the four case studies, I now turn to alternative explanations for the examined outcomes and make the case that the argument I posit is more persuasive. These alternates entail explanations based on partisanship of the governing coalition, strength of labor pressure against privatization, country wealth, and demand-side arguments focusing on the attractiveness of the enterprises, as well as an explanation focusing only on the role of the EU.

The first alternate explanation is the standard partisanship explanation that holds that the right-of-center, pro-market parties are more likely to sell to foreign investors due to their ideological proclivities. Yet, partisan proclivities and market views in postcommunist Europe are complex, and political parties, let alone coalitions, eschew a clear-cut left-right distinction.[3] In any case, with the exception of Slovakia, it was the left-of-center parties (including the communist successor parties in Romania and Poland) that oversaw the privatization deals to foreign investors.

Another explanation posits that strong labor pressure against privatization could have determined the late timing of the Polish privatization, and perhaps also the Czech.[4] Conversely, the co-optation of the Slovak and, to some extent, the Romanian unions inside the biggest enterprises could be expected to have undermined the ability of these actors to resist once the privatization pressure on the government mounted. Yet, the autonomous Polish and Czech unions actually supported privatization to foreign investors, even when the government was ambivalent! By contrast, the Slovak union vehemently opposed privatization, and the Romanian one was moderately opposed to it.

Intuitive explanations based strictly on country wealth in the ECE context also do not provide much analytical leverage. Inspired by the Western European restructuring experience, where the steel sector benefited from substantial state aid, one could posit that the Czech Republic and Poland were more successful than Romania at keeping external pressures at bay simply because they were wealthier and hence better able to support their ailing steel sectors. However, the data on state support for the sector rules this explanation out. In fact, calculations involving levels of state aid administered to the sector and the sector's production output reveal that the average state aid in dollars per ton of steel produced was the lowest in the Czech Republic ($5.4) and Poland ($5.8), followed by Slovakia ($9.1), and Romania ($21.1).[5] Thus, Romania, the poorest country by far, was the most generous to its steelmakers. As chapter 3 argued, this data should be interpreted to reflect the lack of a hard budget constraint rather than a purposeful state-led restructuring effort.

Demand-side arguments suggesting that, by the virtue of their technological and production profile, the Polish and Czech steel producers were simply not as attractive to foreign investors as those in Slovakia and Romania also miss the mark. It is true that the dominant Slovak and Romanian steelworks had more modern basic steel production facilities and overwhelmingly manufactured flat products, which have higher value-added than the long products manufactured by their Czech and Polish counterparts. However, as chapter 6 showed, there was foreign interest in acquiring the Polish and Czech enterprises from the late 1990s through the time of the final sale. Contra demand-side arguments, the data show that these two governments simply did not do much to stoke that interest and facilitate the privatization process.

Finally, one could argue that the EU by itself was a sufficient driving force behind the observed outcomes. However, a straightforward application of Milada Vachudova's notion of the EU's "active leverage," whereby the EU adopted deliberate policies toward accession states and forced them to undertake unpopular and difficult reform measures, cannot explain the surprising timing of the steel-sector privatization.[6] To account for it, one would have to claim that steel-sector reform was somehow less controversial in Romania and Slovakia than it was in Poland and the Czech Republic, which was clearly not the case. Nonetheless, one could relax the assumption of a severe asymmetry of power between the EU and the ECE states and claim that Poland and the Czech Republic were on the forefront of accession and had the political capital to procrastinate on some difficult reforms, against the EU's wishes.[7] By contrast, Slovakia, which was trying to catch up with the first-wave entrants, and Romania, whose credentials were dubious, may have tried to highlight their reformist zeal to be admitted to the EU; a politically difficult (re)privatization would represent a powerful signal of credible commitment to reform. This explanation could potentially hold for Poland and the Czech Republic but is less satisfactory in the Slovak and Romanian cases. In Slovakia, process tracing revealed that the EU considerations played no significant role in the decision to sell VSŽ to U.S. Steel. In Romania, the World Bank was the principal external actor pushing for privatization. In both Slovakia and Romania, however, the EU played an indirect role, strengthening the positions of the international banks and of the IFIs vis-à-vis governments that wanted to demonstrate their reformist credentials. Thus, process tracing the evolution of ownership in the sector has revealed distinct trajectories leading to convergence on transnational capitalism: liberal capitalism dominated by foreign investors. These findings raise the question of how far the convergence argument travels, which I take up next.

Extending the Findings

This book has dealt with the four biggest steel producers among the new EU member states, but the outcome of ownership by foreign investors is broadly shared by the ECE countries. In 2004, Mittal Steel (the predecessor of ArcelorMittal) acquired the biggest steelworks in Bosnia (Mittal Steel Zenica) and in Macedonia (Mittal Steel Skopje). In 2003, U.S. Steel acquired Serbia's Sartid Steelworks (U.S. Steel Serbia). In 2004, Ukrainian Industrial Union of Donbas acquired Hungary's dominant steelworks—Dunaferr—which is responsible for 80% of Hungarian steel production. In 2007, Commercial Metals Company acquired Croatian Sisak Steelworks. In 2003, the Slovenian government sold Slovenian Ironworks to the Russian group Koks (owned by the Industrial Metallurgical Holding), though the Slovenian state continues to hold 25% of shares in the company. In 1999, Bulgaria's largest steelworks, Kremikovtzi, was sold to a domestic company, Daru Metals (later Finmetals Holding), which was purchased by Pramod Mittal's (Lakshmi Mittal's brother) Global Steel Holdings in 2005. The outlier in these countries is Latvia, whose relatively small Liepajas Metalurgs—the only steelworks in the Baltic states—is owned by a group of domestic capitalists and publicly traded on the Riga stock exchange.

Whether the same causal mechanisms as those identified in the country cases studied here explain the privatizations to foreign investors in other ECE countries can only be confirmed through in-depth case study. However, the observed pattern conforms to that seen in the four country cases examined in these chapters. Hungary and Slovenia, two new EU member states that have high state capacity, did not sell their dominant steelworks until around the time of EU accession, as did Poland and the Czech Republic. In line with the argument presented in this book, this was the case not just with the more statist-oriented Slovenia but also with Hungary, which otherwise has been on the forefront of privatization via sales to foreign investors since the very beginning of the transition process (and even before!).[8]

The 1999 decision by Bulgaria—a country where state capacity is slightly higher than in Romania—to pick a questionable (domestic) owner for the country's biggest industrial complex during an attempt to speed up reforms, far in advance of EU accession, is not surprising. After all, without micro-level IFI assistance, Romania made the same kind of mistakes in the late 1990s. It is also not surprising that the subsequent foreign owner did not have the wherewithal to turn the enterprise around or that there were allegations of asset stripping when the enterprise was under foreign ownership.[9] The timing of the privatizations in Serbia, Macedonia, and Bosnia is also not tied to the pressure of EU membership. If anything, these transactions were undertaken as part of a multilateral regional reconstruction effort between the IFIs and the EU.[10] Thus, the timing and context of privatizations in other ECE countries support the argument presented in this book.

The dominance of foreign capital in the "commanding heights" of the

ECE economies earned them the label of liberal but "dependent" capitalism, an economic structure in which the decisions concerning subsidiaries are made in the boardrooms of large, multinational corporations.[11] Indeed, the international strategy of the new owners dictates the business strategy in the various sites, including threats of exit. This raises the question of how this "dependent" transnational capitalism in ECE countries differs from the capitalism of their western and eastern neighbors, both of which have been affected by the global consolidation trends in the sector. Of particular note are the developments in Russia, where the transition process resulted in the emergence of transnational enterprises with global reach.

Turning East: State Capitalism in Russia

Although recent years saw a marked rise in ownership by transnational capital across the entire EU, strikingly different developments have been taking place in Russia and, until recently, in Ukraine. As chapter 6 made clear, unlike in ECE, domestic capitalists in Russia have not only been able to survive but have managed to turn their enterprises into internationally competitive, significant market players. More importantly, the biggest Russian steel producers have themselves started turning into transnational companies, with considerable holdings outside of Russia, not just in ECE, but also in Western Europe and the United States. In these cases, Russian capitalists were often seen as much-needed, even if not fully trusted, strategic partners.

Russia followed a distinct path from its ECE counterparts, which found themselves facing stronger external pressure to sell their enterprises to foreign investors, for several reasons. First, the enterprises in ECE did not have the access, geographical or financial, to natural resources that became the basis for the Russian enterprises' restructuring, including vertical integration. Second, its tremendous endowment of natural resources—namely, oil and gas—provided the Russian state with the financial resources to prop up the sector when needed. Thus, the Russian state could avoid the pressures from both the IFIs and the financial markets as it stepped in to help using the vast resources of the banks it controlled. Third, the Russian state was not constrained by the competition rules of the EU, which forced privatization to foreign investors in those countries—for example, Poland and the Czech Republic—that were able to keep the IFIs and the financial markets at bay.

The development of Russian capitalism in the steel sector epitomizes the processes of economic reform endemic to Russia. In the first phase of transition, the nascent capitalist system, to a far greater extent than in any of the countries considered in this project, brought about the collapse of the Russian economy. For example, while the Russian GDP contracted by 45% in the initial years of economic transition (1989–1998), the economy of the biggest reform laggard considered in this study, Romania, contracted by about half as much: 24%.[12] The resulting devastation for the lives of ordinary Russians had a singularly golden lining for a small group of highly enterprising individuals, often company insiders, who were able to enrich themselves by taking advantage of the rapidity of the unregulated privatization processes.[13] If one examines the evolution of the biggest Russian steelmakers, the familiar picture emerges—enterprise insiders who managed to obtain control over an enterprise in the early to mid-transition years, such as Alexei Mordashov of Severstal or Viktor Rashnikov of the Magnitogorsk Iron and Steel mill, or of oligarchs who made their wealth elsewhere and subsequently entered the steel business, such as Roman Abramovich and Alexander Abramov of Evraz Group or Vladimir Lisin of Novolipetsk.

After the first phase of capital accumulation, the oligarchs had the financial resources and, increasingly, the organizational capability to restructure selected enterprises. Compared to other sectors, such as the lucrative oil and gas industries, the steel industry generally records a relatively small profit margin. It is a testament to the vast natural-resource wealth in Russia, which enabled vertical integration, that the steelworks not only survived the transition but also prospered. The scope of wealth accumulation by the Russian oligarchs is reflected by the 2010 Forbes magazine list of the world's billionaires, which revealed thirteen Russians among the top 100 names (and twenty among the top 150). Of Russia's nine wealthiest men, four were directly involved in steelmaking operations, including Russia's richest man at the time, Vladimir Lisin.[14] As of 2014, the title of Russia's richest man went to Alisher Usmanov, who made his money in steelmaking.[15]

In the second phase of transition, the Russian owners had assets at their disposal that their ECE counterparts could only dream of: not only lower labor costs and subsidized energy prices but also, more importantly, the opportunities for vertical integration of steel operations to include iron ore mines, coal mines, and, sometimes, also the means of shipping their goods. Russia's biggest steel producers, such as Severstal, Magnitogorsk, Evraz Holding, Novolipetsk, and Metalloinvest, all own iron ore mines and, with the exception of Metalloinvest, coal mines. The process of vertical integration was gradual, and the enterprises expanded and consolidated their operations in the early to mid-2000s. The vertical integration of these operations within enterprises led to highly competitive prices in Russia. For example, in the second quarter of 2009, a ton of Russian steel cost $260, compared to a per-ton price of $306 for Brazilian steel, $464 for Chinese steel, and $485 for EU-made steel (average).[16]

With vertical integration, the Russian oligarchs-steelmakers could imitate the "Mittal effect" throughout Russia and beyond: they had the financial resources to restructure and invest in their acquisitions; the production inputs to be self-sufficient; and, increasingly, their own international distribution centers. It is notable that the difficulties and cost of obtaining production inputs—iron ore and coal—represented some of the principal challenges for the ECE steelworks. The parasitic relations between the often politically connected suppliers and intermediaries and the enterprise, which intensified as the latter became financially strapped, are in part what brought the ECE steelmakers to their knees. Although the transparency of the ownership structure of operations in Russia may leave a lot to be desired, in the long run, operating a profitable, vertically integrated enterprise is a more lucrative option than the short-term "tunneling out" of the enterprise via transfer pricing, as happened with VSŽ in Slovakia.

As the Russian steelmakers restructured under domestic ownership—and with the continuing support of the state's resources—they turned their attention to operations beyond Russia's western borders and aspired to become transnational enterprises. In ECE in particular, their knowledge of how to operate on barter-based markets proved to be particularly helpful, especially as they took an interest in Romania. Enterprises in ECE proved to be valuable in that they provided a conduit to the single EU steel market, at least for a fraction of their products.

It was the smaller Russian market players who first entered ECE territory, primarily Romania. Thus, Russia's biggest steel pipe producer, TMK, purchased Artrom Slatina and Reşiţa Steelworks, and Mechel Group/Conares purchased Câmpia Turzii Wire Factory and Târgovişte's Special Steels Company. The first large-scale Russian presence appeared in the Czech Republic, with the purchase by Evraz Holding of the country's third-largest steel producer, Vítkovice Steel. Their Romanian and Czech domestic partners were positive in their assessments of the role of Russian investors, even as the Romanian labor leaders complained about the Russians driving a harder bargain in labor negotiations than the previous managements had done. Western European companies, however, seemed warier of the Russian entry.

In the third, post-2005, phase of transition, after the Russian enterprises integrated vertically and as they started making acquisitions in ECE, the Russian steel magnates strived to gain full respectability on Western markets. This was to be done by being listed on reputable stock exchanges, including the London and New York stock exchanges. Ultimately, Russian steelmakers wanted to be seen as legitimate business partners as they attempted to make the leap from being Russian market players who sell internationally to transnational enterprises.[17]

Severstal took the first steps when it purchased the bankrupt Rouge Steel in Dearborn, Michigan (now Severstal Dearborn). In 2005, it acquired the Italian Lucchini Group, obtaining operations in Italy, France, and the United States. Following these acquisitions, Severstal was producing 40% of its steel outside Russia. Severstal's most ambitious step, however, was to try to play the "white knight" role and prevent Mittal Steel's takeover of Arcelor in 2006.[18] Severstal's owner, Alexei Mordashov, attempted to convince not just Arcelor, Europe's biggest steel producer, but also the business community that Severstal was a suitable partner by arguing that he had all his personal wealth tied up in Severstal and was putting it on the line.[19]

Severstal's efforts were spurned, and Arcelor eventually merged with Mittal Steel to form the biggest steel company in the world, ArcelorMittal, in 2006. Despite this setback, Severstal continued to expand its international operations, going on a shopping spree in the United States, where, in 2008, it purchased the Sparrows Point steel plant in Baltimore, in addition to several smaller plants.

In the meantime, Evraz Group also evolved into a transnational corporation, acquiring plants in Italy and the Czech Republic in 2005 and in South Africa in 2006, as well as in North America in 2007.[20] Evraz Group purchased three plants in the United States and four in Canada. In the midst of the expansion, similarly to Severstal, Evraz engaged in a serious merger discussion with Europe's second largest steel producer—the Anglo-Dutch Corus.[21] The discussions ultimately failed and Corus merged with India's Tata Steel, yet another transnational enterprise originating in the emerging markets.

The failure of these mergers was perceived in Russia as discriminatory against Russian capital. Even as the Russian president, Vladimir Putin, commented that Arcelor's decision to back out of the merger with Severstal "had nothing to do with Russia's purported image as a closed country," members of Russia's upper house of parliament claimed that the decision was political, not economic, in nature. In fact, Western analysts were suspicious of the ties between the Russian capital and the Kremlin. Some analysts went beyond simple wariness. A manager of a New York-based investment firm, and Severstal shareholder, said, "Mordashov is an oligarch, with all that that implies—incestuous relations with politicians, a fast-and-loose attitude to the law and a kind of arrogance that creates risks for shareholders."[22] The Severstal spokeswoman refuted this characterization of Mordashov, saying that it was "entirely wrong" and that "Severstal's management has always respected the law, and our corporate governance fully conforms to the highest international standards."[23]

Other market analysts claimed that it would be "unfathomable" to imagine the merger with Arcelor going through without the approval of the Kremlin.[24] In his letter, published in the *Financial Times*, intended to sway public opinion in favor of the Severstal deal, Mordashov thought it necessary to address these allegations, asserting:

Russian steelmakers generally have an excellent rapport with our government, just as large international companies do in other countries. Contrary to some speculation, we do not have to obtain government approval of our activities as we are public companies. However, as with all other countries, we try to maintain positive relations with our government because we are good corporate citizens.[25]

Yet, despite these denials, there are grounds for concern about the relationship between Russian capital and the Kremlin. Roman Abramovich, who owns a large share in Evraz Group, has been close to the Kremlin. In fact, his involvement in Evraz Group's activities was perceived by analysts as acting on the Kremlin's behalf to forge a Russian steel-industry champion that would rival Arcelor.[26]

Some of the other ties between the political and economic actors are even closer. Alisher Usmanov, for example, is the majority shareholder of Metalloinvest, Russia's third largest steel company and the world's thirty-ninth richest man (Russia's richest).[27] He is also the chairman of Gazprominvestholdings, which manages the foreign investments and operations of Gazprom, Russia's biggest company, which is controlled by the state. Usmanov, whose activities were closely examined in Britain, was certainly able to take advantage of close ties with the government.[28] In 2009, Metalloinvest obtained a $2 billion credit, of which $1 billion was guaranteed by the state, from the state-controlled VTB Bank.[29] Previously, Usmanov had purchased *Kommersant*, Russia's biggest business daily, known for its critical stance vis-à-vis the Kremlin. The move was perceived as an effort by the Kremlin to further clamp down on press freedom in the country, although the newspaper continued to assert its independence.[30]

The Kremlin's relationship with Viktor Rashnikov, the majority stake owner in Magnitogorsk Steel and Iron Company, also raised eyebrows. In selling its remaining 17.8% of shares in the enterprise, in 2004, the state allegedly underpriced its stake while making sure that the current managers emerged as winners. This entailed making a quick announcement of the sale and persuading Rashnikov's biggest rival, Mechel Steel, to drop its bid under a threat of investigation by the tax authorities into its tax payments. According to Mechel's prospectus, published when the company was trying to sell 12% of its stock using American Depositary Shares, Russia's 1999 tax rules "are vaguely drafted, leaving wide scope for interpretation by the Russian tax authorities."[31]

The Yukos case, which was used to bring down Mikhail Khodorkovsky, Russia's richest man at the time and Putin's fierce critic, made the consequences of arbitrary and politicized interpretation of the tax code obvious. The fact that the steel companies paid about 12% of their revenues in taxes in 2004, well below the rate paid by the oil companies, suggests widespread use of transfer pricing between the subsidiaries, in an effort to minimize taxes.[32] Given the ambiguous rules, this could well be used against the companies, should it become politically convenient. However, it is worth noting that the political use of business has its limits. Prime Minister (at the time) Putin's open criticism of Mechel's tax schemes in 2008 prompted a 9% drop in the Russian stock market and a significant decrease in the company's share prices. It subsequently even provoked a minor clash with President (at the time) Medvedev, who said that "law enforcement agencies and government authorities should stop causing nightmares for business."[33]

Other than turning a blind eye to creative accounting schemes by the Russian steelmakers, the Russian state also provides concrete financial help via subsidized loans and administrative measures, such as tariffs and discounts on energy prices. None of these measures are available to Russia's EU counterparts. Given the economic-crisis-related slump in the internal demand for steel in Russia, Putin pledged to boost the construction and automotive industries, with state conglomerates, such as Russian Railways, taking the lead in contracts with domestic suppliers.

The Russian state has also, directly and indirectly, strengthened the enterprises' balance sheets. In 2009, Evraz Group obtained a one-year extension of its $1.8 billion debt facility from the state-owned Vnesheconombank (Russia Development Bank).[34] After overextending itself with a variety of loans to purchase its acquisitions, Mechel Steel was able to pay back some of its short-term debts and to reschedule others using a $1 billion loan from Gazprombank. It was also able to extend a loan from VTB Bank of more than $500 million.[35] Metalloinvest, too, has taken advantage of a VTB Bank loan, in the amount of nearly $2 billion, half of which was guaranteed by the Russian Ministry of Finance.[36] Thus, unlike their ECE competitors, Russian steel companies still have the option of turning to the government for help.

Despite their close relationship with the government, the Russian steel companies have tried to obtain respectability on Western markets by listing some of their shares on the major international stock exchanges, most notably, London's. Thus, in 2010, Evraz Group traded 27% of its shares on the London Stock Exchange. Magnitogorsk did the same with 9.7% of its stock. Severstal's traded stock on the London Stock Exchange was less than 18% of the shares, and Novolipetsk's was 9%, while Metalloinvest planned an initial public offering of about 10% to 20% of its shares, to be listed on the London Stock Exchange.[37] Mechel Steel was a bit of an anomaly since about 15% of its shares were listed on the New York Stock Exchange.

The floating of the shares is intended not only as a vehicle for earning cash for the owners but also as a means of "raising [their] profile with the international investment community."[38] The process forced the companies to become more open about their ownership structure, revenues, and financing schemes. Although the prospectuses left a lot to be desired in terms of transparency, their very existence indicated much progress.[39] For example, in the process of trying to refinance its loans, Mechel Group had to share its reservations with the US Securities and Exchange Commission about whether it could continue operating as a going concern.[40] Although the verdict on Russian capitalism is still out, it is clear that in the steel sector, Russian multinationals have joined the global wave of transnational corporations from emerging market countries that have been revolutionizing the sector.[41]

The case of Ukraine also merits attention. Until recently, it was a hybrid case, sharing similarities with ECE and with Russia; but it seems to have moved decisively in the direction of its ECE neighbors and abandoned the domestic capitalism project.

Ukraine has had its own share of oligarchs with political connections and close relationships between politicians and the business sector. Shortly before the Orange Revolution, in June 2004, the Ukrainian government sold the country's biggest steel mill, Kryvorizhstal Steelworks, to Ukraine's two wealthiest men, Rinat Akhmetov and Viktor Pinchuk, at a highly discounted price. Illustrating the incestuous relations between politics and business, Pinchuk also happened to be the Ukrainian president's son-in-law and a member of parliament. The opposition decried the privatization as an outrage, and Kryvorizhstal Steelworks was renationalized following the Orange Revolution and subsequently put up for auction. Lakshmi Mittal of Mittal Steel was the winner of the 2005 auction, and the reprivatization became a symbol of the pro-Western orientation of the new regime.[42] The $4.8 billion purchase price paid by Mittal Steel was more than five times what the previous owners had paid.

Although the egregiousness of the previous privatization efforts, combined with the initial pro-Western euphoria of the new Ukrainian regime, may account for the Kryvorizhstal developments, it bears highlighting that the new regime remained supportive of the established "national champions." The clearest example was the Industrial Union of Donbas (ISD), a steel company located in the pro-Russia Donetsk region of Ukraine, which voted overwhelmingly against the pro-Western Viktor Yushchenko in the 2004/2005 elections. ISD, like its Russian counterparts, was interested in becoming a transnational enterprise, the first such effort by a Ukrainian steel producer. In 2004, it successfully acquired Hungary's biggest steel producer, Dunaferr. The new post–Orange Revolution government continued to support ISD in its expansion efforts. Viktor Yushchenko, as the new Ukrainian president, and Yulia Tymoshenko, as the prime minister, blatantly lobbied on ISD's behalf as it successfully competed with Mittal Steel for the 2005 acquisition of Częstochowa Steelworks, Poland's second-largest steel producer at the time.[43]

The transnational expansion of Ukraine's steel sector, however, had come to an end, even before the dramatic developments of 2014 that saw Russian annexation of Crimea and put a question mark over the future integrity of the rest of Ukraine's territory. In January 2010, Russian businessman Alexander Katunin, former owner of Evrazholding, purchased a majority stake in ISD in a $2 billion deal. Once again, the state-owned Vnesheconombank provided financing for the deal.[44] This development was foreseen by Anders Åslund, who wrote, in 2009:

> Ukraine is still run by its steel barons, who have rationalized and streamlined their corporate structures . . . but several of them will probably not be able to survive or move to other industries. . . . It is easy to predict that the steel groups that are not shielded from price vagaries through vertical integration or diversification will go under.[45]

Russian investors have also tried to take over the two biggest remaining steel producers under Ukrainian domestic ownership: Ilyich Iron and

Steel Works of Mariupol and Zaporizhstal. Neither company is vertically integrated, making them more susceptible to takeovers.[46] Thus, despite its efforts at transnational expansion, Ukraine seems to be treading the path of its western neighbors, whose biggest steel producers are overwhelmingly under transnational ownership and dependent on their owners' global commercial interests.

Turning West

If the steel sectors of all former communist countries faced the daunting challenge of making the transition from central planning to a market system, the Western European states also faced severe pressure to restructure and rationalize their own steel industries. This section examines the consolidation process in the Western European steel industry, and it compares the resulting capitalist system to that in ECE.

The efforts at steel-sector restructuring in Western Europe began well before there were any signs of transformation taking place in the Eastern bloc. Moreover, the Western European market players also faced the acute globalization pressures and the rise of competition from the new emerging market players in the 1990s. In response to these pressures, the Western European steel industry became dominated by transnational market players. The process, however, differed from the developments in ECE in that the state, with the European Commission's permission, if not blessing, exerted a very strong, and costly, role in the restructuring of the Western European steel industry. Moreover, the resulting Western European steel conglomerates entered into transnational partnerships on a much more equal footing than that of their ECE counterparts.

The EU's insistence on strict implementation of state aid rules in the ECE steel sector stands in stark contrast to the way the acquis was bent to accommodate the economic and political realities of the older member states in prior years. The Commission showed much malleability as it permitted state largesse in the restructuring of the Western European steel industry.

The most visible example was the 1980 crisis on the steel market, which led to a call for the creation of a sectoral policy for the EC steel industry. However, because there were no financial resources to restructure at the Community level, member states were allowed to administer wide-ranging aid, provided it was approved by the European Commission. The aid lasted for twenty years! It was only after July 22, 2002, when the European Coal and Steel Community Treaty was set to expire and when the steel industry was largely restructured, that state aid was prohibited.[47] Thus, the carefully enforced ban on state support for the sector in ECE came into play only *after* the Western European countries had already used substantial state support to restructure their own industry.

The restructuring of the steel sector in Western Europe proceeded in several stages. The first entailed nationally focused, state-led restructuring, either via nationalization or state assistance to private players. With the most difficult restructuring steps completed and state aid no longer administered, there occurred a period of market-based first national and then Europe-wide consolidation. The third and most recent phase entailed the globalization of the sector, as it moved from being a set of European champions to competing transnational players, teaming up with capital originating in emerging markets. Let us examine these three stages in turn.

The first—national—stage of consolidation followed the 1973 oil crisis. The European steel sector found itself in dire straits, caused by both financial constraints and technological changes involving electric arc furnaces, which were more efficient and cost-effective. Although the national governments realized that their steel sectors suffered from both fragmentation and massive overcapacity, they resisted pressures from the European Commission to rationalize production capacity across national lines. The first step was to overcome inefficient domestic competition among the different steel companies.[48]

The various states became directly involved in the steel sector largely via nationalization. The United Kingdom led the way with the 1967 establishment of the British Steel Corporation, which represented an amalgamation of the fourteen most significant steel producers in the country. As part of Margaret Thatcher's revolution, the company was privatized in 1988, and the UK became an ardent supporter of banning all state subsidies in the sector.[49]

The French response to the 1973 oil crisis entailed grants and subsidized loans to the steel sector and the subsequent nationalization of Usinor and Sacilor, major companies in the sector, in 1981. In 1986, as part of the rationalization process, these companies were merged to form a single state-owned enterprise, Usinor Sacilor, which became Europe's largest steel company. State involvement in the sector was substantial, as Usinor Sacilor was recapitalized by the French Treasury in 1984 and 1986. In 1987, the Treasury announced that it would no longer financially assist the enterprise. While the company had to rely on loans from private and state-owned banks, its state-owned status provided an implicit guarantee of debt repayment. Between 1986 and 1995, the company underwent substantial restructuring, increasing productivity and turning a profit in 1994.[50] In 1995, the French state privatized the enterprise.

In the meantime, Spain joined the EU. Membership negotiations provided a spur for the Spanish government to reduce steel output and capacity and to pursue further sectoral consolidation. In 1997, Aceralia Corporación Siderúrgica was formed out of reorganized assets resulting from a merger of Spain's two important steel producers, AHV and Ensidesa, in 1991. The privatization process proceeded later that year, while Aceralia continued to acquire further assets.[51]

At the same time, in Germany, restructuring and consolidation were largely accomplished without state ownership. In 1991, Krupp took over Hoesch, and for more than fifteen years, Thyssen and Krupp discussed a merger proposal. The discussions finally bore fruit in 1999, when ThyssenKrupp became Germany's biggest steel producer.[52]

The second stage in the evolution of the Western European steel sector came at the end of the 1990s. As economic pressures mounted, steel producers engaged in an unprecedented degree of cooperation across national barriers, much in the spirit of the single market. In 1997, Aceralia engaged in a strategic partnership with Luxembourg's Arbed Group. In 1999, the two former primary advocates of cutting state subsidies to the sector—British Steel and Dutch Koninklijke Hoogovens—formed a trans-European merger of unprecedented dimensions, becoming Corus.[53] In the same year, French Usinor (renamed after the privatization of Usinor Sacilor) acquired Belgium's biggest steel producer, Cockerill-Sambre. However, the boldest move came in 2002, with the emergence of Arcelor. A true "European champion," Arcelor resulted from a merger of three European steelmaking giants: the Spanish Aceralia, Luxembourg-based Arbed, and French Usinor.

This second stage of Europeanization in the steel sector quickly morphed into the third one: globalization. This stage was epitomized by the controversial merger of Mittal Steel and Arcelor, announced in June 2006. As discussed earlier, Mittal Steel made a hostile bid for Arcelor, and the management at first tried to use Severstal as a "white knight" to save Arcelor from Mittal Steel's grip. The shareholders—who ultimately prevailed—fiercely resisted the managerial strategy and Severstal's efforts. At the same time, the disturbing rhetoric surrounding Mittal's bid coming from Arcelor's CEO made the difference between Europeanization and globalization explicit. While claiming that "[w]e are a Luxembourg-based company with European cultural values," Arcelor's CEO suggested that Lakshmi Mittal lacked these values, while Alexei Mordashov was a "true European."[54]

The distinction drawn between European and non-European capital, and its appropriateness in the sensitive steel sector, seems to have been rejected as an unfortunate anachronism, to say the least, when Arcelor's shareholders rebelled against the Severstal partnership. This paved the way for the creation of the biggest steel company in the world. The following year, this move was followed by another important merger in the steel sector, involving the Anglo-Dutch Corus, in which a Russian contestant, Evraz, was also rejected, this time in favor of India's biggest steel producer-turned-transnational, Tata Steel.[55]

This brief examination of the economic landscape of Western Europe's steel sector reveals a structure increasingly similar to that of ECE, where the large transnational players deal the cards, and the national champions are the exception, rather than the rule. Although these recent developments reveal a potential for the blurring of differences between the ways in which capitalism operates in ECE and Western Europe/pre-eastern enlargement EU, there are two caveats to keep in mind. First, the remaining domestic-capital-owned steel-making groups in Western Europe, such as the German ThyssenKrupp, the Italian Riva Group, or the Austrian voestalpine Group (formerly Voest-Alpine), are still significant players. The second caveat pertains to the nature of mergers. The companies entering into a merger tend to do so to improve their current market prospects, and not as a dire means of survival, at least not in the short term. Thus, mergers result in shared governance of the new enterprise, as can be seen by examining the composition of the board of directors.

That said, the transnational nature of ownership makes the new owners less beholden to local conditions. Given the market pressures, it is quite likely that we will observe a fuller convergence in European capitalism, albeit with Russian capitalists, like their Western European counterparts, entering the relationship as partners and preserving that status.

From Transitional to Transnational Capitalism: The Role of State Capacity?

This book considered the state as the central and crucial player in the industrial restructuring process, a player that has been overlooked in much of the scholarship on the political economy of transition. At the intersection of national and external pressures, high state capacity is conducive to the autonomy of domestic political actors vis-à-vis external pressures and can lead to maintaining unfinished reform. What the EU provides, at least in the steel and other "sensitive" sectors, is a regulatory framework that promotes economic efficiency by virtually eliminating the possibility of shielding the enterprises from market pressures.[56] In other words, the previously double-edged sword of high state capacity becomes employed in a more unidirectional manner: to implement the rules and regulations guiding the single market, because any deviations from the acquis are closely monitored by the other market players and the EU.

What role does state capacity play in transnational capitalism? High state capacity is crucial for the regulatory function of the state and, more specifically, for its social dimension. After all, market distortions are already closely watched—and reported to the Commission—by fellow market participants. If one looks at the steel sector in ECE, it is clear that the EU has been adamant about inquiring into any perceived vestiges of state aid. In Poland, the Commission conducted a full investigation into allegations of state aid to Częstochowa Steelworks and, in fact, ordered the recovery of €4 million for state aid granted between 1997 and 2002.[57] In the Czech Republic, the Commission investigated illegal aid to Třinecké Steelworks.[58] More recently, the Commission launched an investigation into preferential electricity pricing for ArcelorMittal Galați.[59]

However, the social dimension of the acquis is much weaker than the economic one, and the task of monitoring the enforcement of much of the social legislation falls to the domestic institutions. Thus, the state continues to play a central role in the quality of life of workers of the EU member states from ECE. The differences among countries persist. To illustrate, when asked about the most effective means or institutions labor unions can use to protect workers' interests, a union leader in Poland, without hesitation, named the State Labor Inspectorate. It was, according to the interviewee, the most important and effective "intercessor" in responding to complaints about the employer, "The State Labor Inspectorate comes in, inspects, fines, and demands clarification of irregularities."[60] By contrast, a union leader in Romania claimed that the labor inspectorate there was very weak and that employers were betting on the lack of effectiveness of the authorities. Furthermore, he complained that employees are not interested in legal action or in suing their employers because they expect that the judges will have been bought off by them.[61]

It is possible that the transformation of the state and capacity building on the single-market dimension will create institutional resources that will lead to the eventual strengthening of the state in other areas that are less regulated by the EU. However, the construction of the different dimensions of state capacity will continue to be affected not just by communist—or even precommunist—legacies but, now, also by the legacies of the transition process itself.[62]

Appendixes

Appendix A

List of Interviews

(presented in chronological order)

Industrial policy expert, European Commission, Brussels, Belgium, May 2002

State aid expert, European Commission, Brussels, Belgium, May 2002

Poland Case Study

Henryka Bochniarz, former Minister of Industry and Trade (1991–1992), Warsaw, Poland, July 2001

Civil servant working on the Polish steel-sector restructuring, European Commission, Brussels, Belgium, May 2002

Paweł Ruszkowski, sociologist, labor expert, and adviser, Warsaw, Poland, May 2003

Rafał Towalski, sociologist and labor expert at the Warsaw School of Economics, Warsaw, Poland, May 2003

Journalist covering the steel sector at a major daily, Katowice, Poland, June 2003

Representative #1, Metallurgical Chamber of Industry and Commerce (Hutnicza Izba Przemysłowo-Handlowa), Warsaw, Poland, June 2003; Katowice, Poland, July 2003, June 2008

Michał Górzyński, economist at the CASE Institute, Warsaw, Poland, June 2003

Steel-sector expert, Agency for Industrial Development, Warsaw, Poland, July 2003

Top management member, Częstochowa Steelworks, Częstochowa, Poland, July 2003

Edward Nowak, former Deputy Minister of Economy, Kraków, Poland, July 2003

Trade union leader, Federacja Hutniczych Związków Zawodowych (Federation of Steelworker Trade Unions), Katowice, Poland, July 2003

Trade union leader, Federacja Hutniczych Związków Zawodowych trade union affiliate at Częstochowa Steelworks, Częstochowa, Poland, July 2003

Top management members (4), Florian Steelworks/Polish Steelworks, Świętochłowice, Poland, July 2003

Janusz Steinhoff, former Minister of Economy, Warsaw, Poland, July 2003

Trade union leaders (2), Kadra trade union, Częstochowa, Poland, July 2003

Representative #2, Metallurgical Chamber of Industry and Commerce, Katowice, Poland, July 2003

Civil servant #1, Ministry of Economy, Labor and Social Policy, Industrial Restructuring Division, Warsaw, Poland, July 2003

Civil servant #2, Ministry of Economy, Labor and Social Policy, Industrial Restructuring Division, Warsaw, Poland, July 2003

Civil servant #3, Ministry of Economy, Labor and Social Policy, Industrial Restructuring Division, Warsaw, Poland, July 2003

Civil servant #4 (formerly involved with steel-sector restructuring), Ministry of Economy, Labor and Social Policy, Warsaw, Poland, July 2003

Civil servant #5 (responsible for sectoral social dialogue), Ministry of Economy, Labor and Social Policy, Warsaw, Poland, October 2003

Civil servant #6 (responsible for financial aid from the European Union), Ministry of Economy, Labor and Social Policy, Warsaw, Poland, July 2003

Civil servant #7 (involved in steel-sector oversight), Ministry of State Treasury, Warsaw, Poland, July 2003

Civil servant #8 (involved in steel-sector oversight), Ministry of State Treasury, Warsaw, Poland, July 2003

Legal expert (previously involved in privatization preparations), Ministry of State Treasury, Warsaw, Poland, July 2003

Trade union leader, NSZZ Solidarność Metalworkers' Secretariat,

Katowice, Poland, July 2003; October 2003; June 2008; July 2010

Former trade union leader, NSZZ Solidarność trade union affiliate at Warsaw Steelworks, Katowice, Poland, July 2003

Representative at the Steel Sector Employers' Association, Katowice, Poland, July 2003; June 2008

Emil Wąsacz, former Minister of State Treasury (1997–2000), previously director and earlier trade union leader affiliated with NSZZ Solidarność at Katowice Steelworks, Katowice, Poland, October 2003

Spokesperson, Huta L.W. (formerly Warsaw Steelworks), Warsaw, Poland, October 2003

Manager for Social Affairs, Huta L.W., Warsaw, Poland, October 2003

Trade union leader, NSZZ Solidarność trade union affiliate at Huta L.W., Warsaw, Poland, October 2003

Top management member, Polish Steelworks, Katowice, Poland, October 2003

Trade union leader, Federacja Hutniczych Związków Zawodowych trade union affiliate at Tadeusz Sendzimir Steelworks/Polish Steelworks, Kraków, Poland, November 2003

Trade union leader, NSZZ Solidarność trade union affiliate at Tadeusz Sendzimir Steelworks/Polish Steelworks, Poland, November 2003

Manager #1, Tadeusz Sendzimir Steelworks/Polish Steelworks, Kraków, Poland, November 2003

Manager #2, Tadeusz Sendzimir Steelworks/Polish Steelworks, Kraków, Poland, November 2003

Top management member, Tadeusz Sendzimir Steelworks/Polish Steelworks, Kraków, Poland, November 2003

Steelworks insider, Tadeusz Sendzimir Steelworks/Polish Steelworks, Kraków, Poland, November 2003

Trade union leader responsible for external relations, NSZZ Solidarność (national level), Gdańsk, Poland, June 2008

Trade union leader, Federacja Hutniczych Związków Zawodowych trade union affiliate at ArcelorMittal Poland, Kraków, Poland, July 2010

Trade union leaders (3), NSZZ Solidarność trade union affiliate at ArcelorMittal Poland, Dąbrowa Górnicza, Poland, July 2010

Czech Republic Case Study

Desk officers (3), European Commission, Directorate General for Enlargement, Czech Republic Team, Brussels, Belgium, May 2002

Civil servant, European Commission, Directorate General for Enlargement, Czech Republic Team, Prague, Czech Republic, November 2003

Representatives (2) at the Steel Federation, Inc. (Hutnictví Železa, a.s.), Prague, Czech Republic, November 2003; February 2004

Daniel Munich, Economist, CERGE-EI, Prague, Czech Republic, December 2003

Evžen Kočenda, Economist, CERGE-EI, Prague, Czech Republic, December 2003

Jan Mládek, former Deputy Minister of Finance, Prague, Czech Republic, December 2003

Civil servant #1, Ministry of Industry and Trade, Department of Metallurgy, Prague, Czech Republic, December 2003

Civil servant #2, Ministry of Industry and Trade, Department of Metallurgy, Prague, Czech Republic, December 2003

Civil servant #3, Ministry of Industry and Trade, Department of Metallurgy, Prague, Czech Republic, December 2003; February 2004

Trade union leader, OS KOVO, Prague, Czech Republic, December 2003

Jiří Havel, former National Property Fund chairman (2000–2002), Prague, Czech Republic, March 2004

Top management member, North Moravia steelworks B, North Moravia, Czech Republic, March 2004

Trade union leader, OS KOVO trade union affiliate at Nova Huť, Ostrava, Czech Republic, March 2004

Trade union leader, OS KOVO trade union affiliate at Vítkovice Steel, Ostrava, Czech Republic, March 2004

Consultant, Roland Berger Strategy Consultants, Prague, Czech Republic, March 2004

Vladimír Dlouhý, former Minister of Industry and Trade (1991–1997), Prague, Czech Republic, March 2004

Consultant, Wood and Company, Prague, Czech Republic, March 2004

Top management member #1, North Moravia steelworks A, North Moravia, Czech Republic, April 2004

Top management member #2, North Moravia steelworks A, North Moravia, Czech Republic, April 2004

Former top management member, Revitalization Agency (Revitalizační Agentura), Prague, Czech Republic, June 2008

Václav Novák, former crisis manager, Vítkovice Steelworks, Prague, Czech Republic, June 2008

Slovakia Case Study

Civil servants (3), Ministry of Industry and Trade, Bratislava, Slovakia, April 2004

Enterprise insider #1, U.S. Steel Košice, Košice, Slovakia, April 2004

Enterprise insider #2, U.S. Steel Košice, Košice, Slovakia, April 2004

Enterprise insider #3, U.S. Steel Košice, Košice, Slovakia, April 2004

Expert involved in sale of VSŽ to U.S. Steel, Slovak Government, Bratislava, Slovakia, April 2004

Journalist covering the steel sector at a major daily, Bratislava, Slovakia, April 2004

Priority (the OZ KOVO newspaper) editor, Banská Bystrica, Slovakia, April 2004

Spokesperson, Podbrezová Steelworks, Podbrezová, Slovakia, April 2004

Spokesperson, U.S. Steel Košice, Košice, Slovakia, April 2004

Top management member, U.S. Steel Košice, Košice, Slovakia, April 2004

Trade union leader #1 (national level), OZ KOVO, Bratislava, Slovakia, April 2004

Trade union leader #2 (national level), OZ KOVO, Bratislava, Slovakia, April 2004

Trade union leader (regional level), OZ KOVO, Banská Bystrica, April 2004

Trade union leader, Metalurg trade union affiliate at U.S. Steel Košice, Košice, Slovakia, April 2004

Trade union leader, OZ KOVO trade union affiliate at U.S. Steel Košice, Košice, Slovakia, April 2004

Gabriel Eichler, former crisis manager at VSŽ, Prague, Czech Republic, June 2008

Romania Case Study

Desk officer, European Commission, Directorate General for Enlargement, Romania Team, Brussels, Belgium, May 2002

Civil servant, European Commission, Directorate General for Enlargement, Romania Team, Bucharest, Romania, May 2004

Civil servant #1, Ministry of Industry and Trade, Bucharest, Romania, May 2004

Steel-sector privatization adviser to the Minister of Privatization, Authority for State Assets Recovery (successor to the Authority for Privatization and State Ownership Administration and the State Ownership Fund), Bucharest, Romania, May 2004

Trade union leader, CNSLR Fraţia Trade Union Confederation, Bucharest, Romania, May 2004

Trade union leader, Metarom Federation, Bucharest, Romania, May 2004; June 2004

Trade union leader, Metarom trade union affiliate at Artrom Slatina, Reşiţa, Romania, May 2004

Trade union leader, Metarom trade union affiliate at Combinatul Siderurgic Reşiţa (Reşiţa Steelworks), Reşiţa, Romania, May 2004

Trade union leader, Metarom trade union affiliate at Silcotub Zalău, Reşiţa, Romania, May 2004

Trade union leader, Metarom trade union affiliate at Combinatul de Oţeluri Speciale Târgovişte (Special Steels Enterprise at Târgovişte), Reşiţa, Romania, May 2004

Trade union leader, Metarom trade union affiliate at UBE Târgovişte, Reşiţa, Romania, May 2004

Workers (3) at Reşiţa Steelworks, Reşiţa, Romania, May 2004

Trade union leader, U-Metal Federation, Bucharest, Romania, May 2004

Consultant at the Roland Berger Strategy Consultants, Bucharest, Romania, May 2004

Press Attaché, U.S. Embassy, Bucharest, Romania, May 2004

Expert/Representative at the UniRomSider, Bucharest, Romania, May 2004; June 2004

Civil servant #2, Authority for State Assets Recovery, Bucharest, Romania, June 2004

Civil servant #3, Authority for State Assets Recovery, Bucharest, Romania, June 2004

Civil servant #4, Ministry of Work, Social Solidarity and Family, Bucharest, Romania, June 2004

Journalist covering the steel sector at a major daily, Bucharest, Romania, June 2004

Trade union leaders (2), Federaţia Siderurgistilor din Galaţi (Federatia Sindicală a siderurgiştilor din Galaţi (Galaţi Ferrous Mettallurgy Workers' Trade Union Federation), Galaţi, Romania, June 2004

Top management member, Mittal Steel Galaţi (former Sidex Galaţi and current ArcelorMittal Galaţi), Galaţi, Romania, June 2004

Petru Dandea, Vice President of Cartel Alfa Confederation, Bucharest, Romania, June 2004 (three interviews); July 2010 (two interviews)

Romanian manager working for an international steel distributor, Bucharest, Romania, June 2004

Economic Attaché, US Embassy, Bucharest, Romania, June 2004

Human resources representatives (2), ArcelorMittal Galaţi, Galaţi, Romania, July 2010

Trade union leader, FSS Metarom (Federaţia Sindicatelor din Siderurgie Metarom) , Galaţi, Romania, July 2010

Trade union leader, FSS Metarom trade union affiliate at ArcelorMittal Galaţi, Galaţi, Romania, July 2010

Trade union leader, Solidaritatea trade union affiliate at ArcelorMittal Galaţi, Galaţi, Romania, July 2010

Appendix B

Calculation of State Aid to the Steel Sector

TABLE 8. Calculation of State Aid to the Steel Sector

	Romania	Slovakia	Poland	Czech Republic
Total state aid negotiated with the European Commission, million USD	1,436.5[a]	500.0[b]	871.6[c]	502.6[d]
Total state aid granted, million USD	1,436.5[e]	430.0[f]	707.5[g]	426.8[g]
Crude steel produced, 1993–2004, million tons[h]	68.2	47.3	121.0	79.3
State aid per ton of crude steel produced, 1993–2004, USD[i]	21.1	9.1	5.8	5.4

[a]The amount of maximum allowable state aid approved by the EU in Annex VII to the 2005 Act of Accession is ROL 49,985 billion (June 21, 2005, 2005 O.J. (L 157), 146). This is approximately EUR 1,269 million, as quoted in the European Commission Staff Working Document, *Programme for Restructuring of the Romanian Steel Industry*, Final assessment, COM (2005) 140 final (April 14, 2005). Using the 2003 average USD exchange rate, the amount is $1,436.5 million.

[b]Amount of state aid appearing in Annex XIV to the 2003 Act of Accession (September 23, 2003, 2003 O.J. (L 236), 918). In the case of Slovakia, state aid took the form of tax breaks—not to exceed $500 million—until 2009 for U.S. Steel which took over the accumulated debt of VSŽ in 2000.

[c]The amount of maximum allowable state aid appearing in Protocol 8 to the 2003 EU Act of Accession is PLN 3,387,070,000 (September 23, 2003, 2003 O.J. (L 236), 948). Using the 2003 exchange rate of 1 € = 4.3996 PLN, the European Commission used the amount of €770 million (Max Lienemeyer, "State Aid for Restructuring the Steel Industry in the New Member States," *Competition Policy Newsletter* (Spring 2005), 99, http://ec.europa.eu/competition/publications/cpn/2005_1_94.pdf (accessed April 13, 2013). Using the 2003 average exchange rate of 1 € = $1.132, the amount is $871.6 million.

[d]The amount of maximum allowable state aid appearing in Protocol 2 to the 2003 EU Act of Accession is CZK 14,147,425,201 (2003 O.J. (L 236), 934). Using the 2003 exchange rate of 1 € = CZK 31.846, the European Commission used the amount of €444 million (Lienemeyer, "State Aid for Restructuring," 100). Using the 2003 average exchange rate of 1 € = $1.132, the amount is $502.6 million.

[e]State aid for Romania was negotiated retroactively with the European Commission and entailed, in large part, privatization-related debt write-offs and penalty waivers for late debt payments (Lienemeyer, "State Aid for Restructuring," 102).

[f]Because U.S. Steel Košice did not respect the negotiated caps on production in 2002 and 2003, the European Commission decreased the total permissible state aid to $430 million (Commission decision of September 22, 2004, C (2004) 3496fin in case SK 5/04, Reduction of a Tax Concession Granted by Slovakia to U.S. Steel Kosice).

[g]Both Poland and the Czech Republic were granted a smaller amount of state aid than the maximum permissible amount they negotiated (Poland was awarded 81% and the Czech Republic 85% of the maximum negotiated amount). The actual state aid granted amounted to €625 million ($707.5 million) in the case of Poland and €377 million ($426.8 million) in the case of the Czech Republic (Lienemeyer, "State Aid for Restructuring," 99–100).

[h]My own calculation using statistics from International Iron and Steel Institute, *Steel Statistical Yearbook* (Brussels: IISI, 1991–1994, 2005).

[i]State aid per ton of crude steel produced is calculated using the following formula:

(Total state aid granted, 1993–2004, million USD) / (Crude steel produced, 1993–2004, millions of tons)

For the sake of consistency, the calculation uses crude steel produced over the years 1993–2004 as the denominator because 1993 was the initial year the European Commission identified a country case (Romania) as having granted state aid.

Appendix C

Enterprise List and Ownership Type of Steelworks in the Czech Republic, Poland, Romania, and Slovakia in 1990, 1996, and 2005

TABLE 9. Enterprise List

Enterprise Code	Location	Enterprise Name (English translation, where applicable)
HK	**Poland, Dąbrowa Gór.**	**Katowice Steelworks**
HTS	**Poland, Kraków**	**Tadeusz Sendzimir Steelworks**
HF	Poland, Świętochłowice	Florian Steelworks
HC	Poland, Sosnowiec	Cedler Steelworks
PHS	**Poland, Katowice**	**Polish Steelworks (merger of HK, HTS, HF, and HC)**
HCz	Poland, Częstochowa	Częstochowa Steelworks
HZ	Poland, Zawiercie	Zawiercie Steelworks
HW	Poland, Warszawa	Warsaw Steelworks/Huta L.W.
HO	Poland, Ostrowiec Św.	Ostrowiec Steelworks
HB	Poland, Katowice	Baildon Steelworks
HŁ	Poland, Gliwice	Łabędy Steelworks
NH	**Czech Republic, Ostrava**	**Nová Huť**
VS	Czech Republic, Ostrava	Vítkovice/Vítkovice Steel
TŽ	Czech Republic, Třinec	Třinec Steelworks
Poldi	Czech Republic, Kladno	Poldi
VSŽ	**Slovakia, Košice**	**Eastern Slovak Steelworks**
ŽP	Slovakia, Podbrezová	Podbrezová Steelworks
SG	**Romania, Galați**	**Sidex Galați**
SH	Romania, Hunedoara	Siderurgica Hunedoara
COST	Romania, Târgoviște	Special Steels Enterprise at Târgoviște/ Combinatul de Oțeluri Speciale Târgoviște
ISCT	Romania, Câmpia Turzii	Câmpia Turzii Wire Factory/Industria Sârmei Câmpia Turzii
CSR	Romania, Reșița	Reșița Steelworks
OR	Romania, Oțelu Roșu	Socomet Oțelu Roșu
SC	Romania, Călărași	Siderca Călărași
TI	Romania, Iași	Tepro Iași
PR	Romania, Roman	Petrotub Roman
SZ	Romania, Zalău	Silcotub Zalău
AS	Romania, Slatina	Artrom Slatina

TABLE 10. Ownership Type of Steelworks in the Czech Republic, Poland, Romania, and Slovakia in 1990, 1996, and 2005[a]

Ownership Type	Initial Conditions: 1990			Intermediate (1996) Outcome: Divergence			2005 Outcome: Convergence		
SOEs	**HK**	<u>***NH***</u>	***SG***	**HK**	<u>*VS*</u>*	***SG***			
	HTS	<u>*VS*</u>	*SH*	**HTS**		*SH*			
	HF	<u>*TŽ*</u>	*COST*	HF		*COST*			
	HC	<u>*Poldi*</u>	*ISCT*	HC		*ISCT*			
	HCz		*CSR*	HCz		*CSR*			
	HZ	<u>**VSŽ**</u>	*OR*	HB		*SC*	HŁ		
	HW	<u>ŽP</u>	*SC*	HŁ		*TI*			
	HO		*TI*			*PR*			
	HB		*PR*			*SZ*			
	HŁ		*SZ*			*AS*			
			AS						
Domestic Private				HZ	<u>**VSŽ**</u>	<u>***NH***</u>	<u>*TŽ*</u>		
				HO	<u>ŽP</u>	<u>*TŽ*</u>	<u>ŽP</u>		
						<u>*Poldi*</u>			
FDI				HW			**HK/PHS**		***SG***
							HTS/PHS		*SH*
							HF/PHS		*COST*
							HC/PHS		*ISCT*
							HCz	<u>***NH***</u>	*CSR*
							HW	<u>*VS*</u>	*SC*
							HZ	<u>**VSŽ**</u>	*TI*
							HO		*PR*
									OR
									SZ
									AS

Notes: [a]See Table 9 for enterprise code key.

Enterprise coding by country: Poland (regular font), Czech Republic (***underlined italics***), Slovakia (underlined), Romania (*italics*). The dominant enterprise in the country appears in **boldface** (in Poland, the two largest enterprises are marked).

*VS remained an SOE but managerial privatization devolved oversight to management.

Notes

CHAPTER 1

1. *Dependent market economy* is a new category in the varieties of capitalism literature, which has typically classified economies as liberal or coordinated (or sometimes mixed market). See Andreas Nölke and Arjan Vliegenthart, "Enlarging the Varieties of Capitalism: The Emergence of Dependent Market Economies in East Central Europe," *World Politics* 61 (2009): 670–702. The term dependent is controversial because it evokes dependency and underdevelopment, which many scholars do not regard as an appropriate characterization of the effects of FDI in the region. They point to the developmental potential, existing FDI-imported research and development activities, and the resilience of the transition economies, as shown in the region's weathering of the recent economic crisis. For example, see the discussion by Jan Drahokoupil and Martin Myant in "Putting Comparative Capitalism Research in its Place: Varieties of Capitalism in Transition Economies," in *New Directions in Critical Comparative Capitalisms Research*, ed. Matthias Ebenau, Ian Bruff, and Christian May (London: Palgrave Macmillan, 2015). I use the term *transnational capitalism* throughout the book, which has more neutral developmental connotations.

2. For an argument highlighting how the EU actors protected their own interests during the ECE enlargement process, see Wade Jacoby, "Managing Globalization by Managing Central and Eastern Europe: The EU's Backyard as Threat and Opportunity," *Journal of European Public Policy* 17 (2010): 416–432; David L. Ellison, "Divide and Conquer: The European Union Enlargement's Successful Conclusion?" *International Studies Review* 8 (2006): 150–165.

3. For a discussion of investment opportunities in the steel sector from the investors' perspective, see T. Daly and E. Dalli, "Central and Eastern European Steel after EU Accession," *Steel Times International* (November–December 2005): 41–58.

215

4. Saul Estrin, Kirsty Hughes, and Sarah Todd, *Foreign Direct Investment in Central and Eastern Europe: Multinationals in Transition* (London: Royal Institute of International Affairs, 1997), 34; Gábor Hunya, "Recent Impacts of FDI on Growth and Restructuring in Central and Eastern European Transition Countries" (WIIW Research Report No. 284, Vienna Institute for International Economic Studies, 2002); Miklós Szanyi, "Experiences with Foreign Direct Investment in Eastern Europe," *Eastern European Economics* 36 (1998): 28–48.

5. According to the ratings by the European Bank for Reconstruction and Development (EBRD), published annually in *Transition Report*, the Czech Republic had 3.62 points out of a possible 4.33, closely followed by Poland (3.58), then by Slovakia (3.46), and finally, by Romania (3.00). These scores represent a mean of ratings on each of the eight dimensions of economic reform assessed by the EBRD: large-scale privatization, small-scale privatization, governance and enterprise restructuring, price liberalization, trade and foreign exchange system, competition policy, banking reform, and interest rate liberalization, as well as the reform of securities markets and nonbank financial institutions. Each country is annually assigned a grade ranging from 1 to 4+, or 4.33, in increments of .33 (+ or –), with 1 indicating a complete lack of a given reform, and 4.33 demonstrating the extent of reform encountered in advanced market democracies. See European Bank for Reconstruction and Development, *Transition Report* (London: EBRD, 1994–2010).

6. European Bank for Reconstruction and Development, *Transition Report* (London: EBRD, 2000), 74. Accounting for the disparity in population size, the Czech Republic led the way with $1,447 per capita in FDI, but Poland surpassed the other two countries, with $518 in FDI per capita, compared to Slovakia's $391 and Romania's $252.

7. European Commission desk officer responsible for Romania, interview with the author, Brussels, Belgium, May 2002.

8. František Bouc, "Steel's Time Running Short," *The Prague Post*, April 10, 2002.

9. Civil servant closely involved in the privatization process at the Ministry of State Treasury, interview with the author, Warsaw, Poland, July 2003.

10. Poland's two instances of domestic ownership resulted from debt workouts.

11. I borrow Anna Grzymała-Busse's definition of the state as the set of formal institutions that implement policy and enforce legal sanctions. Anna Grzymała-Busse, *Rebuilding Leviathan: Party Competition and State Exploitation in Post-Communist Democracies* (Cambridge: Cambridge University Press, 2007), 3.

12. Michael Mann, "The Autonomous Power of the State: Its Origins, Mechanisms and Results," *European Journal of Sociology* 25 (1984): 189. I thank Stephen Hanson for a useful discussion on this topic. See also Jessica Fortin, "Is There a Necessary Condition for Democracy? The Role of State Capacity in Postcommunist Countries," *Comparative Political Studies* 45 (2012): 903–930.

13. For analysis of the politics of the state reform process in East Central Europe, see Grzymała-Busse, *Rebuilding Leviathan.* Also, see a comparative discussion of the postcommunist state building from a broader regional perspective in Anna Grzymała-Busse and Pauline Jones Luong, "Reconceptualizing the State: Lessons from Postcommunism," *Politics and Society* 30 (2002): 529–554. For empirical analysis of state capacity in postcommunist countries over time, see Jessica For-

tin, "A Tool to Evaluate State Capacity in Post-Communist States, 1989–2006," *European Journal of Political Research* 49 (2010): 654–686.

14. World Bank, *World Development Report: The State in a Changing World* (New York: Oxford University Press, 1997). Some of the works that highlight the role of the state in development include Meredith Woo-Cumings, *The Developmental State* (Ithaca, NY: Cornell University Press, 1999); Peter Evans, *Embedded Autonomy: States and Industrial Transformation* (Princeton, NJ: Princeton University Press, 1995); Sylvia Maxfield and Ben Ross Schneider, eds., *Business and the State in Developing Countries* (Ithaca, NY: Cornell University Press, 1997); and Ben Ross Schneider, *Business and the State in Twentieth Century Latin America* (Cambridge: Cambridge University Press, 2004).

15. I thank Jennifer Pribble for a useful discussion.

16. Luigi Manzetti, *Privatization South American Style* (Oxford: Oxford University Press, 2000); Hector Schamis, *Re-Forming the State: The Politics of Privatization in Latin America and Europe* (Ann Arbor: University of Michigan Press, 2002).

17. Schamis, *Re-Forming the State*.

18. See, for example, critiques of the neostatist project by Gabriel Almond and Myron Weiner in Myron Weiner and Samuel P. Huntington, *Understanding Political Development* (Boston: Little, Brown, 1987), 476–477 and xxii, respectively. For the neostatist perspective, see the chapters by Peter B. Evans, Eric A. Nordlinger, and Joel S. Migdal in the same volume, as well as Peter B. Evans, Dietrich Rueschemeyer, and Theda Skocpol, *Bringing the State Back In* (New York: Cambridge University Press, 1985).

19. For an early discussion of the insufficient scholarly and policy focus on the state as both the object and the subject of restructuring, see Alice Amsden, Jacek Kochanowicz, and Lance Taylor, *The Market Meets Its Match* (Cambridge, MA: Harvard University Press, 1994). For existing studies of the role of the state, see Grzegorz Ekiert, "The State after Socialism: Poland in Comparative Perspective," in *Nation-State in Question*, ed. T. V. Paul, G. John Ikenberry, and John Hall (Princeton, NJ: Princeton University Press, 2003), 291–320; Jadwiga Staniszkis, *Postkomunizm—Próba Opisu* (Gdańsk: Wydawnictwo Słowo/Obraz Terytoria, 2001); Lawrence King and Aleksandra Sznajder, "The State-Led Transition to Liberal Capitalism," *American Journal of Sociology* 112 (2006): 751–801; Dorothee Bohle and Béla Greskovits, "The State, Internationalization, and Capitalist Diversity in Eastern Europe," *Competition and Change* 11 (2007): 443–466; and Dorothee Bohle and Béla Greskovits, *Capitalist Diversity on Europe's Periphery* (Ithaca, NY: Cornell University Press, 2012). For an economic sociological perspective on the role of the state in attracting foreign investment, see Nina Bandelj, *From Communists to Foreign Capitalists: The Social Foundations of Foreign Direct Investment in Postsocialist Europe* (Princeton, NJ: Princeton University Press, 2008). Foremost examples of studies focusing on the state as the dependent variable include Grzymała-Busse, *Rebuilding Leviathan*; Conor O'Dwyer, *Runaway State Building* (Baltimore: Johns Hopkins University Press, 2006); Venelin Ganev, *Preying on the State* (Ithaca, NY: Cornell University Press, 2007); Gerald M. Easter, *Capital, Coercion, and Postcommunist States* (Ithaca, NY: Cornell University Press, 2012).

20. For a thorough discussion of communist and precommunist legacies, see the essays in Grzegorz Ekiert and Stephen E. Hanson, eds., *Capitalism and Democracy*

in Central and Eastern Europe: Assessing the Legacy of Communist Rule (Cambridge: Cambridge University Press, 2003); Herbert Kitschelt et al., *Post-Communist Party Systems: Competition, Representation and Inter-Party Cooperation* (Cambridge: Cambridge University Press, 1999).

21. David Stark and László Bruszt, *Postsocialist Pathways* (Cambridge: Cambridge University Press, 1998).

22. Roman Frydman and Andrzej Rapaczynski, *Privatization in Eastern Europe: Is the State Withering Away?* (Budapest: Central European University Press, 1994). See the discussion in Stark and Bruszt, *Postsocialist Pathways.*

23. Peter A. Hall and David Soskice, *Varieties of Capitalism: The Institutional Foundations of Comparative Advantage* (Oxford: Oxford University Press, 2001). *Varieties of Capitalism* spurred a vigorous debate. See, for example, Bob Hancké, Martin Rhodes, and Mark Thatcher, *Beyond Varieties of Capitalism: Conflict, Contradictions, and Complementarities in the European Economy* (Oxford: Oxford University Press, 2007); and Dorothee Bohle and Béla Greskovits, "Varieties of Capitalism and Capitalism 'tout court,'" *European Journal of Sociology* 50 (2009): 355–386. The limits of this framework in accounting for institutional change and the political process, as well as for the role of external pressures, were particularly relevant objections to its application to the transition economies. For excellent reviews of the literature on varieties of capitalism in the postcommunist world, see Nölke and Vliegenthart, "Enlarging the Varieties of Capitalism"; Bohle and Greskovits, *Capitalist Diversity*, 9–13; and Drahokoupil and Myant, "Putting Comparative Capitalism Research in Its Place."

24. See, respectively, Nölke and Vliegenthart, "Enlarging the Varieties of Capitalism"; Drahokoupil and Myant, "Putting Comparative Capitalism Research in Its Place"; and Bohle and Greskovits, *Capitalist Diversity.*

25. Nölke and Vliegenthart, "Enlarging the Varieties of Capitalism," 672.

26. Ibid., 694.

27. Bandelj, *From Communists to Foreign Capitalists*; Jan Drahokoupil, *Globalization and the State in Central and Eastern Europe: The Politics of Foreign Direct Investment*. Routledge Series on Russian and East European Studies (London: Routledge, 2008); Rachel Epstein, *In Pursuit of Liberalism: International Institutions in Postcommunist Europe* (Baltimore: Johns Hopkins University Press, 2007).

28. For an interesting discussion of the literature on the EU's role in privatization and encouraging FDI, see Jacoby, "Managing Globalization," 419. See also Gergö Medve-Bálint, "The Role of the EU in Shaping FDI Flows to East Central Europe," *Journal of Common Market Studies* 52 (2014): 35–51.

29. This body of literature is too voluminous to summarize here. For a review that highlights the various directions (and inconsistent findings) of EU studies research, see Rachel E. Epstein and Wade Jacoby, "Eastern Enlargement Ten Years On: Transcending the East-West Divide?" *Journal of Common Market Studies* 52 (2014): 1–16. For an exemplar of comparative analysis showing differing impact of the EU depending on policy area, see Wade Jacoby, *The Enlargement of the European Union and NATO: Ordering from the Menu in Central Europe* (New York: Cambridge University Press, 2004).

30. Mitchell Orenstein, Stephen Bloom, and Nicole Lindstrom, *Transnational Actors in Central and East European Transitions* (Pittsburgh: University of Pittsburgh Press, 2008).

31. David R. Cameron powerfully illustrates the enormity of the burden the *acquis communautaire* imposed on transition states already preoccupied with the multifaceted reform processes in "The Challenges of EU Accession," *East European Politics and Societies* 17 (2003): 24–41.

32. Or, to borrow a term from Milada Vachudova's work, this was the EU's "active leverage" at its finest. See, especially, chapters 5 to 7 in *Europe Undivided: Democracy, Leverage, and Integration after Communism* (Oxford: Oxford University Press, 2005).

33. Frydman and Rapaczynski, *Privatization in Eastern Europe*, 143.

34. Michael Shafir, *Romania: Politics, Economics and Society* (London: Pinter, 1985), 46–47.

35. Vladimír Dostál, *Dějiny hutnictí železa v Československu*, vol. 3 (Prague: Academia, 1988), 69.

36. Ibid., 57.

37. Michael D. Shafer, *Winners and Losers: How Sectors Shape the Developmental Prospects of States* (Ithaca, NY: Cornell University Press, 1994), 10. See also Béla Greskovits, "Leading Sectors and Variety of Capitalism in Eastern Europe," *Actes du Gerpisa* 39 (2005): 113–128.

38. The data for the Czech Republic, Poland, and Romania are from 1990. In the case of Slovakia, the reference year is 1993 since data using current prices was not available for earlier years. WIIW Industrial Database on Central and Eastern Europe (Vienna: Vienna Institute for International Economic Studies, 2005).

39. Ibid. The data are from 1992, the first year for which comparable data was available.

40. The 1999 raw minerals figure, as percentage of exports, was also similar, ranging from 2.1% in the Czech Republic to 5.3% in Romania. World Bank, *World Development Indicators* (Washington, DC: World Bank, 2009).

CHAPTER 2

1. Interview with Edward Nowak, Kraków, July 2003.

2. Herbert Kitschelt et al., *Post-Communist Party Systems* (Cambridge: Cambridge University Press, 1999), 39.

3. The Europe Agreements were signed with Poland and Czechoslovakia in 1991 and with Romania in 1993. Following the split of Czechoslovakia in 1993, both successor states concluded separate Europe Agreements. For a comprehensive, in-depth examination of the EU effects on democratic and economic reform, see Milada Anna Vachudova, *Europe Undivided* (Oxford: Oxford University Press, 2005).

4. Each duty was to be reduced to 80% of the basic duty on the date of the agreement's effective date. This meant January 1, 1992, in the case of Czech Republic, Poland, and Slovakia, and January 1, 1993, in the case of Romania. Further reductions in 20% increments were planned for the subsequent years. As a result of the 1993 Copenhagen Summit, however, the date of liberalization was moved up one year. See Hutnictví Železa, "Úroveň celních tarifů s EU a ostatními zeměmi," unpublished manuscript, photocopy, Prague, 2004.

5. In Poland, the trade liberalization schedule was uniform, and, pursuant to

Article 10(3) of the Europe Agreement, trade was to be liberalized seven years after the Agreement's effective date. In the Czech Republic, Slovakia, and Romania, the Europe Agreements established separate schedules for less and more sensitive products. Trade in less sensitive products was to be completely liberalized after five years (Article 11(2) of the three Europe Agreements). Free trade in the more sensitive products was to be achieved after nine years, although the tariff reduction schedules differed slightly in Romania (Article 11(3) in Czech and Slovak Europe Agreements and Article 11(4) in the Romanian one). See Europe Agreement establishing an association between the European Communities and their Member States, of the one part, and the Republic of Poland, of the other part, December 31, 1993, 1993 O.J. (L 348); Europe Agreement establishing an association between the European Communities and their Member States, of the one part, and the Czech Republic, of the other part, December 31, 1994, 1994 O.J. (L 360).

6. Only in the case of Poland did the Europe Agreement leave the question of further state aid open, by stating, "The Association Council shall, taking into account the economic situation of Poland, decide whether the period of five years could be extended." See Europe Agreement, protocol 2, chapter III, article 8(4).

7. Commission of the European Communities. *Industrial Forum on Enlargement. Industrial Restructuring Report*, September, 2000.

8. In the case of Romania, the deadline was December 2004.

9. The viability criteria entailed gross operating profit (EBITDA) of at least 13.5% of steel sales revenue, and a minimum return (EBIT) of at least 1.5% of steel sales revenue. Before the viability test was to be carried out, depreciation and financial charges as a percentage of steel sales revenue in each plan year had to satisfy the minimum accounting conditions (depreciation of not less than 7% of turnover, financial charges of not less than 3.5%, and a 2.5% price-cost squeeze applied). See Commission of the European Communities, European Commission Staff Working Document, *Programme for Restructuring of the Romanian Steel Industry*, Final Assessment, COM (2005) 140 final (April 14, 2005).

10. The literature also makes a distinction between "defensive" and "strategic" restructuring. Defensive restructuring refers to activities undertaken in order to reduce costs and scale down enterprise activity, such as laying off workers, cutting obsolete production lines, and getting rid of nonproductive assets. Its goal is to ensure immediate survival of the enterprise and does not mean that a long-term strategy necessarily exists. By contrast, strategic restructuring is based on a long-term business strategy that entails "profound redeployment of assets" and introduces new product lines and processes, technologies and investments. See Irena Grosfeld and Gérard Roland, "Defensive and Strategic Restructuring in Central European Enterprises," *Emergo* 3 (1996): 21–46. Although this distinction is useful for highlighting the disparity of capital investment necessary for the two types of restructuring, it obscures the politics of restructuring. Defensive restructuring is at least as, if not more, politically contentious than strategic restructuring.

11. Barbara Pytel, "Sector Study of the Iron and Steel Industry in Poland," *Emergo* 2 (1995): 32.

12. Raj Desai, "Organizing Markets: Property Rights, Governance, and the Politics of Industrial Privatization in a Post-Communist Economy" (PhD diss., Harvard University, 1996), 171.

13. The year 1987 is used as the reference because it was the year of the highest consumption in the three countries during the 1985–1989 time period.

14. International Iron and Steel Institute, *Steel Statistical Yearbook,* 1991, 2005 (Brussels: IISI).

15. International Iron and Steel Institute, *Steel Statistical Yearbook,* 2009.

16. My own calculations using national publications: FNS Metarom, *InfoMetal.* "A XIII-a Conferinţă," October 22–24, 2003, 6; Hutní Projekt Ostrava and Vysoká Škola Báňská. *Studie Plánu Restrukturalizace Českého Ocelářského Průmyslu,* Ostrava, 1999, 47; Desai, "Organizing Markets," 170; International Iron and Steel Institute, *Steel Statistical Yearbook,* 1991.

17. Commission of the European Communities, *Steel industry country fiches,* June 23, 2004.

18. The open-hearth process, which is energy and labor-intensive and harmful to the environment, dominated crude steel production in developed market economies until the 1960s, when it started to be replaced by basic oxygen furnaces and electric arc furnaces. Continuous casting is a technology that eliminates the intermediate step of casting ingots, which require reheating before they can be rolled into slabs suitable for use in a rolling mill. Instead, the molten steel is poured and pulled at high speed through a curved mold and is then cut into slabs that can immediately be used in the rolling mill, saving energy and time, and increasing the yield of finished steel product from each ton of liquid steel by at least 5%. See Christopher G. L. Hall, *Steel Phoenix: The Fall and Rise of the U.S. Steel Industry* (New York: St. Martin's, 1997), 6–9.

19. International Iron and Steel Institute, *Steel Statistical Yearbook,* 1991–1994; Hutnictví Železa, *Hutnická Ročenka 1994* (Prague: Ocelot, 1994), 69.

20. Hutnictví Železa, *Hutnická Ročenka.*

21. The distinction is not domestic versus foreign buyers, but strategic versus nonstrategic buyers. There is no a priori assumption that foreign owners are better.

22. Incomplete reform, which gives beneficiaries of the status quo an incentive to resist full reform, is similar to Joel Hellman's idea of partial reform. However, in Hellman's conceptualization, the winners of the first reform wave use economic influence to capture the state and prevent it from continuing economic reform in *other domains.* See Joel Hellman, "Winners Take All: The Politics of Partial Reform in Postcommunist Transitions," *World Politics* 50 (1998): 203–234. I thank Juliet Johnson for emphasizing this difference.

23. "Governing coalition" is used interchangeably with "government."

24. Coalition members may have dissimilar preferences, driven by ideology or by material incentives. This book remains agnostic as to the extent either prevails. For a critique of the materialist and power-based approaches to privatization policy development and a discussion of ways in which ideology shapes privatization policy, see Hilary Appel, "The Ideological Determinants of Liberal Economic Reform: The Case of Privatization," *World Politics* 52 (2000): 520–549.

25. For a discussion of the managerial pressures, see Hilary Appel and John Gould, "Identity Politics and Economic Reform: Examining Industrial-State Relations in the Czech and Slovak Republics," *Europe-Asia Studies* 52 (2000): 111–131.

26. Michael Mann, "The Autonomous Power of the State: Its Origins, Mechanisms and Results," *European Journal of Sociology* 25 (1984): 189.

27. See a discussion by Peter Evans in *Embedded Autonomy: States and Industrial Transformation* (Princeton, NJ: Princeton University Press, 1995), 19.

28. Alice Amsden, Jacek Kochanowicz, and Lance Taylor, *The Market Meets Its Match* (Cambridge, MA: Harvard University Press, 1994).

29. Jessica Fortin, "A Tool to Evaluate State Capacity in Post-Communist States, 1989–2006," *European Journal of Political Research* 49 (2010): 654. For an in-depth analysis of the outcomes of state reform in postcommunist countries and the political processes underpinning these reforms, see Anna Grzymała-Busse, *Rebuilding Leviathan: Party Competition and State Exploitation in Post-Communist Democracies* (Cambridge: Cambridge University Press, 2007).

30. Kitschelt et al., *Post-Communist Party Systems*, 35–41.

31. Ibid.

32. Ibid.

33. Martin Dangerfield, "Ideology and the Czech Transformation: Neoliberal Rhetoric or Neoliberal Reality," *East European Politics and Societies 11* (1997): 441.

34. Kitschelt et al., *Post-Communist Party Systems*, 23.

35. See also the discussion by Lawrence King and Iván Szelényi about the effects of whether the technocratic faction (allied with dissident intellectuals) or the bureaucratic faction of the communist elite prevailed at the outset of the transition process. Lawrence King and Iván Szelényi, "The New Capitalism of Eastern Europe: Towards a Comparative Political Economy of Post-Communism," in *Handbook of Economic Sociology*, ed. Neil J. Smelser and Richard Swedberg (Princeton, NJ: Princeton University Press, 2005): 205–229. See also Gil Eyal, Iván Szelényi, and Eleanor Townsley, *Making Capitalism without Capitalists: The New Ruling Elites in Eastern Europe* (New York: Verso, 2000).

36. In the case of the Czech Republic and Poland, there is a trade-off between the greater technocratic resources of the Polish state and the greater control over society in the Czech Republic.

37. This kind of logic and behavior has certainly been observed in economies undergoing reform. See Edward S. Steinfeld, *Forging Reform in China: The Fate of State-Owned Industry* (Cambridge: Cambridge University Press, 1998).

38. Fortin, "A Tool to Evaluate State Capacity."

39. By focusing on indicators specific to the economic domain, the measure operationalizes state capacity in a context-appropriate way. For a discussion of operationalizing state infrastructural power, see Hillel Soifer, "State Infrastructural Power," *Studies in Comparative International Development* 43 (2008): 245.

40. The data for the four countries for both time periods are correlated at .93 (Pearson's *r*). I thank Jessica Fortin for making the country-year data available.

41. The correlation between the State Capacity Index and the Euromoney Country Risk rankings in 2000 was .97 (significant at .05 level, two-tailed test). The correlation between state aid in the steel sector and the State Capacity Index was –.99; and between state aid and the Euromoney Country Risk rankings, it was –.98 (both significant at .01 level, two-tailed test). For Euromoney Country Risk rankings, see Keri Geiger, "Waiting for the Dust to Settle," *Euromoney* 377 (2000): 214–224.

42. Given that the high correlation between the internal and external assessments of state capacity could be driven by the "expert opinion" nature of the

indexes that are used to create both measures, I used another, domestically derived, proxy for state capacity: a measure of corruption provided by the World Bank's survey of domestic businesses. In line with the data presented above, the corruption measure taken from the 1999 Business Environment and Enterprise Performance Survey correlates with state capacity measure at –.99. See Joel S. Hellman, Geraint Jones, and Daniel Kaufmann, "Seize the State, Seize the Day: State Capture, Corruption, and Influence in Transition" (Policy Research Working Paper 2444, World Bank, Washington, DC, 2000); World Bank, *Business Environment and Enterprise Performance Survey* (Washington, DC: World Bank, 1999), http://data.worldbank.org/data-catalog/BEEPS (accessed October 10, 2014).

43. The steps following the initial policy choice shown in figure 2 are italicized in figure 3.

44. Mark Granovetter, "Economic Action and Social Structure: The Problem of Embeddedness," *American Journal of Sociology* 91 (1985): 481–510.

45. In Poland (World Bank funding): Hatch Associates et al., *Raport Końcowy*, June 1992; in the Czech Republic (PHARE funding): *Hutnické listy*, "Restrukturalizace československého hutnictví," no. 7–8 (1992): 84–89, and no. 9 (1992): 43–45; in Romania (PHARE funding): Ministère de l'Industrie—Roumanie, Sofres Conseil, Sofresid, Cedres, *Programme de Restructuration de la Siderurgie Roumaine*, 1993, 7.

46. Coordination of capacity cuts should not be confused with "central planning" or oligopolistic tendencies. The problem here is that the creation of additional capacity would exacerbate the already existing problem of excess capacity. Progressive liberalization of trade would certainly enlarge the size of the market, increasing the exposure of these enterprises to competition. Moreover, given the scope of desired investments, there was the inherent danger that they would not be finished before the competitive external pressures increased, leading to enterprises' collapse.

47. Mark Granovetter, "Problems of Explanation in Economic Sociology," in *Networks and Organizations*, ed. Nitin Nohria and Robert Eccles (Boston: Harvard Business School Press, 1992), 25.

48. Individual prestige becomes weighted by the prestige of the enterprise in the sector.

49. David Stark and László Bruszt, *Postsocialist Pathways* (Cambridge: Cambridge University Press, 1998), 132, 164. Others, notably Gerald A. McDermott, do not dispute the capacity of the managers to engage in restructuring, but they disagree with Stark and Bruszt's characterization of networks. According to McDermott, the institutional ties that underpinned networks dating from the communist period have undergone substantial transformation during transition, making "the loyalties, interests, and goals of both the network as a whole and each member a priori indeterminate" (19). Therefore, the government has a unique role as a monitor, mediator, and negotiator with network actors in an effort to create effective governance rules (169). Gerald A. McDermott, *Embedded Politics: Industrial Networks and Institutional Change in Postcommunism* (Ann Arbor: University of Michigan Press, 2002).

50. Although deeply influenced by the communist legacy of investment hunger and desire to augment professional prestige (see János Kornai, *Economics of Shortage*

[Amsterdam: North-Holland, 1980], 191–193), managerial behavior in ECE cannot be solely explained by it, as it shares similarities with managerial attitudes noted in research on corporate behavior in advanced industrialized countries. Excessive optimism contributed to the crisis of the Western European steel industry in the 1970s, and subsequently marred its restructuring. See Yves Mény and Vincent Wright, "State and Steel in Western Europe," in *The Politics of Steel: Western Europe and the Steel Industry in the Crisis Years (1974–1984)*, ed. Yves Mény and Vincent Wright (Berlin: Walter de Gruyter, 1986), 13, 16. The phenomenon has also been noted in the literature on takeovers. Research has shown that the profitability of mergers often fell short of pre-merger expectations. See Ellen B. Magenheim and Dennis C. Mueller, "Are Acquiring-Firm Shareholders Better Off after an Acquisition?," in *Knights, Raiders, and Targets: The Impact of the Hostile Takeover*, ed. John C. Coffee Jr., Louis Lowenstein, and Susan Rose-Ackerman (New York: Oxford University Press, 1988). Moreover, the "hubris hypothesis" of takeovers posits that managers seek to acquire firms, not out of economic motives, but to further their personal goals. See Richard Roll, "The Hubris Hypothesis of Corporate Takeovers," *Journal of Business* 59 (1986): 197–216; Patrick A. Gaughan, *Mergers, Acquisitions, and Corporate Restructurings* (New York: Wiley, 1996), 136–138.

51. McDermott, *Embedded Politics*, 169.

52. Grzegorz Ekiert and Jan Kubik, "Contentious Politics in New Democracies: East Germany, Hungary, Poland, and Slovakia, 1989–93," *World Politics* (1998): 547–581; Béla Greskovits, *The Political Economy of Protest and Patience: East European and Latin American Transformations Compared* (New York: Central European University Press, 1998); Stephen Crowley and David Ost, eds., *Workers after Workers' States* (Lanham, MD: Rowman and Littlefield, 2001).

53. Juliusz Gardawski et al., *Rozpad Bastionu? Związki zawodowe w gospodarce prywatyzowanej* (Warsaw: Instytut Spraw Publicznych, 1999); Juliusz Gardawski, *Związki zawodowe na rozdrożu* (Warsaw: Instytut Spraw Publicznych, 2001).

54. For a discussion of "time of extraordinary politics," see Leszek Balcerowicz, *Socialism, Capitalism, Transformation* (Budapest: Central European University Press, 1995).

55. John Williamson, ed., *The Political Economy of Policy Reform* (Washington, DC: Institute for International Economics, 1994); Joan Nelson, ed., *Fragile Coalitions: The Politics of Economic Adjustment* (New Brunswick, NJ: Transaction, 1989).

56. Jeffrey Sachs, *Poland's Jump to the Market Economy* (Cambridge, MA: MIT Press, 1993), 81.

57. Roman Frydman and Andrzej Rapaczynski, *Privatization in Eastern Europe: Is the State Withering Away?* (Budapest: Central European University Press, 1994), 143.

58. Stark and Bruszt, *Postsocialist Pathways*, 188–191.

59. Depending on the country, umbrella organizations differ. In some, there is a single umbrella organization for the sector as a whole, while in others there are different umbrella organizations that coordinate their demands. The umbrella organizations are held accountable by the enterprise-level unions. As for the role of the state in the early transition period, it was important even in bipartite negotiations because the employers represented SOEs.

60. Mancur Olson, *The Rise and Decline of Nations: Economic Growth, Stagflation, and Social Rigidities* (New Haven, CT: Yale University Press, 1982), 47.

61. Ibid., 48.

62. At times, Olson considers a single enterprise-level union to qualify as encompassing, at others, he claims that only peak associations that cover several industries can credibly result in increased efficiency gains for society. Ibid., 52–53.

63. Ibid.

64. The option of bailing enterprises out becomes severely restricted once the EU competition policy rules become binding.

CHAPTER 3

1. Ministerul Economiei şi Comerţului, "Strategia de Restructurare a Industriei Siderurgice din Romania Pentru Perioada 2004–2010—Actualizare 2004—Sinteză," Bucharest, March 2004, 13.

2. Ibid. In 1989, Siderurgica Hunedoara produced 3.2 million tons of steel. By 2002, that number had dwindled to 297,000 tons. At the same time, Sidex's decline was from nearly 7.7 million tons to 4.5 million tons.

3. The total 1989 employment estimates for the Romanian steel sector range between 136,000 and 207,856 people, despite coming from three credible sources. In 1998, the employers' association put the 1989 employment figure in the sector at 188,920 workers. In the same year, a group of engineering and research bureaus put the figure at 136,000 workers. See ICEM, IPROMET, IPROLAM, IEI, IEM, "Strategia Sectorială în Industria Siderurgică din România. Sinteză," Bucharest, October 1998, 75. The sectoral trade union organization, Metarom, put the 1989 figure at 207,856 workers. See FNS Metarom, *InfoMetal*, "A XIII-a Conferinţă," October 22–24, 2003, 6. The latter two sources agree on the number of employees in 1997, putting it at 122,290. I decided to use the trade union data, as the organization would be expected to have the best information about the level of employment, and their figure is relatively close to the figure given by the employers' association.

4. Raj Desai, "Organizing Markets: Property Rights, Governance, and the Politics of Industrial Privatization in a Post-Communist Economy" (PhD diss., Harvard University, 1996), 170; Hutní Projekt Ostrava and Vysoká Škola Báňská, *Studie Plánu Restrukturalizace Českého Ocelářského Průmyslu*, Ostrava, 1999, 47.

5. Hutní Projekt Ostrava and Vysoká Škola Báňská, *Studie Plánu*, 17; International Iron and Steel Institute, *Steel Statistical Yearbook 2005* (Brussels: IISI, 2005), 47–51.

6. Desai, "Organizing Markets," 170; Hutní Projekt Ostrava and Vysoká Škola Báňská, *Studie Plánu*, 47.

7. Desai, "Organizing Markets," 164.

8. Vítkovice is located in the city center and Nová Huť is built on the city outskirts, in the town of Kunčice.

9. Najwyższa Izba Kontroli, "Informacja o wynikach kontroli restrukturyzacji przekształceń własnościowych w hutnictwie żelaza i stali" (Warsaw: NIK, February 2003), 22.

10. Florian Steelworks, Kościuszko Steelworks, Ostrowiec Steelworks, Baildon Steelworks, Bankowa Steelworks, and Łabędy Steelworks were all founded in the first half of the nineteenth century; while Batory Steelworks, Ferrum Steelworks, and Zawiercie Steelworks were founded in the second half. Cedler Steelworks was

founded at the turn of the century. See Eugeniusz Rączka, *Hutnictwo w Polsce na początku XXI wieku* (Katowice: SITPH, 2003).

11. Given the economic problems Poland faced in the late 1970s, the equipment for the enormous rolling mill envisioned by Katowice Steelworks was shipped off and subsequently installed in Russia's Magnitogorsk Steelworks. Source: numerous interviews in the Polish steel sector, 2003–2004.

12. International Iron and Steel Institute, *Steel Statistical Yearbook 2005*, 47–51, my own calculation.

13. Ministère de l'Industrie—Roumanie, Sofres Conseil, Sofresid, Cedres, "Programme de Restructuration de la Siderurgie Roumaine," Bucharest, 1993, 7; Ministerul Industriilor, Departamentul Industriei Metalurgice, "Strategia de Restructurare a Siderurgiei Românești, Sinteză," Bucharest, November, 1993, 15; interview with a Romanian manager working for an international steel distributor, Bucharest, June 2004.

14. PHARE stands for "Poland and Hungary Assistance for the Restructuring of the Economy," and was one of the primary pre-accession instruments financed by the European Community.

15. Interview with UniRomSider (Association of Steel Producers of Romania) expert, Bucharest, May 2004.

16. Ministerul Industriilor, "Strategia de Restructurare," 2. For the individual mills to be profitable, their capacity utilization needs to be at least 80%.

17. Ministère de l'Industrie—Roumanie, Sofres Conseil, Sofresid, Cedres, "Programme de Restructuration," 7; Ministerul Industriilor, "Strategia de Restructurare," 15; Romanian manager working for an international steel distributor, interview.

18. The two smaller enterprises were Oțelinox and Lamdro.

19. These were Republica, Artrom, Petrotub, and Silcotub.

20. Ministère de l'Industrie—Roumanie, Sofres Conseil, Sofresid, Cedres, "Programme de Restructuration," 6–8.

21. Ministerul Industriilor, "Strategia de Restructurare," 14. Note that the 1989 estimates of employment in the sector oscillated between 136,000 and around 207,000 people. See note 3.

22. The 1989 employment level for Czechoslovakia was 163,000 people. See note 6.

23. The study expected the years 1991–1996 to be devoted to restructuring and privatization. It noted that the privatization process would influence restructuring. Privatization, in turn, would take several years and depend on the investment potential of foreign partners, who were experiencing adverse economic conditions. "Restrukturalizace československého hutnictví"—informace SPHŽ, *Hutnické listy*, č. 7–8/1992, 84–89 and č. 9/1992, 43–45; "Nákladná restrukturalizace hutnictví," *Hospodářské noviny*, May 21, 1992; Desai, "Organizing Markets," 172.

24. Andrzej Reczyński, "Omówienie i ocena studium restrukturyzacji hutnictwa wykonanego przez konsorcjum kanadyjskie," *Problemy Projektowe* (January–March 1993): 1–11; interview with Henryka Bochniarz, former minister of industry and trade, Warsaw, July 2001.

25. The most pressing technological problems were the widespread use of the

open-hearth process and the paucity of continuous casting technology. Reczyński, "Omówienie i ocena studium."

26. In 1980, Polish production of raw steel hit a record 19.5 million tons.

27. Reczyński, "Omówienie i ocena studium."

28. Barbara Pytel, "Sector Study of the Iron and Steel Industry in Poland," *Emergo* 2 (1995): 23; Dorota Sajnug and Danuta Zagrodzka, "Co naprawić, a co zamknąć," *Gazeta Wyborcza*, March 12, 1992.

29. The initial draft recommended immediate closure of crude steel production facilities at Sendzimir Steelworks; the eventual proposal represented a compromise brokered by the government with the managers and unions in the plant. See Pytel, "Sector Study of the Iron and Steel Industry"; Dorota Sajnug, "Jak hartuje się stal," *Gazeta Wyborcza* (Warsaw edition), November 17, 1992.

30. These were Batory Steelworks, Bobrek Steelworks, Bankowa Steelworks, Buczek Steelworks, Jedność Steelworks, Szczecin Steelworks, and Warsaw Steelworks.

31. Najwyższa Izba Kontroli, "Informacja o wynikach kontroli," 25.

32. Reczyński, "Omówienie i ocena studium"; Hatch Associates et al., "Raport Końcowy," June, 1992; Najwyższa Izba Kontroli, "Informacja o wynikach kontroli"; Jan Macieja, "Restrukturyzacja hutnictwa żelaza i stali," in *Studia nad restrukturyzacją sektorów przemysłowych w Polsce*, ed. Adam Lipowski and Jan Macieja (Warsaw: Polish Academy of Sciences, Institute of Economics), 65.

33. Rifts within the Solidarity camp were first noted in a press article published on November 10, 1989. Antoni Dudek, *Pierwsze Lata III Rzeczypospolitej* (Kraków: Arcana, 2002), 118. The political scene in Poland was fragmented and volatile, and, following the first fully free elections in November 1991, no fewer than twenty-four different parties and groups were represented in the parliament. The biggest parties representing different camps were as follows: neoliberals were overwhelmingly associated with Kongres Liberalno-Demokratyczny (Liberal Democratic Congress); social liberals, with Unia Demokratyczna (Democratic Union); conservatives/nationalists, with either Wyborcza Akcja Katolicka (Catholic Electoral Action), Konfederacja Polski Niepodległej (Confederation of Independent Poland), or Porozumienie Obywatelskie Centrum (Center Civic Agreement). Finally, worker self-management was most strongly espoused by NSZZ "Solidarność."

34. Dudek, *Pierwsze Lata*, 118–133; Maciej Bałtowski, *Przekształcenia Własnościowe Przedsiębiorstw Państwowych* (Warsaw: Wydawnictwo Naukowe PWN, 2002); Mitchell Orenstein, *Out of the Red: Building Capitalism and Democracy in Postcommunist Europe* (Ann Arbor: University of Michigan Press, 2001), 112–115; Connie Squires Meaney, "Foreign Experts, Capitalists, and Competing Agendas: Privatization in Poland, the Czech Republic, and Hungary," in *Liberalization and Leninist Legacies*, ed. Beverly Crawford and Arend Lijphart (Berkeley: University of California Press, 1997).

35. Andrew Berg, Olivier Jean Blanchard, "Stabilization and Transition: Poland, 1990–91," in *The Transition in Eastern Europe*, vol. 1, ed. Oliver Jean Blanchard, Kenneth Froot, and Jeffrey Sachs (Chicago: University of Chicago Press, 1994), 52, quoted in Orenstein, *Out of the Red*, 35.

36. Bałtowski, *Przekształcenia Własnościowe*, 95.

37. Ibid., 88–89; Orenstein, *Out of the Red*, 112–115.

38. Polish mass privatization never gave the citizens an opportunity to purchase enterprises directly. Rather, they owned a piece of largely foreign-consultancy-managed investment funds that managed the enterprise shares. See Maciej Bałtowski and Tomasz Mickiewicz, "Privatization in Poland: Ten Years After," *Post-Communist Economies* 12 (2000): 425–443.

39. Given the legacy of decentralization and self-management among the Polish enterprises, unless an enterprise was commercialized, its worker council had to agree to any changes of ownership. The government could only overcome a worker council veto by issuing a specific decision to do so. By the end of 1995, 788 enterprises had been privatized using MEBOs; Orenstein, *Out of the Red*, 118.

40. For a discussion of the various means by which this was done, see Jadwiga Staniszkis, "Political Capitalism in Poland," *East European Politics and Societies* 5 (1991): 127–141; Jacek Kuroń and Jacek Żakowski, *Siedmiolatka czyli kto ukradł Polskę?* (Wrocław: Wydawnictwo Dolnośląskie, 1997), 116–121.

41. Bałtowski, *Przekształcenia Własnościowe*, 77.

42. Robert E. Kennedy, "A Tale of Two Economies: Economic Restructuring in Post-Socialist Poland," *World Development* 25 (1997): 841–865.

43. Janusz Lewandowski, "Spawozdanie Komisji Przekształceń Własnościowych oraz Komisji Polityki Gospodarczej, Budżetu i Finansów o rządowym projekcie ustawy o narodowych funduszach inwestycyjnych i ich prywatyzacji" (druki nr 861, 860 i 860-A), Posiedzenia Sejmu, sprawozdania stenograficzne. 1 kadencja, 43 posiedzenie, 1 dzień (April 29, 1993).

44. Bałtowski, *Przekształcenia Własnościowe*, 18; Maria Jarosz, "Oblicza prywatyzacji," in *Manowce Polskiej Prywatyzacji*, ed. Maria Jarosz (Warsaw: Wydawnictwo Naukowe PWN SA, 2001), 24.

45. Interview with Edward Nowak, former deputy minister of economy, Kraków, July 2003.

46. Ibid.

47. Ryszard Stelmaszczyk, "Mniejsza produkcja i zużycie," *Rzeczpospolita*, February 14, 1997; Irena Bytomska, "Gorszy rok Hutnictwa," *Nowe Życie Gospodarcze*, April 13, 1997; Jan Dziadul, "Stalowy uchwyt," *Polityka*, May 16, 1998; interviews with industry insiders, Poland, June–November 2003. Polskie Stronnictwo Ludowe has traditionally been translated as the "Polish Peasants' Party."

48. Jerzy Sadecki and Danuta Walewska, "Granice otworzymy później," *Rzeczpospolita*, May 28, 1996.

49. An interviewed representative of the Metallurgical Chamber of Industry and Commerce claimed that there was no "political will" to privatize the mills in the mid-1990s, Warsaw, June 2003.

50. Najwyższa Izba Kontroli, "Informacja o wynikach kontroli," 47.

51. Freedom Union was a 1994 merger of the neoliberal Liberal Democratic Congress (Kongres Liberalno-Demokratyczny) and the social liberal Democratic Union (Unia Demokratyczna).

52. Antoni Kowalik, "Bez generalnych zmian," *Rzeczpospolita*, November 26, 1997, emphasis added.

53. Ryszard Stelmaszczyk, "Koncern albo problemy," *Rzeczpospolita*, February 7, 1998, emphasis added.

54. The original document was called "Stanowisko rządu Rzeczypospolitej Polskiej w sprawie restrukturyzacji hutnictwa żelaza i stali."

55. Macieja, "Restrukturyzacja hutnictwa," 64.

56. Pytel, "Sector Study of the Iron," 21.

57. Bochniarz, interview.

58. Interview with a Sendzimir Steelworks insider and longtime Solidarity activist, Kraków, November 2003.

59. Interview with Janusz Steinhoff, former minister of economy, Warsaw, July 2003. The neoliberals' reservation was largely unfounded. First, at the time, the structure of the two biggest enterprises was complementary. Second, to be able to compete with each other, the enterprises would have needed to make redundant investments in a market already suffering from overcapacity. Nowak, interview.

60. Lawrence P. King and Aleksandra J. Sznajder, "The State-Led Transition to Liberal Capitalism: Neoclassical, Organizational, World Systems, and Social Structural Explanations of Poland's Economic Success," *American Journal of Sociology* 112 (2006): 771.

61. Lidia Oktaba and Tomasz Świderek, "Jeszcze więcej holdingów," *Rzeczpospolita*, April 20, 1995. See Macieja, "Restrukturyzacja hutnictwa żelaza i stali," 67, for a discussion of the proposal.

62. Barbara Cieszewska, "Razem czyli osobno," *Rzeczpospolita*, September 19, 1996.

63. Jerzy Sadecki and Danuta Walewska, "Granice otworzymy później," *Rzeczpospolita*, May 28, 1996.

64. Ryszard Czarnecki, "Cztery programy dla hutniczych sektorów," *Rzeczpospolita*, July 11, 1996.

65. Krzysztof Fronczak, "Nie Zatrzymujmy Się," *Nowe Życie Gospodarcze*, September 7, 1997.

66. Jerzy Sadecki, "Sprzeciw Huty Sendzimira," *Rzeczpospolita*, March 11, 1998.

67. Interview with Emil Wąsacz, former minister of state treasury, Katowice, Poland, October 2003.

68. Interview with a civil servant at the Ministry of Economy, Warsaw, July 2003 (translated by the author).

69. Interviews with civil servants (nos. 1–6) at the Ministry of Economy, Warsaw, July 2003.

70. The ODA was similar to the ODS in policy terms, though it was more open to foreign investment.

71. Orenstein, *Out of the Red*; Martin Myant, *Rise and Fall of Czech Capitalism* (Northampton, MA: Edward Elgar, 2003); Jan Mládek, "Národní kapitalismus v Čechách: A co s ním dál?" *Respekt*, September 10, 2001.

72. Being very politically savvy, the Czech reformers also reluctantly included clearly specified social policies in an effort to make their initial reform package more popular. See Orenstein, *Out of the Red*, 71–72.

73. Various interviews in Prague, December 2003, March 2004.

74. Orenstein, *Out of the Red*, 76–79; Quentin Reed, "Corruption in Czech Privatization: The Dangers of 'Neo-Liberal' Privatization," in *Political Corruption in Transition: A Sceptic's Handbook*, ed. Stephen Kotkin and András Sajó (Buda-

pest: Central European University Press, 2002); Myant, *Rise and Fall*, 14; Mládek, "Národní kapitalismus."

75. Gerald A. McDermott, *Embedded Politics: Industrial Networks and Institutional Change in Postcommunism* (Ann Arbor: University of Michigan Press, 2002), 90–92. The main Czech candidates to establish joint ventures were Škoda Plzeň and ČKD Praha in engineering; Tatra and Liaz in trucks; Aero in aircraft; Chemapol in chemopetrols; Poldi in high-end steel; and Vitkovice, Třinecké Železarny, and Nová Huť in medium- and low-end steel. Desai, "Organizing Markets."

76. See Myant, *Rise and Fall*, for analysis of the privatization process into the hands of domestic investors. Also, interview with a consultant from Wood and Company, Prague, March 2004.

77. Myant, *Rise and Fall*, 194.

78. For an argument concerning pragmatism of the Czech neoliberal reformers, see Orenstein, *Out of the Red*.

79. Interview with Vladimír Dlouhý, former minister of industry and trade, Prague, Czech Republic, March 2004.

80. Ibid.

81. Desai, "Organizing Markets," 210.

82. McDermott, *Embedded Politics*, 92; *Hospodářské noviny*, "Poznámky k českému hutnictví," May 13, 1997.

83. McDermott, *Embedded Politics*, 91. The joint ventures that were created largely pertained to activities supporting steel production, such as generator production, liquefied gas, valves, ceramics, etc. See Desai, "Organizing Markets," 211.

84. Poldi Ocel represented 76% of Poldi's assets, and its steelmaking facilities were known as Poldi 1, producing special steel, and Poldi 2, specializing in construction steel.

85. FNM was the state agency responsible for financial restructuring of enterprises, negotiating privatization deals, and exercising property rights over shares of firms permanently or temporarily owned by the state.

86. Dlouhý, interview; see also Myant, *Rise and Fall*, 195. Liabilities reached CZK 5.1 billion in 1992. Desai, "Organizing Markets," 241–242.

87. The remaining shares were to be acquired by the banks as part of debt-equity swaps. Myant, *Rise and Fall*, 195.

88. Ibid.

89. Dlouhý, interview.

90. Myant, *Rise and Fall*, 196.

91. *Hospodářské noviny*, "Fond národního majetku oznámí další postup vůči firmě Bohemia Art," February 15, 1996. The court dismissed the argument that the transfer of shares should have been invalidated because the buyer had not met the payment deadlines and because the transfer had not been properly approved by the FNM. Myant, *Rise and Fall*, 197; *Hospodářské noviny*, "Případ Poldi Ocel už projednává obchodní soud," March 12, 1996.

92. Třinecké Železárny, *Presentation*, photocopy, March 2004.

93. *Hospodářské noviny*, "FNM diskutuje o postupu privatizace železáren," April 9, 1995.

94. Eva Kijonková, "DB Invest sází na dohodu Třince s Novou hutí," *Hospodářské noviny*, October 20, 1995; Eva Kijonková and Ivana Kněžínková, "Majoritu v

Třineckých železárnách má získat společnost Moravia Steel," *Hospodářské noviny*, October 26, 1995.

95. *Hospodářské noviny*, "Úvěr na získání majority v Třineckých železárnách poskytnou české banky," November 2, 1995; Pavel Šmíd, "Moravia Steel mění vlastnickou strukturu," *Hospodářské noviny*, May 30, 2002.

96. See Myant, *Rise and Fall*, 200; Tomáš Němeček, "Mečbol za čtrnáct milionů," *Respekt*, June 12, 2000. Šrejber had also been accused of illegally siphoning off CZK 13 million into the Šrejber Tennis Investing company in 1995 and 1996. Leah Bower, "First Convicted Tunneler Gets 10-Year-Term," *The Prague Post*, January 17, 2001.

97. James Drake, "Campaign-Financing Headache Hits ODS," *The Prague Post*, November 26, 1997. See also Orenstein, *Out of the Red*; and Abby Innes, *Czechoslovakia: The Short Goodbye* (New Haven, CT: Yale University Press, 2001).

98. Dlouhý, interview.

99. Ibid.

100. Interviews with representatives of Steel Federation, Inc. (the sectoral-level steel producers' and employers' organization), Prague, February 2004.

101. Interview with Jiří Havel, former FNM chairman, Prague, March 2004. See also Zuzana Kubátová and Tereza Zavadilová, "Hutě tě nevolají," *Týden*, August 16, 1999; Jaroslav Pětroš, "Manažerská Odyssea," *Nová Huť*, April 25, 2000.

102. Resolution of the Government of the Czech Republic Nr. 485 from September 18, 1996.

103. Markéta Huleová, "Díky amatérsky sepsané smlouvě drží Petrcíle stát pod krkem," *Právo*, May 5, 2000; Markéta Huleová, "Češka svedl vinu na špatnou smlouvu s Petrcíle na Klause," *Právo*, May 6, 2000; *Severočeské noviny*, "Ukončení smlouvy s FNM o Nové huti přinese firmě Petrcíle stamilióny," October 2, 2000.

104. Dlouhý, interview; Robert Čásenský and Petr Šimůnek, "Byznys je lepší než politika," *Mladá fronta Dnes*, July 27, 2001; Reed, "Corruption in Czech Privatization."

105. In the case of Nová Huť, the share amounted to 18.25%; and in the case of Vítkovice, 17.30%.

106. International Finance Corporation, "1997 Annual Report" (Washington, DC: IFC, 1998).

107. Ministry of Industry and Trade of the Czech Republic (Ministerstvo průmyslu a obchodu České republiky), *Pro Schůzi Vlády, Věc: Postup Programu restrukturalizace ocelařského průmyslu v České republice, Příloha č.1 k Předkládací zprávě*, Prague, January 14, 2002, 1.

108. The loan agreement with the IFC was signed on June 27, 1997, but concrete "privatization" measures did not follow until the Government Resolution Nr. 608 of October 1, 1997. *Nová Huť*, "Výstavba minihutě může pokračovat," July 11, 1997.

109. The $250 million loan entailed a $75 million loan from IFC's own account and a $175 million syndicated loan. IFC press release, "IFC Signs US $250 Million Financing Agreement for Privatized Czech Steel Company," Prague, June 27, 1997.

110. Slovak autonomy within Czechoslovakia increased during post–Prague Spring "normalization" period, starting in 1969. The fact that national sovereignty was not at the forefront of Slovak concerns after the collapse of communism was reflected by the initial low level of public support for the separation of the two

states. Miroslav Kusý, "Slovak Exceptionalism," in *The End of Czechoslovakia*, ed. Jiří Musil (Budapest: Central European University Press, 1997), 145; Sharon L. Wolchik, "The Politics of Ethnicity in Post-Communist Czechoslovakia," *East European Politics and Societies* 8 (1994): 153–188.

111. Innes, *Czechoslovakia*, 147–167.

112. Ibid., 162.

113. Ibid., 159–161.

114. Ibid., 159.

115. See essays in Musil, *The End of Czechoslovakia*, and Innes, *Czechoslovakia*, for a detailed analysis of the "velvet divorce."

116. Begging the question of its legitimacy, the decision was never voted on in a referendum, despite earlier promises that one would be held.

117. In the second round, 3.4 million people, or almost all adult citizens, registered to participate. See Ivan Mikloš, "Economic Transition and the Emergence of Clientalist Structures in Slovakia," in *Slovakia: Problems of Democratic Consolidation*, ed. Soňa Szomolányi and John A. Gould (Bratislava: Slovak Political Science Association, 1997), 62.

118. Mikloš, "Economic Transition," 64; Grigorij Mesežnikov, "Political Parties as Actors of Reforms in Slovakia," in *Slovakia: Ten Years of Independence and a Year of Reforms*, ed. Grigorij Mesežnikov and Oľga Gyárfášová (Bratislava: Institute for Public Affairs, 2004).

119. Marian Leško, *Mečiar a mečiarizmus* (Bratislava: VMV, 1994), 215.

120. Slovak Government Information Service, *Analysis of the Inherited State of the Economy and Society*, 1999.

121. *Trend*, "Desať rokov VSŽ," October 11, 2000.

122. Marían Hutňan, "Chronológia krízy vo VSŽ, kedysi vlajkovej lode Slovenska (I)," *Hospodarské noviny*, August 26, 1999; Anton Marcinčin, "Reštrukturalizácia podnikov," in *Hospodárska Politika na Slovensku 1990–1999*, ed. Anton Marcinčin and Miroslav Beblavý (Bratislava: Slovak Foreign Policy Association, 2000), 329.

123. Gabriel Beer, "Dcéra musí matku počuvať," *Trend*, September 29, 1993.

124. Róbert Matejovič, "VSŽ a ich príbuzní," *Slovenský Profit*, December 17, 1996.

125. Ibid.

126. Gabriel Beer, "Mimoriadne valné zhromaždenia VSŽ a Priemyselnej banky v tónine tímu V. Mečiara," *Trend*, March 23, 1994.

127. Matejovič, "VSŽ a ich príbuzní."

128. Michael J. Kopanic, "Stealing the Eastern Slovak Steelworks," *Central Europe Review* 2 (2000), http://www.ce-review.org/00/1/kopanic1_steel.html (accessed October 7, 2014); Karol Wolf, "Před velkou inventurou," *Euro*, November 9, 1998.

129. Daniel Tácha, "VSŽ: Manažeři z východu naivní nejsou," *Týden*, July 29, 1996.

130. Marcinčin, "Reštrukturalizácia podnikov," 329.

131. *Trend*, "Desať Rokov VSŽ." There were other examples of the intertwined relationship between VSŽ and HZDS, as in the construction of HZDS's new headquarters in Bratislava by VSŽ. *Trend*, "HZDS hľadá nové sídlo, VSŽ jedno stavia," January 3, 1998. See also *Právo*, "HZDS dostalo výpověď," June 4, 1999.

132. Vladimir Tismaneanu, *Stalinism for All Seasons* (Berkeley: University of California Press, 2003), 4.

133. Ibid., 242.

134. Vladimir Pasti, *The Challenges of Transition: Romania in Transition* (Boulder, CO: East European Monographs), 270. Much in the same spirit, pluralism received very mixed, if not hostile, reaction within the FSN; the movement's main ideologue, Silviu Brucan, insisted that internal factionalism within the FSN could replace regular political party competition. Tismaneanu, *Stalinism for All Seasons*, 241.

135. In a bout of nationalist pride, Ceauşescu strived—and succeeded—to repay Romania's foreign debt to Western countries in the 1980s, leading to far-reaching misery among the Romanian population, which was subjected to meager food rations and rampant power cuts.

136. Pasti, *The Challenges of Transition*; Tom Gallagher, *Modern Romania* (New York: New York University Press, 2005), 90.

137. The legitimacy of the new leaders was questioned soon after the revolution, with student unrest erupting as early as February of 1990 and continuing until it was contained, with no casualties, by the army. This was the first time that the Jiu Valley miners had visited Bucharest to show their support for those in power, above all, Iliescu. Pasti, *The Challenges of Transition*, 276; Stelian Tanase, "Reform and Counter-Reform in Romanian Political Life," *Romanian Journal of Sociology* 8, no. 2 (1997): 76.

138. To place Romania in a comparative perspective, in the post-1989 period in Poland and Czechoslovakia, inventories accounted for -3% to 3% of GDP. In Romania, the proportion reached 10.5% in 1990, 13.7% in 1991, and 16.2% by 1992. Foreign currency reserves stood at $1.7 billion at the end of 1989, but they decreased to under $400 million at the end of 1990. Daniel Daianu, "Macro-Economic Stabilization in Post-Communist Romania," in *Romania in Transition*, ed. Lavinia Stan (Brookfield, VT: Dartmouth, 1997), 101.

139. Subsequently, a contingent of the Jiu Valley miners arrived in Bucharest, and in coordinated actions brutalized students and others near the University Square, as well as devastated opposition party headquarters. Iliescu's decision to thank the miners for their support tarnished Romania's image abroad and led to the suspension of trade agreement negotiations with the European Community. See Pasti, *The Challenges of Transition*, 277–279, for a discussion of these events.

140. Another plan, implemented late in 1993, proved more successful in fighting inflation, in part by dramatically raising interest rates. Daianu, "Macro-Economic Stabilization," 102–108.

141. Dragos Negrescu, "A Decade of Privatization in Romania," in *Economic Transition in Romania*: Past, Present, and Future; Proceedings of the Conference "Romania 2000—10 Years of Transition," ed. Christof Rühl and Daniel Daianu (Bucharest: World Bank and Romanian Center for Economic Policies, 1999), 377.

142. These included armament, energy, mining and natural gas, and post and railway transportation, as well as some areas belonging to other branches, as decided by the government. Ibid., 379–380.

143. Only in 1997 did the *regies autonomes* enter the privatization process, following a law stipulating that they be either commercialized or converted into public institutions with regulatory functions.

144. The mission of the FPS was to be accomplished within seven years. John S. Earle and Dana Săpătoru, "Privatization in a Hypercentralized Economy: Case of Romania," in *Privatization in the Transtition to a Market Economy*, ed. John S. Earle, Roman Frydman, and Andrzej Rapaczynski (London: Pinter, 1993), 154–160.

145. In 1994, the amount that could be obtained by the private ownership funds was raised to 60% of the capital of selected enterprises.

146. The MEBO law (no. 77/1994) provided very favorable conditions for acquiring the remaining shares from the FPS. Negrescu, "A Decade of Privatization," 401–403.

147. Following the distribution of coupons in 1992, the program came to a standstill; only about 20% of the total number of certificates had been swapped for shares by the beginning of 1995. Under IFI pressure, the mass privatization process was resurrected in 1995–1996. Negrescu, "A Decade of Privatization," 403–406; Lavinia Stan, "Romanian Privatization Program: Catching Up with the East," in *Romania in Transition*, ed. Lavinia Stan (Brookfield, VT: Dartmouth, 1997), 137–141.

148. Negrescu, "A Decade of Privatization," 386–387.

149. *Mediafax*, "Ministerul Industriei și Comerțului a nominalizat 30 de societăți comerciale din industrie care trebuie privatizate rapid," Bucharest, March 5, 1997.

150. *Mediafax*, "Oferta de privatizare pentru 1998 cuprinde un număr de 2,745 societăți comerciale," Bucharest, February 25, 1998.

151. According to the European Bank for Reconstruction and Development, the enterprise restructuring score remained at the modest 2 (out of a best possible score of "4+") throughout the 1994–2004 period. European Bank for Reconstruction and Development, *Transition Report* (London: EBRD, 1994–2004).

152. Ministerul Industriilor, "Strategia de Restructurare," 15–17.

153. Based on the information corroborated during three interviews: at the Ministry of Economy and Trade, with a UniRomSider expert, and with the vice president of Cartel Alfa Confederation, Bucharest, May–June 2004.

154. Interview with Petru Dandea, vice president, Cartel Alfa Confederation, Bucharest, June 2004.

155. UniRomSider expert, interview.

156. Parliamentary debate statement made by Mircea Nicu Toader (PD), Ședința Camerei Deputaților din 6 iunie 2002, "Dezbaterea și adoptarea Proiectului de Lege privind unele măsuri pentru întărirea disciplinei contractuale" (translated by the author), http://www.cdep.ro/pls/steno/steno.stenograma?ids=5298&idm=6&idl=1 (accessed October 25, 2014).

157. See the parliamentary statement of PNȚCD, Partidul Național Țărănesc Creștin și Democrat (Christian-Democratic National Peasants' Party) deputy, Vasile Pavel: Ședința Camerei Deputaților din 17 martie 1998. "Vasile Pavel—comentarii ocazionate de afirmații ale secretarului general al PDSR referitoare la implicarea PNȚCD în procesul de privatizare, făcute cu prilejul unei vizite la Bacău" (translated by the author), http://www.cdep.ro/pls/steno/steno.stenogram a?ids=24&idm=1,06&idl=1 (accessed October 25, 2014); numerous interviews in Romania, May–June 2004.

158. *Mediafax*, "Mugur Isărescu comentează dificultățile întîmpinate de procesul de privatizare," Mangalia, November 11, 2000 (translated by the author).

159. *Ziua*, "Rechizitoriul lui Sorin Dimitriu," September 3, 1998.

160. *Mediafax*, "FPS vinde 90 de întreprinderi prin intermediul băncilor străine de investiţii," Bucharest, July 21, 1998.

CHAPTER 4

1. David Stark and László Bruszt, *Postsocialist Pathways: Transforming Politics and Property in East Central Europe* (Cambridge: Cambridge University Press, 1998), 194.

2. Interview with a civil servant responsible for the steel sector at the Ministry of Economy and Trade (formerly the Ministry of Industry and Trade), Bucharest, May 2004 (translated by the author).

3. Interview with a journalist covering the steel sector at a major daily, Bucharest, June 2004.

4. Interview with UniRomSider expert, Bucharest, May 2004.

5. Interview with the leader of Metarom Federation, Bucharest, June 2004 (translated by the author).

6. Usinor Consultants, "Asistenţa Tehnică Pentru Restructurarea Industriei Siderurgice Româneşti, Raportul Final Nr.0. Sinteza," Bucharest, July 2000, 6 (translated by the author).

7. Ibid.

8. See data provided by the European Commission in 2004: http://europa. eu.int/comm/enterprise/steel/restructuring-steel/fiche_2003_ro.pdf   (accessed April 20, 2006).

9. Interview with Petru Dandea, vice president, Cartel Alfa Confederation, Bucharest, June 2004.

10. Information corroborated in separate interviews with Metarom, Cartel Alfa, and UniRomSider representatives. The UniRomSider expert claimed that the investment was funded from private sources. However, state aid given to the enterprises, as appearing in the 2004 sectoral restructuring program notes, included a $8.8 million direct transfer from the state budget, as well as total aid of about $26.5 million. See "Strategia de Restructurare a Industriei Siderurgice din Româ-nia," approved by H.G. nr. 655/29.04.2004, published in M.O. (Official Journal) nr. 425/12.05.2004, 97. This supports the information given by the trade union representatives. Moreover, the 1993 program draft envisaged large "capital infu-sion" *(aport de capital)* in the enterprise, by far the largest such single transfer in the entire sector ($172 million), claiming that "some other enterprises, and, especially, Siderca Călăraşi, cannot support the necessary investment effort [using their own resources or bank credits]." See Ministerul Industriilor, Departamentul Industriei Metalurgice, "Strategia de Restructurare a Siderurgiei Romaneşti. Sinteză," Bucha-rest, November, 1993, 13 (translated by the author).

11. Anton Marcinčin, "Reštrukturalizácia podnikov," in *Hospodárska Politika na Slovensku 1990–1999*, ed. Anton Marcinčin and Miroslav Beblavý (Bratislava: Slo-vak Foreign Policy Association, 2000), 331. Diósgyör Steelworks went bankrupt in March 2000.

12. Marían Hutňan, "Chronológia krízy vo VSŽ, kedysi vlajkovej lode Slovenska (II)," *Hospodarské noviny*, August 27, 1999.

13. VSŽ had controlling interest in significant banks, such as the Premyselná Banka Košice, Dopravná Banka, Postová Banka, as well as a significant share in mid-sized banks, such as Poľnobanka and Istrobanka. VSŽ also owned a significant share of Slovenská Poist'ovňa, which in 1995 controlled 78% of all insurance sold in Slovakia. See Michael J. Kopanic, "Stealing the Eastern Slovak Steelworks," *Central Europe Review* 2 (2000), http://www.ce-review.org/00/1/kopanic1_steel.html (accessed October 7, 2014); see also *Trend*, "VSŽ Košice získali pätinu akcií Slovenskej poist'ovne," July 24, 1996.

14. Marcinčin, "Reštrukturalizácia podnikov," 330.

15. Newspapers controlled by VSŽ included *Národná obroda*, *Večer*,and *Lúč*. Also, the enterprise had influence over the state-owned radio Fun as well as the local TV station, TV Naša. Daniel Tácha, "VSŽ: Manažeři z východu naivní nejsou," *Týden*, July 29, 1996.

16. The Czechoslovak sectoral organizations, or the *výrobní hospodařskě jednotky*, were dismantled only in 1987–1988, under the perestroika-inspired decentralization reforms. Raj Desai, "Organizing Markets: Property Rights, Governance, and the Politics of Industrial Privatization in a Post-Communist Economy" (PhD diss., Harvard University, 1996), 179; interview with representatives at Steel Federation, Inc. (Hutnictví Železa, a.s.), Prague, November 2003. At the same time, one should not overestimate the rigidity of these structures, which had been undergoing relatively frequent and nontrivial changes during the communist period. See Vladimír Dostál and Jaroslav Purš, *Dějiny hutnictí železa v Československu*, vol. 3 (Prague: Academia, 1988).

17. *Zrzeszenie* was never endowed with whatever diminished power it was supposed to have over the individual enterprises, as written into its statute. See Barbara Pytel, "Sector Study of the Iron and Steel Industry in Poland," *Emergo* 2 (1995): 19. In any event, the two largest steelworks, Katowice Steelworks and Sendzimir Steelworks, had always enjoyed special treatment in that, although they were dependent on the Iron and Steel United for allocation of inputs, they answered directly to the minister overseeing the steel industry. Interview with the steel-sector expert at the Agency for Industrial Development, Warsaw, July 2003.

18. Jan Sztobryn, "Hutnictwo Żelaza i Stali w Polsce Niepodległej" (unpublished manuscript, photocopy. Warsaw: Agencja Rozwoju Przemysłu, 2001); interviews at the Agency for Industrial Development, Warsaw, July 2003, and at the Metallurgical Chamber of Industry and Commerce, Katowice, June 2003.

19. Interview with a civil servant at the Ministry of Economy, Warsaw, July 2003; interview with a top management member at the Polish Steelworks (PHS), formerly at the Ministry of Economy, Katowice, October 2003.

20. Pytel, "Sector Study," 19. These views were also voiced by the current president of the Metallurgical Chamber of Industry and Commerce, both in numerous press interviews and in an interview with the author.

21. Interview with a representative at the Metallurgical Chamber of Industry and Commerce, Katowice, July 2003 (translated by the author).

22. Ibid.

23. Interview with Emil Wąsacz, Katowice, October 2003.

24. Pytel, "Sector Study," 22–24; interview with a civil servant at the Ministry of State Treasury, Warsaw, July 2003.

25. Aleksandra Sznajder, "Still Restructuring? The Politics of the Steel Sector Restructuring in Poland and the Effects of EU Accession" (paper presented at the 2003 Annual Meeting of the American Political Science Association, Philadelphia, 2003).

26. Jan Dziadul, "Stal na złom," *Polityka*, February 13, 1999, 63 (translated by the author).

27. Dziadul "Stal na złom," 62; Małgorzata Bos-Korzewska and Artur Morka, "Pytania o restrukturyzację hutnictwa," *Rzeczpospolita*, November 14, 1998; Artur Morka, "Ostrzeżenia Austriaków," *Rzeczpospolita*, November 23, 1998.

28. Jan Dziadul, "Jak rdzewieje stal," *Polityka*, August 11, 2001, 58. Other reasons were the ongoing merger between British Steel and Hoogovens to form Corus and the company's poor financial results at the time. See Jan Dziadul, "Jak rozhartowała się stal," *Polityka*, November 25, 2000, 68–71.

29. Wąsacz, interview. In three separate interviews with the author, both the management and the leaders of two significant unions in the steelworks talked with pride about the technological investments made in the mill, Częstochowa, July 2003.

30. For a detailed coverage of the investments, see Eugeniusz Rączka, *Hutnictwo w Polsce na początku XXI wieku* (Katowice: SITPH, 2003), 87–90; Andrzej Wypych et al., "Modernizacja i restrukturyzacja huty 'Częstochowa': Stan obecny i kierunki dalszych przemian," *Problemy Projektowe* 3 (1996), 66–71.

31. Najwyższa Izba Kontroli, "Informacja o wynikach kontroli restrukturyzacji przekształceń własnościowych w hutnictwie żelaza i stali" (Warsaw: NIK, February 2003), 49; Barbara Cieszewska et al., "Nie tylko zmiana nazwy," *Rzeczpospolita*, October 7, 2002.

32. Tomasz Wolf, "Wyścig kosztownych liderów," *Nowe Życie Gospodarcze*, January 3, 1999.

33. Dorota Margas, "Trzy kroki do przodu," *Rzeczpospolita*, December 13, 1993.

34. Wolf, "Wyścig kosztownych liderów."

35. Dariusz Wieczorek, "Impexmetal dostanie 200 mln złotych," *Parkiet*, July 24, 2003; Marcin Góralewski, "Impexmetal sprzedaje Zawiercie za 200 mln zł," *Puls Biznesu*, July 24, 2003.

36. Rączka, *Hutnictwo w Polsce*, 106.

37. Najwyższa Izba Kontroli, "Informacja o wynikach kontroli," 40–41.

38. Jan Dziadul, "Stalowa wola," *Polityka*, no. 20, May 18, 2002, 68.

39. Najwyższa Izba Kontroli, "Informacja o wynikach kontroli," 41.

40. Dziadul, "Stalowa wola," 68–69.

41. Interview with Edward Nowak, Kraków, July 2003.

42. Interview with a representative of the Steel Sector Employers' Association, Katowice, July 2003; Wąsacz, interview; Barbara Cieszewska, "Walcownia zamknięta na kłódkę," *Rzeczpospolita*, July 1, 2002; Jacek Czarnecki, "Modernizacja na kredyt," *Rzeczpospolita*, January 4, 1996.

43. Najwyższa Izba Kontroli, "Informacja o wynikach kontroli," 59–60.

44. Cieszewska, "Walcownia zamknięta na kłódkę"; Witold Gadomski, "Z tego pieca będzie awantura," *Rzeczpospolita*, May 23, 2003.

45. Desai, "Organizing Markets," 179; Steel Federation representatives, interview. For a discussion of the transformation of the different types of production networks, or *Výrobní Hospodařské Jednotky* (VHJs), and their implications for

restructuring, see Gerald McDermott, *Embedded Politics: Industrial Networks and Institutional Change in Postcommunism* (Ann Arbor: University of Michigan Press, 2002).

46. Desai, "Organizing Markets," 178; Steel Federation representatives, interview.

47. Steel Federation representatives, interview.

48. "Restrukturalizace československého hutnictví"—informace SPHŽ, *Hutnické listy*, č. 9/1992, 45; in interviews with the author, the top managers of two North Moravia steel plants confirmed this sentiment (March and April 2004).

49. Interview with a civil servant at the Ministry of Industry and Trade, Department of Metallurgy, Prague, December 2003.

50. Martin Myant, *Rise and Fall of Czech Capitalism* (Northampton, MA: Edward Elgar, 2003), 194.

51. Interview with top management member #1, at North Moravia steel works A, April 2004 (translated by the author).

52. Myant, *Rise and Fall*, 194; Eva Kijonkova, "Spory hutní lobby," *Hospodářské noviny*, January 26, 1994.

53. *Hospodářské noviny*, "Záruku za úvěr na výstavbu kontilití nejspíše obdrží všichni tři žadatelé," January 25, 1994.

54. *Hospodářské noviny*, "Třinecké železárny se záměru postavit moderní minihuť ještě nevzdávají," February 24, 1994; *Hospodářské noviny*, "Třinecké železárny nemají žádnou náhradu za koncepci vybudování vlastní minihutě," September 16, 1994.

55. "Nová huť odmítá spojení s Vítkovicemi," *Hospodářské Noviny*, October 30, 1995; "Valná hromada v Nové huti, a.s., Ostrava," *Hutnické listy* č. 12/1995, 35.

56. Ivo Psota, "Vysoké Pece Ostrava, a.s., správné rozhodnutí?" *Nová Huť*, September 26, 2000, 9; Steel Federation representatives, interview (2004); Desai, "Organizing Markets," 229.

57. Kijonkova, "Spory hutní lobby."

58. Desai, "Organizing Markets," 229; *Hospodářské noviny*, "Nová huť nechce válku hutníků," February 25, 1994.

59. Šárka Swiderová, "Premiér Zeman posvětil spojení tří největších hutí, *Moravskoslezský den*, October 26, 1999; *Hospodářské noviny*, "Vznik hutního konsorcia se posouvá," December 15, 1999.

60. Interview with the top management member at North Moravia steelworks B, March 2004. This sentiment was echoed by top management members #1 and #2 at North Moravia steelworks A, April 2004 (translated by the author).

61. Jana Paštiková and Karin Pelikánová, "Gross se příjel podívat, jak by mohlo fungovat spojení hutí," *Moravskoslezský den*, August 9, 1999.

62. Interview with a consultant at the Roland Berger Strategy Consultants, Prague, March, 2004.

63. UniRomSider expert, interview.

64. *Mediafax*, "Combinatul siderurgic SIDEX Galaţi îşi va reduce personalul cu 10,000–10,200 de personae pînă în anul 2000," Bucharest, February 22, 1998.

65. Ibid.

66. *Mediafax*, "Sidex va fi privatizat prin intermediul băncii de investiţii Flemings," Bucharest, January 25, 1998.

67. *Mediafax*, "Ministerul Industriei şi Comerţului a întocmit o listă cu mari

societăţi aflate în pericol de faliment," Bucharest, September 10, 1997; *Mediafax*, "Conducerea FPS este nemulţumită de managementul Combinatului Siderurgic Galaţi," Bucharest, August 11, 1997.

68. *Mediafax*, "Acţionarii Sidex au schimbat conducerea combinatului," Galaţi, March 5, 1999.

69. *Mediafax*, "Administratorul interimar al combinatului Sidex Galaţi, Ionel Borş, a declarat că principalul său obiectiv este menţinerea în stare de funcţionare a combinatului," March 7, 1999.

70. Aurel Stancu, "Balli santajeaza SIDEX-ul," *Viaţa Liberă*, April 17, 2000; *Mediafax*, "Programul de reeşalonare a datoriilor Sidex ar putea fin semnat în această săptămînă," August 24, 1999; *Mediafax*, "Consilierul guvernamental Petrişor Peiu va fi noul preşedinte al Consiliul de Administraţie al combinatului Sidex," July 29, 1999.

71. *Adevărul*, "Ani grei de puscarie pentru fosta conducere de la SIDEX," December 17, 2005.

72. *Economist Intelligence Unit*, "Romania Industry: Survey of the Metals Sector," August 31, 2001.

73. *Mediafax*, "Ovidiu Muşetescu: Cei care lucrează la combinatul Sidex trebuie să înţeleagă că simpla dorinţă de a turna oţel nu mai stă în picioare," Bucharest, February 14, 2001.

74. *Mediafax*, "Departamentul de Control al Guvernului a definitivat raportul controlului efectuat la combinatul Sidex," Bucureşti, November 1, 1999.

75. *Ziua*, "Rechizitoriul lui Sorin Dimitriu," September 3, 1998; *Evenimentul Zilei*, "Balli a pus mina sip e SIDEX Galati," March 10, 1998.

76. Interview with the steel-sector analyst at the Roland Berger Strategy Consultants, Bucharest, May 12, 2004; interview with the steel-sector privatization adviser to the minister of privatization at the AVAS, Bucharest, May 2004.

77. "Strategia de Restructurare a Industriei Siderurgice din România," approved by H.G. no. 655/29.04.2004, published in M.O. no. 425/12.05.2004, 94–101. The actual number was $592.4 million.

78. Total state aid anticipated by the program entailed $2,399.7 million, of which $1,069.7 was to be granted in 2003–2010. However, the final total figure to be granted over the years 2003–2004, approved by the European Commission, was $1,436.5 million (see Appendix B).

79. For example, out of CONEL customers' total debts of ROL 4,108 billion in 1999 ($275 million, using the exchange rate on April 15, 1999, including ROL 1,281 billion in interest [$85.8 million]), Sidex's debts reached ROL 707 billion ($47.4 million); those of COS Târgovişte, ROL 295 billion ($19.8 million); and Siderurgica Hunedoara's debts stood at ROL 250 billion ($16.7 million). See *Mediafax*, "CONEL va reduce cantitatea de energie electrică furnizată către Sidex Galaţi," Bucharest, May 3, 1999.

80. Interview with a COS Târgovişte union leader, Reşiţa, May 2004 (translated by the author).

81. Ibid.

82. *Mediafax*, "Combinatul siderurgic Sidex Galaţi a ocupat primul loc," Bucharest, October 14, 1998; *Mediafax*, "SNCFR conduce în topul marilor datornici la bugetul asigurărilor sociale de stat," Bucharest, August 10, 1998.

83. *Mediafax*, "Banca de investiţii Flemings acordă consultanţă FPS în vederea privatizării Sidex," Bucharest, November 19, 1998.

84. *Mediafax*, "Preşedintele FPS recomendă conducerii Sidex să utilizeze banii din depozitul constituit de FPS la BCR pentru a achita salariile restante," Bucharest, March 3, 1999 (translated by the author).

85. *Mediafax*, "Fostul preşedinte al FPS susţine că Sidex Galaţi este controlată de firmele căpuşă ale foştilor ofiţeri ai serviciilor de informaţii," Cluj-Napoca, May 4, 2000.

86. For example, during an Ordinary General Shareholders Assembly in 1999, "the representatives of the State Ownership Fund proposed that the penalties arising from tax-payment default be included in the loss account. Such penalties amount to 2,237 billion, while the share capital is 2,586 billion lei. The proposal was rejected, since otherwise the conditions for commencement of liquidation proceedings would have been met." See "48 billion lei losses last year for Sidex-Galati," 1999, www.romanian-daily.ro/ARHTR2/RED101.html (accessed on April 20, 2006).

87. *Mediafax*, "Oferta de privatizare pentru 1998 cuprinde un număr de 2,745 societăţi comerciale," Bucharest, February 25, 1998.

88. *Mediafax*, "FPS vinde 90 de întreprinderi prin intermediul băncilor străine de investiţii," Bucharest, July 21, 1998.

89. The table also notes problems connected to post-privatization labor conflict, which will be discussed in chapter 5.

90. *Mediafax*, "Departamentul de Control al Guvernului propune FPS analizarea modului de respectare a contractului de privatizare a societăţii Tepro de către Zelezarny Veseli," Bucharest, July 20, 1999; *Mediafax*, "Liderul grupului parlementar din Senat, Dan Vasiliu, senator de Iaşi, a cerut, luni, demisia preşedintelui FPS Radu Sîrbu, pentru privatizarea în condiţii 'neclare' a societăţii Tepro din Iaşi," Bucharest, June 7, 1999.

91. *Mediafax*, "Preşidentele FPS consideră privatizarea Tepro Iaşi perfect legală," Bucharest, July 21, 1999; *Mediafax*, "FPS a decis să acţioneze în instanţă firma Zelezarny Veseli în vederea rezoluţionării contractului de privatizare a societăţii Tepro Iaşi," Bucharest, May 10, 2000.

92. UniRomSider expert, interview (May 2004). These allegations were made in a 2001 report issued by the government's Office of Control and Anti-Corruption. See *Mediafax*, "Mai mult de jumătate din societăţile comerciale privatizate de fostul FPS au înregistrat rezultate economice mai slabe după privatizare," Bucharest, March 2, 2001.

93. Florin Ghetau and Razvan Voican, "The Czechs of Tepro Iasi Get Ready to Pack," *Ziarul Financiar*, June 28, 2000.

94. *Mediafax*, "FPS a decis să acţioneze în instanţă firma Zelezarny Veseli în vederea rezoluţionării contractului de privatizare a societăţii Tepro Iaşi," Bucharest, May 10, 2000.

95. *Mediafax*, "Directorul firmei Zelezarny Veseli respinge acuzaţiile privind implicarea sa în asasinarea liderului sindical de la Tepro Iaşi," Bucharest, September 19, 2000; Ruxandra Popescu, "Fostul director al Tepro Iaşi Victor Bălan a ajuns de la balamuc, după graţii," *Adevărul*, August 10, 2005. *Mediafax*, "Reprezentanţii FPS

şi cei ai firmei Zelezarny Veseli vor discuta posibilitatea rezilierii contractului de privatizare a societăţii Tepro Iaşi," September 12, 2000.

96. *Mediafax*, "Curtea Supremă de Justiţie a anulat contractul de privatizare a Tepro Iaşi îcheiat între FPS şi Zelezarny Veseli," Bucharest, October 20, 2000. The attorney general prepared an appeal against that decision, presumably to defend against allegations about the insecurity of property rights in Romania, but he later backed out of this position. See *Mediafax*, "Procurolul general al României a promovat recurs în anulare împotriva deciziei judecătoreşri de anulare a contractului de privatizare a Tepro Iaşi," Bucharest, May 7, 2001; *Mediafax*, "Contractul de vînzare-cumpărare a pachetului majoritar al Tepro Iaşi este nul," Bucharest, October 8, 2001.

97. *Mediafax*, "Senatorul PDSR de Neamţ Radu Timofte l-ar putea acţiona în judecată pe Radu Sîrbu care l-acuzat de implicare în protestul angajaţilor de la Petrotub," Piatra Neamţ, November 11, 1999.

98. *Mediafax*, "Comisia pentru industrii a Camerei Deputaţilor îi cere preşedintelui FPS să prezinte contractile de privatizare încheiate pentru societăţile Petrotub Roman şi Silcotub Zalău," Bucharest, Novemebr 17, 1999; *Mediafax*, "Derularea procesului de privatizare a societăţii Petrotub Roman va fi analizată de Guvern," Bucharest, November 17, 1999.

99. *Mediafax*, "Laminoarele societăţii Petrotub SA Roman au fost închise din cauza lipsei de comenzi," Piatra Neamţ, February 4, 2000.

100. *Mediafax*, "Consiliul Concurenţeu a anulat privatizarea societăţii Petrotub SA Roman cu firma britanică Tubman International," Piatra Neamţ, June 15, 2000.

101. *Mediafax*, "Petrotub Roman ar putea fi liquidată dacă va fi transferată la FPS," Bucharest, June 26, 2000.

102. *Mediafax*, "Deputatul PD Cristian Rădulescu consideră îndreptăţită moţinea simplă depusă de PDSR privind activitatea FPS," Bucharest May 17, 2000.

103. *Mediafax*, "Societatea Noble Ventures a preluat efectiv controlul Combinatului Siderurgic Reşiţa," Reşiţa, September 8, 2000.

104. *Mediafax*, "This Year's Largest American Investment Was Inaugurated in Reşiţa," Reşiţa, October 4, 2000.

105. Dandea, interview.

106. UniRomSider expert, interview.

107. According to the privatization contract, Noble Ventures agreed to make a $6 million capital infusion by the end of December 2000, pay an installment of a state-guaranteed $35 million loan to a Spanish bank consortium by mid-October 2000, and to start the $26 million investments expected in the first year after the take-over. *Mediafax*, "Angajaţii Combinatului Siderurgic Reşiţa continuă mitingul de protest," Reşiţa, January 15, 2001.

108. See, for example, *Mediafax*, "Combinatul Siderurgic Reşiţa a fost rebranşat la sistemul energetic," Reşiţa, February 4, 2001; *Mediafax*, "Combinatul Siderurgic Reşiţa a fost din nou debranşat de la reţeaua de energie electrică," Reşiţa, February 8, 2001.

109. *Mediafax*, "Ovidiu Muşetescu: Executivul nu va acorda garanţii de stat pentru creditul solicitat de CSR," Bucharest, February 5, 2002.

110. *Mediafax*, "Guvernul a aprobat eşalonarea datoriilor către stat şi ştergerea

penalităţilor şi a majorărilor de întîrziere pentru Combinatul Siderurgic Reşiţa," Bucharest, May 30, 2001; *Mediafax*, "Contractul de privatizare a combinatului de la Reşiţa va fi rezoluţionat, dacă partenerii nu îşi respectă angajamentele, afirmă premierul Adrian Năstase," Bucharest, June 7, 2001.

111. *Mediafax*, "Ovidiu Muşetescu despre investiţia anunţată recent de Noble Ventures la CSR: 'Pentru cine e oxigenul, pentru muncitori sau pentru oţel?'" Reşiţa, June 15, 2001; *Mediafax*, "Contractul de privatizare a CSR a fost desfiinţat," January 15, 2003; *Mediafax*, "Ovidiu Muşetescu: Statul se poate angaja în luarea unor măsuri de asistenţă financiară la CSR," Bucharest, August 21, 2002.

112. International Centre for Settlement of Investment Disputes, Washington, DC. In the proceedings between Noble Ventures, Inc. (claimant) and Romania (respondent), ICSID Case No. ARB/01/11, October 12, 2005.

113. Sorin Elisei, "Rusi de la Conares ar vrea si Otelu Rosu," *Ziarul Financiar*, November 19, 2002.

114. *Mediafax*, "Combinatul metalurgic din Oţelu Roşu nu-şi poate relua activitatea," Reşiţa, January 12, 2001.

115. *Mediafax*, "Guvernul va încercasă creeze 'un instrument' care să permită statului să intervină în cazul în care clauzele contractelor de privatizare nu sînt respectate," Bucharest, February 1, 2001.

116. *Mediafax*, "Gavazzi Steel SA Oţelu Roşu va beneficia de eşalonarea obligaţiilor bugetare restante pe o perioadă de cinci ani," Bucharest, July 12, 2001; *Mediafax*, "Situaţia de la Oţelu Roşu este urmarea nerespectării obligaţiilor investiţionale, afirmă APAPS," Bucharest, July 14, 2002.

117. *Ziarul Financiar*, "Activele Gavazzi Steel, scoase la licitatie," January 18, 2005.

118. VSŽ became an active participant in the privatization process by creating or co-creating four investment funds, which gave it shares in as many as 126 enterprises. Marían Hutňan, "Chronológia krízy vo VSŽ, kedysi vlajkovej lode Slovenska (I)," *Hospodarské noviny*, August 26, 1999.

119. *Trend*, "Desať rokov VSŽ," October 11, 2000.

120. Róbert Matejovič, "VSŽ a ich príbuzní," *Slovenský Profit*, December 17, 1996.

121. Tácha, "VSŽ: Manažeři z východu."

122. *Hospodarské noviny*, "Ivan Lacko: Pod výrazný pokles zisku VSŽ sa podpísala najmä chaotická obchodná politika kombinátu," October 22, 1997; Marcinčin, "Reštrukturalizácia podnikov," 331.

123. Hutňan, "Chronológia krízy . . . (II)," August 27, 1999.

124. Marcinčin, "Reštrukturalizácia podnikov," 333.

125. *Trend*, "VSŽ Košice získali syndikátny úver 125 miliónov dolárov," April 3, 1996. The syndicated loan by ING Bank N.V. involved eighteen different banks and was the first such loan given to a private entity without government guarantees. It was to be used to restructure VSŽ's existing debt by changing its structure from short term to long term, as well as for new investments.

126. Peter Laca, "VSŽ Sacks Leaders, Rezeš and Son Take Over," *The Slovak Spectator*, March 12, 1998.

127. *Hospodarské noviny*, "Chronológia spolupráce U.S. Steel Group a VSŽ, a.s.," June 9–11, 2000.

128. Kopanic, "Stealing the Eastern Slovak Steelworks."

129. Marcinčin, "Reštrukturalizácia podnikov," 332; *Financial Times*, "Banks aim for VSZ $450m debt standstill," December 28, 1998.

130. *Podbrezovan*, "Výročná Správa akciovej spoločnosti Železiarne Podbrezová," Nr 11/93; interview with the Podbrezová Steelworks spokesperson, Podbrezová, April 2004.

131. Podbrezová Steelworks spokesperson, interview.

132. Železiarne Podbrezová, a.s. 1992 *Annual Report* and 2005 *Annual Report*.

133. Podbrezová Steelworks spokesperson, interview. Železiarne Podbrezová, a.s. 2010 *Annual Report*, http://wwwlds.zelpo.sk/e-brochure/zelpo/rocna_sprava/2010/en/index.html (accessed October 30, 2014).

134. Podbrezová Steelworks spokesperson, interview; interview with *Priority* editor, Banská Bystrica, Slovakia, April 2004.

135. *Priority* editor, interview.

136. In 1989, the enterprise hired about 5,000 workers. Jacek Bołdok, "Zamiast wesołego miasteczka," *Polityka*, May 4, 1996, 59–60.

137. In fact, economists were surprised by how well the SOEs could adapt to the new economic reality. This showed that it was the hard budget constraint and not ownership type that drove economic performance. See Brian Pinto, Marek Belka, and Stefan Krajewski, "Transforming State Enterprises in Poland: Evidence on Adjustment by Manufacturing Firms," *Brookings Papers on Economic Activity* 24 (1993): 213–270.

138. It was a rare occurrence in the region that the local government and population were not supportive of the local steelworks and can probably be explained by the job-generating rapid development of Warsaw. In fact, the local council even floated the idea of liquidating the enterprise and turning its grounds into an amusement park. See Bołdok, "Zamiast wesołego miasteczka."

139. In the third quarter of 1991, the workers voted to accept a temporary 30% salary cut to help overcome the enterprise's financial catastrophe. Ibid.; interview with the Huta L.W. (formerly, Warsaw Steelworks) spokesperson, Warsaw, October 2003.

140. Capital-Accounting Agency (Agencja Kapitałowo-Rozliczeniowa, S.A.) was supposed to arrange for the payment of debts and/or reach agreement with the lenders, as well as manage or liquidate the unused capital. See Parliamentary statement of Andrzej Żebrowski, Undersecretary of State in the Ministry of Ownership Transformation, February 4, 1994. Posiedzenia Sejmu, sprawozdania stenograficzne. 2 kadencja, 11 posiedzenie, 2 dzień (February 4, 1994).

141. Huta L.W. spokesperson, interview; interview with the Solidarity trade union leader at Huta L.W. (formerly, Warsaw Steelworks), Warsaw, October 2003.

142. Severo Bocchio, "Historia straconych okazji, przyszłość pełna nadziei," *Magazyn Huty L.W.*, September 2002.

143. Marek Pol, the minister of industry and trade, during parliamentary debate on the Warsaw Steelworks strike: Posiedzenia Sejmu, sprawozdania stenograficzne. 2 kadencja, 25 posiedzenie, 2 dzień (July 7, 1994); 17 punkt porządku dziennego: Informacja rządu o sytuacji strajkowej w Hucie Lucchini-Warszawa. Also speech by Wiesław Kaczmarek, the minister of ownership transformation, same source.

144. Najwyższa Izba Kontroli, "Informacja o wynikach kontroli," 62. Also con-

firmed by Jan Macieja in "Restrukturyzacja hutnictwa żelaza i stali," in *Studia nad restrukturyzacją sektorów przemysłowych w Polsce*, ed. Adam Lipowski and Jan Macieja (Warsaw: Polish Academy of Sciences, Institute of Economics, 1998), 72. The guarantees were for up to 60% of the value of the loan.

145. This probably explains why the European Commission took 1997 as the starting point for calculating state aid accorded to the sector.

146. Jędrzej Bielecki, "Półtora roku na prywatyzację," *Rzeczpospolita*, July 25, 2000.

147. Disregarding the better technological state of VSŽ, it is important to keep in mind that Slovak crude steel production for that year was 41% of the Polish output. Moreover, let us recall that the Romanian write-off of tax arrears and penalties for late payments for 1993–2002 was equal to nearly $600 million. According to an audit by the Polish Supreme Audit Office, state aid for the years 1998–2001 was evaluated at $209 million (PLN 849.7 million). However, tax and social-security-contribution arrears represented only about 27% of that sum (12% and 15%, respectively). Far greater sums were expended on employment restructuring and environmental protection categories. Moreover, no social insurance payments and only 18% of tax arrears were forgiven—a total of $4.5 million (PLN 18.3 million).

148. See Cieszewska, "Walcownia zamknięta na kłódkę"; Witold Gadomski, "Z tego pieca." The state-of-the-art rolling mill was eventually purchased, in 2004, by the German BGH Edelstahl for PLN 27 million. Its construction back in 1998 cost PLN 220 million. *Rzeczpospolita*, Dział Ekonomiczny: Firmy, April 3, 2004; Polish Press Agency, "Syndyk czeka na oferty na Baildon," *Puls Biznesu*, July 21, 2003.

149. For an overview, see Maciej Bałtowski, *Przekształcenia Własnościowe Przedsiębiorstw Państwowych* (Warsaw: Wydawnictwo Naukowe PWN, 2002), 109–111, and Anna Krajewska, "Efektywność Restrukturyzacji Finansowej Przedsiębiorstw. Aspekty Oddłużeniowe" in *Restrukturyzacja Przedsiębiorstw w Procesie Tranformacji Gospodarki Polskiej*, ed. Elżbieta Mączyńska (Warsaw: Wydawnictwo DiG, 2001), 317; Ewa Zychowicz, "Metoda dla zadłużonych kolosów," *Rzeczpospolita*, March 23, 1998.

150. Najwyższa Izba Kontroli, "Informacja o wynikach kontroli."

151. Witold Gadomski, "Stal się leje, moc truchleje," *Gazeta Wyborcza* (Warsaw edition), December 12, 1997.

152. Krzysztof Mering, "Najpierw prywatyzacja," *Nowe Życie Gospodarcze*, August 2, 1998; Jarosław Zwoliński, "Huta Katowice zmieniła oblicze," *Nowe Życie Gospodarcze*, March 12, 2000.

153. As will be apparent from the ensuing discussion of employment restructuring in the sector, many of these workers found jobs in the spun-off enterprises rather than leaving the industry altogether.

154. Zwoliński, "Huta Katowice zmieniła oblicze," 43.

155. Witold Gadomski, "Cena przetrwania," *Gazeta Wyborcza*, December 22, 1999.

156. Jerzy Sadecki, "Huta im. Sendzimira układa się z wierzycielami," *Rzeczpospolita*, September 12, 2001.

157. Najwyższa Izba Kontroli, "Informacja o wynikach kontroli," 35; Irena Dryll, "Dramat polskiej stali," *Nowe Życie Gospodarcze*, April 22, 2001, 39.

158. Katarzyna Jędrzejewska, "Zaciągane, potem umarzane," *Rzeczpospolita*, December 22, 2000.

159. Nowak, interview.

160. Civil servant at the Ministry of Industry and Trade, interview.

161. Ministry of Industry and Trade of the Czech Republic and Eurostrategy Consultants, "Restructuring of the Steel Industry in the Czech Republic: Industry Restructuring Plan," Prague, May 2001, 56.

162. Myant, *Rise and Fall*, 197; *Hospodářské noviny*, "Případ Poldi Ocel už projednává obchodní soud," March 12, 1996.

163. *Hospodářské noviny*, "Kladenské Ušlechtilé a Konstrukční oceli jsou v konkursu," July 25, 1997. In a very controversial move, within six months of the commencement of bankruptcy proceedings, Stehlík created a new company, Poldi Steel, into which he transferred assets of Poldi Ocel and its three subsidiaries: Poldi Foundry (Huť Poldi), Poldi Specialty Steel (Ušlechtilé oceli Poldi) and Poldi Construction Steel (Konstrukční oceli Poldi). See *Ekonom*, "Rozparcelovaná POLDI," May 25, 2000, 12; Myant, *Rise and Fall*, 198.

164. The sale of Poldi Foundry and Poldi Specialty Steel to Ferra, a company owned by domestic capitalist Zdenek Zemek of Z-Group was unsuccessful and production ground to a halt. The assets were eventually sold to a German company, which restarted production, albeit in a much scaled-down form. *Ekonom*, "Rozparcelovaná POLDI"; Zuzana Kubátová and Tereza Zavadilová, "Hutě tě nevolají," *Týden*, August 16, 1999; *Hospodářské noviny*, "Poldovka hledá nového majitele," July 29, 1999; *Svět hospodaářství*, "Komu patříš Poldinko," December 1, 1999.

165. *Ekonom*, "Rozparcelovaná POLDI"; Pavel Šmíd, "Kladenskou válcovnu získají Třinecké železárny," *Hospodářské noviny*, February 27, 2002.

166. Pavel Šmíd, "Moravia Steel mění vlastnickou strukturu," *Hospodářské noviny*, May 30, 2002.

167. Jaroslav Kmenta, "Bývalí ministři ODS jsou podezřelí z korupce," *Mladá fronta Dnes*, January 18, 1999; Jiří Leschtina and Jiří Kubík, "Všechno jsme dělali metodou zkoušky a omylu," *Mladá fronta Dnes*, June 17, 2000. The enterprise was originally supposed to cover environmental liability using its own means. For backing out of the investments plans, see *Severočeské noviny*, "Měla ODS prsty v privatizaci Třineckých železáren?" November 29, 1997.

168. Jakub Vosátka, "Pomohou nám Američané?" *Ekonom*, April 6, 2000.

169. Civil servant at the Ministry of Industry and Trade, interview.

170. Jaroslav Baďura and Petr Žižka, "Vítkovice, dříve všemocný gigant, jsou na kolenou," *Mladá fronta Dnes*, January 26, 2000; Marcela Šperkerová, "Dostali šanci," *Euro*, November 6, 2000.

171. Kubátová and Zavadilová, "Hutě tě nevolají."

172. Marek Pražák, "Hutě platily velké sumy za rady, které zřejmě nikdy nedostaly," *Mladá fronta Dnes*, May 9, 2001.

173. Interview with Jiří Havel, former FNM chairman, Prague, March 2004.

174. Havel, interview; Marcela Šperkerová and Pavel Páral, "Čistotný Novák," *Euro*, February 26, 2001.

175. Steel Federation representatives, interviews (2003 and 2004); Břetislav Meca, "Osinek zahájil," *Magazin Vítkovice*, April 27, 2000.

176. Gabriel Beer, "České huty neskoro reagujú na oživenie trhu a pýtajú štátnu pomoc," *Trend*, August 30, 2000.

177. *Magazín Vítkovice*, "Konkurs už nehrozí," September 29, 2001; Mar-

cela Šperkerová, "Dostali šanci"; Anna Bortlíčková, "Soud potvrdil Vítkovicím vyrovnání," *Právo*, August 7, 2001.

178. Marek Pražák and Pavla Nováková, "Vítkovice Steel zřejmě získa stat," *Mladá fronta Dnes*, March 28, 2002.

179. Havel, interview; Steel Federation representatives, interview; *Nová Huť*, "Nová Huť očekává ztrátu za rok 1999 ve výši 2.9 mld. Kč," March 14, 2000; Libuše Bautzová, "NH: Zablokovaná partie," *Ekonom*, August 3, 2000.

180. International Iron and Steel Institute, *Steel Statistical Yearbook*, 1991–1994, 2005 (Brussels: IISI).

181. Hutnictví Železa, a.s., *Hutnická Ročenka 1994* (Prague: Ocelot, 1994); International Iron and Steel Institute, *Steel Statistical Yearbook 2005*, 23.

CHAPTER 5

1. See John Williamson, ed., *The Political Economy of Policy Reform* (Washington, DC: Institute for International Economics, 1994); Joan Nelson, ed., *Fragile Coalitions: The Politics of Economic Adjustment* (New Brunswick, NJ: Transaction, 1989).

2. Interview with Metarom leader, Bucharest, June 2004; interview with Petru Dandea, Cartel Alfa vice president, Bucharest, June 2004. By contrast with Cartel Alfa, CNSLR-Frația, the successor to the communist-era union organization, was closely associated with the communist successor Social Democratic Party (PSD). See David A. Kideckel, "Winning the Battles, Losing the War: Contradictions of Romanian Labor in the Postcommunist Transformation," in *Workers after Workers' States*, ed. Stephen Crowley and David Ost (Lanham, MD: Rowman and Littlefield, 2001), 105.

3. Metarom leader, interview.

4. Dandea, interview, quoted in Aleksandra Sznajder Lee, "Between Apprehension and Support: Social Dialogue, Democracy, and Industrial Restructuring in Central and Eastern Europe," *Studies in Comparative International Development* 45 (2010): 48.

5. Metarom leader, interview; interview with civil servant at the Ministry of Work, Social Solidarity and Family, Bucharest, May 2004.

6. Interview with COST union leader, Reşiţa, May 2004; Dandea, interview; Metarom leader, interview.

7. Metarom leader, interview.

8. Serghei Mesaros, "The Evolving Structure of Collective Bargaining in Europe 1990–2004. National Report. Romania," University of Florence/European Commission, 2005. The initial law was Law no. 13/1991, which was replaced by Law no. 130/1996.

9. Civil servant at the Ministry of Work, Social Solidarity and Family, interview.

10. Dandea, interview, quoted in Sznajder Lee, "Between Apprehension and Support," 48.

11. Interview with Silcotub union leader, Reşiţa, May 2004, quoted in Sznajder Lee, "Between Apprehension and Support," 49.

12. COST union leader, interview.

13. UniRomSider represents employers in the iron and steel industry and is part

of the Metalurgia Employers' Federation, which also represents steel pipe manufacturers (Unitub) and nonferrous metal industry manufacturers (Argos). http://www.conpirom.ro/membri.php?id=1 (accessed November 1, 2014).

14. The severance payments were provided for in the (non-steel-sector-specific) Government Ordinance of 1997 (Guvenul României, "Ordonanță de urgență nr. 9 din 14/04/1997").

15. *Mediafax*, "Federația Metarom cere, printr-o campanie de proteste, ca restructurarea metalurgiei să fie accelerată," Bucharest, July 20, 1998.

16. The two relevant documents were the government decree of 1999 (Guvenul României, "Ordonanță de Urgență [nr.146] privind unele măsuri de acompaniament social pentru salarații din industria metalurgică ale căror contracte individuale de muncă vor fi desfăcute ca urmare a concedierilor collective," Monitorul Oficial al României, Partea I, Nr. 281/5/X/1999) and the 2001 law approving the government decree (Parlamentul României, Lege nr. 145 din 3 aprilie 2001 pentru aprobarea Ordonanței de urgență a Guvernului nr. 146/1999); interview with Petru Dandea, Bucharest, July 2010.

17. *Info Metal International* (Bucharest: Metarom, 2003), 16.

18. *Mediafax*, "Sindicaliştii din industria siderurgică vor declanşa acțiuni de protest față de privatizarea anumitor societăți din acest sector," Bucharest, October 23, 2000.

19. *Mediafax*, "Tepro Iaşi ar putea intra în faliment dacă activitatea societății nu va fi reluată, spune preşedintele societății," Bucharest, June 24, 1999.

20. In July 2000, a district court decided that Tepro had unlawfully fired 593 workers, who were subsequently supposed to be reinstated with compensation for the thirteen months that had passed since the layoff. See *Mediafax*, "Revista Presei—Cu doar cinci săptămîni înainte de scadență, 'Fierarii Veseli' investesc 150,000 de dolari la Tepro Iaşi, titrează 'Adevărul,'" July 25, 2000.

21. *Mediafax*, "Radu Sîrbu speră că, în urma îtîlnirii cu parlamentarii, problemele referitoare la situația Tepro Iaşi se vor clarifica," Bucharest, June 18, 1999.

22. *Mediafax*, "Senatorul PD de Iaşi Dan Vasiliu cere FPS să rezilieze, dacă este cazul, contractul de vînzare a societății Tepro," Bucharest, May 8, 2000.

23. Interview with UniRomSider expert, Bucharest, May 2004. Suspicions were quickly directed at the Czech owner, who denied the allegations. According to media reports of the subsequent criminal investigation, however, the decision to murder Săhleanu was taken during an August 2000 meeting at which Zemek was present. Although Romanian participants in the crime, as well as a Czech accomplice, were given up to twenty-three-year prison terms, Zemek did not bear legal responsibility for the crime. *Mediafax*, "Directorul firmei Zelezarny Veseli respinge acuzațiile privind implicarea sa în asasinarea liderului sindical de la Tepro Iaşi," Bucharest, September 19, 2000; Ruxandra Popescu, "Fostul director al Tepro Iaşi Victor Bălan a ajuns de la balamuc, după grații," *Adevărul*, August 10, 2005.

24. *Mediafax*, "Petrotub Roman a înregistrat, în primele nouă luni ale anului, pierderi de 99.2 miliarde lei şi datorii de 670 miliarde de lei," Bucharest, November 11, 1999.

25. *Mediafax*, "Singura şansă pentru Petrotub Roman este includerea acesteia într-un grup cu acoperire mondială, spune preşedintele FPS," Bucharest, November 11, 1999.

26. *Mediafax*, "Conducerea societății Petrotub a desfăcut contractul de muncă al liderului de sindicat, Marius Cornenco," Piatra Neamț, January 20, 2000.

27. In part, the union's militancy can be explained by the initial plans to close the enterprise. The union took it upon itself to defend the steelworks against closure. In fact, after the protest meetings of 1994, involving thousands of workers and local inhabitants of this monoindustrial town, Reşiţa Steelworks obtained a government-guaranteed $60 million loan for the installation of an electric furnace with an excessive production capacity of 100,000 tons, as opposed to 50,000 tons, which would have been more appropriate given Romania's market conditions. Metarom leader, interview; *Mediafax*, "Combinatul Siderurgic Reşiţa scoate la vînzare 51 la sută din capitalul social," Bucharest, May 28, 1997.

28. UniRomSider expert, interview.

29. Interview with the Reşiţa union leader, Reşiţa, May 2004.

30. According to an interview with the Reşiţa Steelworks union leader, the unionists found out through the internet that Noble Ventures, Inc., was a consulting company, and they were afraid that it did not have the financial resources to invest in the steelworks.

31. Interview with a group of Reşiţa Steelworks workers, some of whom participated in the hunger strike against Noble Ventures, Reşiţa, May 2004.

32. *Mediafax*, "Angajaţii Combinatului Siderurgic Reşiţa continuă mitingul de protest," Reşiţa, January 15, 2001.

33. Reşiţa union leader, interview; ibid.

34. Constantin Mănăilă, "VATRA Reşiţa după greva foamei," *InfoMetal*, Nr. 48, July 2001; *Mediafax*, "Directorul general al Combinatului Siderurgic Reşiţa intenţionează să-l acţioneze în instanţă pe liderul sindical Iancu Muhu," Reşiţa, April 12, 2001.

35. *Mediafax*, "Sindicaliştii de la Reşiţa şi Oţelu Roşu au cerut Autorităţii pentru Privatizare să trimită o echipă de control la cele două combinate siderurgice," January 16, 2001; *Mediafax*, "Circa 1.000 de angajaţi de la Reşiţa şi Oţelu Roşu au blocat circulaţia la Caransebeş," Caransebeş, August 23, 2002; *Mediafax*, "Un număr de 30 de angajaţi de la Gavazzi Steel Oţelu Roşu au intrat în greva foamei," Reşiţa, July 8, 2002.

36. *Mediafax*, "În prima jumătate a anului FPS a privatizat 1,106 societăţi comerciale," Bucharest, July 28, 1999.

37. Silcotub union leader, interview.

38. *Mediafax*, "Programul de restructurare a societăţii Silcotub Zalău prevede disponibilizarea a 257 angajaţi," Bucharest, March 19, 1999.

39. *Mediafax*, "Premerul Radu Vasile a cerut investitorilor de la Petrotub să renunţe la disponibilizări, susţin sindicaliştii de la Metarom," Bucharest, November 23, 1999.

40. Ibid.

41. For example, over the years 2001–2002, the enterprise invested $15 million to develop three new production lines. *Mediafax*, "Silcotub Zalău va investi 15 milioane dolari pentru dezvoltarea capacităţii de producţie," May 29, 2001.

42. *Mediafax*, "FPS a vîndut firmei Mixasset din Anglia 58% din acţiunile societăţii comerciale Artrom Slatina," Bucharest, March 31, 1998.

43. *Mediafax*, "Artrom Slatina a înregistrat, în primele nouă luni ale anului, o pierdere de 27,6 miliarde lei," Bucharest, November 13, 1998.

44. Interview with the Artrom Slatina union leader, Reșița, May 2004 (translated by the author).

45. *Mediafax*, "O firmă austriacă a depus o ofertă pentru preluarea pachetului de acțiuni deținut de FPS la Artrom Slatina," January 21, 1999.

46. Artrom Slatina union leader, interview.

47. Ibid.

48. Ibid.

49. According to interviewed union leaders, the wages at Sidex were double the average salary for the branch. Dandea, interview; COST union leader, interview.

50. Dandea, interview.

51. COST union leader, interview.

52. *Mediafax*, "Federația Sindicală a Siderurgiștilor de la Sidex solicită elaborarea unei hotărîri speciale pentru privatizarea Combinatului," October 26, 2000.

53. *Mediafax*, "Liderul sindicaliștilor de la Sidex respinge afirmația lui Radu Sîrbu potrivit căreia sindicaliștii sînt de acord cu conținutul notei de privatizare a combinatului," October 29, 2000.

54. Interview with union leaders at Mittal Steel Sidex, Galați, June 2004.

55. *Mediafax*, "Circa 60 muncitori de la Sidex Galați au blocat intrarea în sala de ședințe a reprezentantului FPS în Adunarea Generală a Acționarilor," November 2, 2000.

56. *Mediafax*, "Ovidiu Mușetescu: Cei care lucrează la combinatul Sidex trebuie să înțeleagă că simpla dorință de a turna oțel nu mai stă în picioare," Bucharest, February 14, 2001.

57. *Economist Intelligence Unit*, "LNM acquires Sidex," August 3, 2001.

58. Dandea, interview; Luminița Chivu, "Representativeness of the European Social Partner Organisations: Steel Industry—Romania," EIRO Briefing Paper, September 2009, www.eurofound.europa.eu/eiro/studies/tn0811027s/ro0811029q.htm (accessed November 1, 2014).

59. Interview with AVAS/APAPS representative, Bucharest, May 2004.

60. Perhaps as a token gesture, in June 2002, the government agreed that the contents of the social package would be a binding part of the privatization contract, under the penalty of canceling the contract, but it withdrew the proposal in response to IMF pressure. Federația Națională Sindicală METAROM, "Raport de Activitate, Conferința a XIII-a, Băile Herculane, 22–24 October 2003," 4–5.

61. Ibid., 3–4; interview with Metarom union leader, Bucharest, May 2004.

62. Civil servant at the Ministry of Work, Social Solidarity and Family, interview.

63. Metarom leader, interview.

64. Reșița union leader, interview.

65. COST union leader, interview.

66. *Mediafax*, "Proprietarul combinatului Sidex a preluat Tepro Iași," Bucharest, May 30, 2003.

67. Dandea, interview.

68. Interview with a top management member at Mittal Steel Galați (now ArcelorMittal Galați), Galați, June 2004.

69. Ibid.

70. *Mediafax*, "A început întocmirea listerol cu disponibilizații de la Siderurgica," Deva, April 6, 2003.

71. Petru Vaidoş, "Siderurgica în faţa privatizării," *InfoMetal*, December 2003; *Mediafax*, "Marş de protest al siderurgiştilor pe străzile Hunedoarei," Deva, October 30, 2003; *Mediafax*, "Întîlnire între sindicalişti din Siderurgica şi reprezentanţii LNM Holding," Deva, Novemebr 3, 2003.

72. Petru Vaidoş, "Siderurgica în faţa privatizării."

73. Dandea, interview. The unions also resented the fact that the tender was drawn up so as to make LNM the only real bidder fitting the description, specifying a five-year experience in steel production and over $1 billion turnover. See Marius Cornenco, "Privatizarea PETROTUB Roman—o şansă sau o eroare?" *InfoMetal*, December 2003.

74. Cornenco, "Privatizarea PETROTUB."

75. COST union leader, interview.

76. Marían Hutňan, "Chronológia krízy vo VSŽ, kedysi vlajkovej lode Slovenska (II)," *Hospodarské noviny*, August 27, 1999.

77. Anton Marcinčin, "Reštrukturalizácia podnikov," in *Hospodárska Politika na Slovensku 1990–1999*, ed. Anton Marcinčin and Miroslav Beblavý (Bratislava: Slovak Foreign Policy Association, 2000), 329.

78. Interview with OZ KOVO leader, Bratislava, April 2004; *Trend*, "Odborári z VSŽ sa podujali oslabiť najsilnejší odborový zväz," November 27, 1996, reprinted in *Priority*, December 16, 1996.

79. Interview with OZ KOVO leadership, Bratislava, April 2004. See a series of articles in *Priority*, the OZ KOVO publication, from December 16, 1996.

80. Interview with Metalurg leader, Košice, April 2004; OZ KOVO leadership, interview; Hutňan, "Chronológia krízy. . . ."

81. The relationship quickly soured, as KOZ claimed that its demands were being ignored by the new government. By 2004, the unions perceived that the government was approaching social dialogue in the same high-handed manner as Mečiar. For early signs of dissatisfaction, see, for example, Alena Gottweisová, "Odbory žiadajú účasť na privatizácii," *Hospodarské noviny*, September 3, 1999; Dana Švecová, "Krik a odborárske analýzy," *Profit*, February 16, 1999.

82. Interview with a VSŽ insider, Košice, April 2004.

83. *Hospodarské noviny*, "Odborári chcú odvolať G. Eichlera z funkcií," August 26, 1999.

84. *Trend*, "Desať rokov VSŽ," October 11, 2000.

85. Vladimír Duduk, "Útek z hrobárovej lopaty," *Profit*, June 9, 2000.

86. *Trend*, "Desať rokov VSŽ."

87. Marián Hutňan, "Boj o Východoslovesnké železiarne vrcholí," *Hospodarské noviny*, September 20, 1999.

88. The official reasons for the motion to fire Eichler were his use of costly consultancies, uncoordinated manner of leading the restructuring process within VSŽ, and failure to act in VSŽ's best interests. *Hospodarské noviny*, "Odborári chcú."

89. *Hospodarské noviny*, "Súčasná akcionarská štruktúra VSŽ Košice," September 22, 1999.

90. Dagmar Truncová, "Byl to recept na krach," *Euro*, September 27, 1999; *Hospodářské noviny*, "Pozice manažerů zesílily, odborář Gruber rezignoval," September 23, 1999.

91. Gabriel Beer, "Hrozba bankrotu vo VSŽ stále nie je zažehnaná," *Trend*, September 29, 1999.

92. VSŽ insider, interview.

93. *Hospodárské noviny*, "Chronológia spolupráce U.S. Steel Group a VSŽ, a.s.," June 9–11, 2000.

94. Reuters Company News, "Slovak VSZ Shareholders Approve U.S. Steel Takeover," October 12, 2000.

95. Interview with a top management member at U.S. Steel Košice, Košice, May 2004; interview with the OZ KOVO union leader at U.S. Steel Košice, Košice, May 2004; Metalurg leader, interview.

96. OZ KOVO leadership, interview; Metalurg leader, interview.

97. Based on nine interviews with current and past union leaders, a representative of employers' association in the sector, and a civil servant at the Ministry of Work and Social Policy, July and October 2003. See also Rafał Towalski, "Związki zawodowe w procesie restrukturyzacji sektora hutnictwa żelaza i stali," in *Związki Zawodowe a Restrukturyzacja—Bariery czy Kompromis*, ed. Leszek Gilejko (Warsaw: SGH, 2003).

98. Interview with a Sendzimir Steelworks insider, Kraków, November 2003; Krzysztof Banasiak, "Solidarni z hutą," *Gazeta Wyborcza*, November 30, 1998.

99. Banasiak, "Solidarni z hutą."

100. Ibid.; interview with Edward Nowak, former Deputy Minister of Economy, Kraków, July 2003.

101. In the third quarter of 1991, the workers voted for a temporary 30% salary cut to help overcome the enterprise's financial catastrophe. Jacek Bołdok, "Zamiast wesołego miasteczka," *Polityka*, May 4, 1996, 59–60.

102. Interview with a Huta L.W. (formerly, Warsaw Steelworks) spokesperson, Warsaw, October 2003.

103. Interview with the former Solidarity leader at Huta L.W. (formerly, Warsaw Steelworks), Katowice, July 2003.

104. Leszek Gilejko, Paweł Gieorgica, and Paweł Ruszkowski. *Społeczni Aktorzy Restrukturyzacji* (Warsaw: Centrum Partnerstwa Społecznego 'Dialog,' 1997), 81–97; Towalski, "Związki zawodowe," 140.

105. Former Solidarity leader at Huta L.W., interview; Huta L.W. spokesperson, interview; also, see Gilejko, Gieorgica, and Ruszkowski, *Społeczni Aktorzy Restrukturyzacji*; and Towalski, "Związki zawodowe." The strike's spectacular nature, and its eventual resolution using high-profile mediation, led to the establishment of excellent social dialogue at the plant level, later emulated in other enterprises. Bołdok, "Zamiast wesołego miasteczka"; confirmed by interviews with a Solidarity leader and a Huta L.W. spokesperson in 2003.

106. Nowak, interview.

107. According to Jan Macieja, wage strikes represented a "significant cause of problems" for the Polish steelworks in the early years of transition, as they increased the costs and uncertainty of delivery. See Jan Macieja, "Restrukturyzacja hutnictwa żelaza i stali," in *Studia nad restrukturyzacją sektorów przemysłowych w Polsce*, ed. Adam Lipowski and Jan Macieja (Warsaw: Polish Academy of Sciences, Institute of Economics, 1998), 61.

108. Towalski, "Związki zawodowe," 138.

109. Interview with a representative of the Steel Sector Employers' Association, Katowice, July 2003. The most famous wage strike occurred in 1994 in Katowice Steelworks and lasted eighteen days. It resulted in enormous losses for the enterprise, including departure of several very valuable clients. The unions were not able to obtain much as far as the benefits were concerned, compared to what the management offered just prior to the beginning of the strike. Towalski, "Związki zawodowe," 139–140; Witold Gadomski, "Stal się leje, moc truchleje," *Gazeta Wyborcza* (Warsaw edition), December 12, 1997.

110. Gadomski, "Stal się leje."

111. Interview with a civil servant at the Ministry of Economy, Labor and Social Policy, Warsaw, July 2003; Mitchell Orenstein, *Out of the Red: Building Capitalism and Democracy in Postcommunist Europe* (Ann Arbor: University of Michigan Press, 2001).

112. Civil servant at the Ministry of Economy, Labor and Social Policy, interview; Towalski, "Związki zawodowe."

113. Towalski, "Związki zawodowe," 135; representative of the Steel Sector Employers' Association, interview.

114. Representative of the Steel Sector Employers' Association, interview.

115. Krzysztof Mering, "Najważniejszy jest czas," *Nowe Życie Gospodarcze*, August 8, 1998, 14–16.

116. Gadomski, "Stal się leje."

117. Interview with Solidarity Metalworkers' Secretariat leader, Katowice, July 2003.

118. Representative of the Steel Sector Employers' Association, interview.

119. Vera Trappmann, *Fallen Heroes in Global Capitalism: Workers and the Restructuring of the Polish Steel Industry* (Houndsmill, UK: Palgrave Macmillan, 2013).

120. Solidarity Metalworkers' Secretariat leader, interview.

121. Multi-Union Committee of PHS, S.A. Fax sent to Piotr Czyżewski, the Minister of State Treasury on June 13, 2003, photocopy.

122. Ibid.

123. Ibid.

124. Although labor power increased following the 1998 victory of the Czech Social Democratic Party, the unions tended to stay out of the political arena and to maintain a strictly social-partner role.

125. Anna Pollert, *Transformation at Work* (London: Sage, 1999); David Ost, "Illusory Corporatism in Eastern Europe: Neoliberal Tripartism and Postcommunist Class Identities," *Politics and Society* 28 (2000): 503–530; Orenstein, *Out of the Red*; Mitchell Orenstein and Henry Hale, "Corporatist Renaissance in Post-communist Central Europe?" in *The Politics of Labor in a Global Age*, ed. Christopher Candland and Rudra Sil (Oxford: Oxford University Press, 2001); Elena Iankova, *Eastern European Capitalism in the Making* (Cambridge: Cambridge University Press, 2002).

126. Interview with OS KOVO leader, Prague, March 2003.

127. OS KOVO leader, interview; interview with representatives at the Steel Federation, Prague, November 2003; interview with a civil servant at the Ministry of Industry and Trade, Prague, December 2003.

128. Interview with Vladimír Dlouhý, former minister of industry and trade, Prague, March 2004.

129. This was in addition to the corruption-tainted sale of Třinecké Železárny in 1996, which resulted in the Civic Democratic Party (ODS) scandal in 1997.

130. The account of the rally came from the Stehlík-controlled newspaper, *Práce* [Work]: "Stehlík: Vlastní vláda likviduje českého podnikatele," May 29, 1996.

131. *Pražské noviny*, "Odboráři Stehlíka nechtějí," December 14, 1996; *Hospodářské noviny*, "Obdoráři vyzvali V. Stehlíka k předání hutí," December 16, 1996; *Hospodářské noviny*, "V kladenské Poldi je stávková pohotovost," February 24, 1997.

132. *Hospodářské noviny*, "Odboráři kladenské Poldi nehodlají bránit generálního ředitele Stehlíka," February 28, 1996.

133. *Hospodářské noviny*, "Odbory se stavějí proti Vladimíru Stehlíkovi, May 23, 1996.

134. *Hospodářské noviny*, "Odboráři kladenské Poldi."

135. OS KOVO leader, interview; civil servant at the Ministry of Industry and Trade, interview.

136. Civil servant at the Ministry of Industry and Trade, interview; Jároslav Marek, "Hra s čísly o zaměstnanosti v Nové huti," *Kovák*, July 2002. Over the three-year period, however, 17,000 workers left the sector; most of them, not covered by the Social Program, subsequently continued working in the spun-off enterprises.

137. Interview with Václav Novák, Prague, June 2008, quoted in Sznajder Lee, "Between Apprehension and Support," 41.

138. Ibid.

139. Pavel Šmíd, "Odboráři Nové huti vyhlásili protestní demonstrace," *Hospodářské noviny*, February 1, 2002. Alena Adámková, "Hutníci hrozí vládě stávkou," *Večerník Praha*, January 22, 2002.

140. *Právo*, "Po výstražné stávce se otřáslo vedení Nové huti," April 28, 2000; Hana Čápová, "Tuneláři neprojdou!" *Respekt*, May 2, 2000.

141. Aleksandra Sznajder Lee and Vera Trappmann, "Von der Avantgarde zu den Verlierern der Transformation: Gewerkschaften im Prozess der Restrukturierung der Stahlindustrie in Mittel- und Osteuropa," *Industrielle Beziehungen* 17 (2010): 170–191.

142. At the national level, 54% of polled workers felt that nobody represented their interests well, compared with 12% who felt that Solidarity did it well and 9% who felt that OPZZ did it well. Leszek Gilejko, *Związki Zawodowe a Restrukturyzacja—Bariery czy Kompromis* (Warsaw: Oficyna Wydawnicza SGH, 2003), 54–55. See also Vera Trappmann, *Fallen Heroes in Global Capitalism*.

143. Aleksandra Sznajder Lee and Vera Trappmann, "Overcoming Postcommunist Labour Weakness: Attritional and Enabling Effects of MNCs in Central and Eastern Europe," *European Journal of Industrial Relations* 20 (2014): 111–127.

CHAPTER 6

1. *Economist Intelligence Unit*, "Romania: Country Briefing," February 3, 2004.

2. World Bank, *Romania Country Brief 2004* (Washington, DC: World Bank, 2004).

3. The State Property Fund (FPS), which was charged with both privatization and enterprise restructuring, became known for extending soft money to the enterprises without requiring concomitant market adjustment. Until 1997, such support was described as "widespread" and, later, as "discrete." Vladimir Pasti, *The Challenges of Transition—Romania in Transition* (Boulder, CO: East European Monographs), 338. See also Dragoş Negrescu, "A Decade of Privatization in Romania," in *Economic Transition in Romania. Proceedings of the Conference "Romania 2000—10 Years of Transition*, ed. Christof Rühl and Daniel Daianu (Bucharest: World Bank and Romanian Center for Economic Policies, 1999), 377. During the initial five years of transition, there were three cases of general bailouts of the enterprises, intended to eliminate the accumulated interenterprise debt. See Pasti, *The Challenges of Transition*, 319. The Romanian record should be contrasted with Poland's 1993 Enterprise and Bank Restructuring Act that placed the Polish banking system on a sound financial footing and avoided the moral hazard problem.

4. Pasti, *The Challenges of Transition*, 338–339.

5. For a discussion of the program, see Simeon Djankov, "The Enterprise Isolation Program in Romania," *Journal of Comparative Economics* 27 (1999): 281–293.

6. Interview with a civil servant at the Authority for State Assets Recovery, Bucharest, June 2004.

7. *Financial Times*, "Fitch IBCA downgrades Slovakia and Romania," December 24, 1998, 22.

8. Ibid.

9. World Bank, "Romania Country Assistance Evaluation" (Report No. 32452, Washington, DC: World Bank, 2005), 11.

10. Other areas included the reform of the banking sector; social security reform; and promotion of the business environment. *Mediafax*, "World Bank: Interview," October 27, 2000; civil servant at the Authority for State Assets Recovery, interview.

11. *Mediafax*, "FPS a ales 64 de societăţi comerciale cu capital de stat care să fie privatizate, restructurate sau lichidate potrivit acordului PSAL," Bucharest, May 27, 1999; World Bank, "Romania Country Assistance Evaluation," 10.

12. *Mediafax*, "Radu Sîrbu: '2000 va fin anul în care vom oferta tot,'" Bucharest, December 21, 1999.

13. *Mediafax*, "Au fost semnate contractele de consultanţă pentru privatizarea a 44 de societăţi comerciale cu capital de stat," Bucharest, March 14, 1999.

14. World Bank, "Romania Country Assistance Evaluation," vi. These were the Private Sector Institution Building Loan, extended in June 1999 ($25 million), and the Private and Public Sector Institution Building Loan, extended in September 2002 ($18.6 million).

15. *Mediafax*, "Autoritatea pentru Privatizare şi-ar putea retrage plîngerea în justiţie împotriva Noble Ventures dacă investitorul American îşi onorează investiţiile asumate," Bucharest, June 28, 2001.

16. World Bank, "Memorandum of the President of the International Bank for Reconstruction and Development and the International Finance Corporation to the Executive Directors on a Country Assistance Strategy for Romania" (Report No. 22180-RO, Washington, DC, May 22, 2001), I (emphasis added).

17. Interview with the European Commission desk officer for Romania, Directorate-General for Enlargement, Bucharest, May, 2004.

18. Interview with UniRomSider expert, Bucharest, May, 2004.

19. Letter from Enrico Grillo Pasquarelli (European Commission) to Petru Ianc (responsible for the preparation of the steel-sector restructuring program at the Ministry of Economy and Trade), dated January 20, 2000.

20. For a discussion of this particular conflict, see *Mediafax*, "Sorin Dimitriu critică programul de restructurare a siderurgiei propus de ministerul Industriei și Comerțului," Bucharest, November 21, 1997.

21. Usinor Consultants, "Asistența Tehnică Pentru Restructurarea Industriei Siderurgice Românești, Raportul Final Nr. 0, Sinteza," Bucharest, July 2000.

22. *Mediafax*, "Ministry of Industry Proposes That Five Iron-and-Steel Plants Be Excluded from PSAL," Bucharest, October 19, 2000.

23. *Mediafax*, "FPS Chairman Does Not Accept Decision of Ministry of Industry for Restructuring Iron and Steel Industry," Bucharest, October 19, 2000.

24. *Mediafax*, "Government Approves Privatization Strategies for Three Iron-and-Steel Plants," Bucharest, October 20, 2000.

25. The FPS retained an almost 87% stake in the enterprise. According to the sale announcement, the remaining shares were distributed among two privatization funds (a total of 11%) and a nearly 2% stake was held by "others." State Ownership Fund, "Advertisement for the Sale of Shares through Negotiation: Sidex S.A," photocopy.

26. *Mediafax*, "Necesarul fondurilor pentru restructurarea industriei metalurgice depășește un miliard și jumătate de dolari," Bucharest, October 24, 1997; *Mediafax*, "Sidex va fi privatizat prin intermediul băncii de investiții Flemings," Bucharest, January 25, 1998; *Mediafax*, "Guvernul va discuta strategia de privatizare a societăților comerciale în 1998," Bucharest, February 24, 1998; *Mediafax*, "Banca de investiții Flemings acordă consultanță FPS în vederea privatizării Sidex," Bucharest, November 19, 1998.

27. *Economist Intelligence Unit*, "Romania Industry: Survey of the Metals Sector," Country Briefing, August 31, 2001.

28. *Mediafax*, "Principalele prevederi ale scrisorii de intenție a guvernului român pentru prelungirea acordului stand-by cu FMI," Bucharest, May 3, 2000.

29. *Mediafax*, "Consortium Including Fieldstone Private Capital Group and Deloitte & Touche to Grant Consulting for Privatizing Sidex Galați," June 29, 2000.

30. *Mediafax*, "Președintele FPS răspunde acuzațiilor ministrului Transporturilor," Suceava, May 29, 2000.

31. *Mediafax*, "FPS va scoate la vînzare 70% din acțiunile combinatului Sidex," Bucharest, November 9, 2000.

32. The Party of Social Democracy of Romania (PDSR) actually won a plurality of the votes in the 2000 elections. In 2001, it merged with the Romanian Social Democratic Party to form the Social Democratic Party (PSD).

33. Interview with the steel-sector privatization adviser to the minister of privatization at the Authority for State Assets Recovery, Bucharest, May 2004.

34. *Mediafax*, "AGA Combinatului Siderurgic Sidex Galați a desemnat noul Consiliu de Administrație al combinatului," Galați, February 9, 2001.

35. *Mediafax*, "Romanian Companies That Want to Buy Sidex Do Not Understand the Attitude of PM," Bucharest, March 15, 2001.

36. Ibid.

37. *Mediafax*, "Usinor Tries to Persuade the Government to Resume Privatization of Sidex Galati," Bucharest, July 18, 2001.

38. The letter caused scandal in Britain. First, it was perceived as an example of influence peddling, since Lakshmi Mittal had previously made a financial contribution to the Labour Party. Secondly, Blair presented the privatization as an effort to support the British steel industry. This attracted biting criticism from the British steel lobby, which represented competitors of the Dutch Antilles-based LNM (later Netherlands-based Mittal Steel) was a competitor. *Economist Intelligence Unit*, "Sidex Scandal Could Have Local Repercussions," April 5, 2002.

39. Ibid.

40. *Economist Intelligence Unit*, "LNM Acquires Sidex," August 3, 2001.

41. Interview with the European Commission desk officer responsible for Romania, Brussels, May 2002.

42. *Mediafax*, "Litigiu între proprietarul COS Tîrgoviște și o firmă de prestări servicii," June 17, 2003. In the case of COS Tîrgoviște, Conares was the only party interested in purchasing the enterprise, with the privatization contract worth $36 million, including $10 million in tax arrears, 14 million dollars'worth of capital investments, and a further $7.3 million in environmental protection investments. *Mediafax*, "Capitalul social al COS Tîrgoviște ar putea crește cu 2.8 milioane dolari," Bucharest, April 3, 2003; *Mediafax*, "Conares Trading alocă peste 36 milioane dolari pentru COS Tîrgoviște," May 22, 2003. In the case of Industria Sârmei Câmpia Turzii, the long products producer specializing in wire production, Conares competed with LNM Holdings for exclusivity in the negotiations. The eventual contract was worth €27.2 million, with €16.5 million to be invested in the enterprise's assets, and €3.3 million to be invested in environmental protection. The share packet was sold for €2.5 million. *Mediafax*, "APAPS a vîndut cinci societăți pentru 37 milioane de euro," Bucharest, March 14, 2003.

43. *Mediafax*, "Grupul LNM, proprietarul Sidex, este singurul oferant pentru Tepro Iași," Bucharest, March 19, 2003.

44. *Mediafax*, "Proprietarul combinatului Sidex a preluat Tepro Iași," Bucharest, May 30, 2003.

45. *Mediafax*, "Petrotub va intra în administrare specială," Bucharest, September 18, 2002.

46. Interview with a top management member at Mittal Steel Galați (former Sidex Galați and current ArcerlorMittal Galați), Galați, June 2004.

47. Ibid.

48. In the case of Siderurgica, LNM was to pay $21 million in commercial debts, $12 million for modernization investments, $4 million for environmental protection, and $5.4 million for working capital. The shares were sold for $1 million. Petrotub's $83 million package entailed $6 million for the shares, $40 million for commercial debts, $18 million for technological investments, and $13 million for environmental investments. *Mediafax*, "LNM Holdings NV a preluat Siderurgica Hunedoara și Petrotub Roman," Bucharest, October 28, 2003.

49. "Situatia privatizarilor societatilor din portfoliul APAPS, incluse in PSAL I si PSAL II—28.05.2004." APAPS mimeo (photocopy), May 28, 2004, 2. The sev-

enth stand-by agreement with the IMF was approved on July 7, 2004. International Monetary Fund, *IMF Country Report No. 04/319* (Washington, DC: IMF, 2004), 34.

50. *Mediafax*, "Datoriile bugetare acumulate de CS Reşiţa vor fi anulate," Bucharest, April 1, 2004; *Mediafax*, "Aproape 20 de societăţi vor efectua concedieri colective," November 18, 2003; steel-sector privatization adviser to the minister of privatization, interview. According to the deal, the new investor was to allocate $10 million for new investments and $3 million for environmental protection, in addition to giving $6 million in working capital and taking on (commercial) debts in the amount of €10 million ($12.5 million). *Mediafax*, "Sinara Handel a preluat de la APAPS Combinatul Siderurgic Reşiţa," Bucharest, February 11, 2004.

51. "Strategia de Restructurare a Industriei Siderurgice din România," approved by H.G. nr. 655/29.04.2004, published in M.O. nr. 425/12.05.2004, 97.

52. Commission of the European Communities, European Commission Staff Working Document, "Programme for Restructuring of the Romanian Steel Industry, Final Assessment," COM (2005) 140 final (April 14, 2005), 9.

53. Ibid.

54. Zdeněk Lukas, "Slovakia: Challenges on the Path towards Integration" (WIIW Research Report No. 261, Vienna Institute for International Economic Studies, no. 261, December 1999), 10.

55. Slovak Government Information Service, "Analysis of the Inherited State of the Economy and Society," Bratislava, 1999.

56. World Bank, "Implementation Completion Report on a Loan in the Amount of Euro 200 Million to the Slovak Republic for an Enterprise and Financial Sector Adjustment Loan" (Report No. 31073, Washington, DC: World Bank, 2004), 5; World Bank, *Slovak Republic: Enterprise and Financial Sector Adjustment Loan Project* (Washington, DC: World Bank, 2000).

57. Slovak Government Information Service, "Analysis of the Inherited State."

58. *Financial Times*, "Fitch IBCA Downgrades."

59. The government actually requested more-stringent tranche release conditions than initially expected, "with a view to use the EFSAL as a vehicle to support the domestic reform momentum." World Bank, *Implementation Completion Report*, 4.

60. *Financial Times*, "Banks Aim for VSZ $450 Million Standstill," December 28, 1998, 18.

61. Interview with Gabriel Eichler, Prague, June 2008.

62. Ibid.

63. Ibid.

64. Ibid.

65. Ibid.

66. Michael Kopanic, "Stealing the Eastern Slovak Steelworks," *Central Europe Review* 2 (2000), http://www.ce-review.org/00/1/kopanic1_steel.html (accessed October 7, 2014).

67. *Trend*, "Štát kúpil súhlas pre U.S. Steel," May 24, 2000.

68. According to the deputy prime minister, Ľubomir Fogaš, the Priemyselná Banka debacle highlighted the government's deficiency as far as decision-making and communication were concerned. *Hospodarské noviny*, "Prezident VSŽ má podporu vlády SR—Mimoriadne rokovanie o Priemyselnej banke," September 22, 1999.

69. Vladimír Duduc, "Ako Penta k miliónom prišla," *Profit,* June 9, 2000.

70. *Hospodarské noviny,* "Prezident R. Schuster sa stretol s J. Rezešom," September 8, 1999.

71. Reuters Company News, "Slovak VSZ Shareholders Approve U.S. Steel Takeover," October 12, 2000.

72. Reuters Company News, "U.S. Steel Takes Over Slovak Steel Mill VSZ," October 12, 2000.

73. Milada Anna Vachudova, *Europe Undivided* (Oxford: Oxford University Press, 2005), 200–202.

74. *Košický Korzár,* "Dispute with US Steel Košice Is Resolved, Mr. Verheugen Claims," March 20, 2004; "Documents concerning the accession of the Czech Republic, the Republic of Estonia, the Republic of Cyprus, the Republic of Latvia, the Republic of Lithuania, the Republic of Hungary, the Republic of Malta, the Republic of Poland, the Republic of Slovenia and the Slovak Republic to the European Union," Annex XIV: List referred to in Article 24 of the Act of Accession: Slovakia, September 23, 2003, 2003 O.J. (L 236), 918.

75. Burcu Eke and Ali M. Kutan, "IMF-Supported Programmes in Transition Economies: Are They Effective?" *Comparative Economic Studies* 47 (2005): 23–40. Poland concluded a standby agreement with the IMF in February 1990, followed by an extended fund facility loan in 1991. Both agreements were suspended. The first suspension was due to a quick rise in unemployment and steep decline of output. The second one took place after both sides realized that the agreed-upon targets were overly optimistic. Poland subsequently signed standby agreements with the IMF in 1993 and in 1994, tied to requests made by Paris and London Clubs in connection with their reduction of the Polish debt. In addition to macroeconomic policies, the last agreement also provided for the speeding up of the mass privatization project.

76. World Bank, "Memorandum of the President of the International Bank for Reconstruction and Development and the International Finance Corporation to the Executive Directors on a Country Assistance Strategy for the Republic of Poland" (Report No. 24783-POL, November 13, 2002, Washington, DC: World Bank, 2002), ii.

77. In 1997, the Polish government negotiated retaining the 9%, instead of the 6%, duty on the EU steel products. In 1998, instead of the uniform 3% rate, a 6% duty was in place for the first half of the year, followed by 4% in the second. Finally, the 3% rate was retained in 1999 when trade in steel products was to be liberalized fully. Andrzej Muńko, "Ocena i Stanowisko Unii Europejskiej w Odniesieniu do Procesów Restrukturyzacji Przedsiębiorstw w Polskim Przemyśle Hutnictwa Żelaza i Stali (1990–1998)" (Warsaw, INE PAN Working Paper No. 13, 1999), 34–36; Jerzy Sadecki, "Będą negocjacje z Unią Europejską," *Rzeczpospolita,* June 21, 1996; Jędrzej Bielecki, "Polska odmawia otwarcia rynku hutnictwa," *Rzeczpospolita,* November 15, 1997; Jacek Safuta, "Stalą w stal," *Polityka,* October 21, 1998; Jędrzej Bielecki, "Coraz dłużej, coraz trudniej," *Rzeczpospolita,* February 4, 1999.

78. Barbara Cieszewska, "Program restrukturyzacji powinien być wspólny," *Rzeczpospolita,* January 7, 1998.

79. Jędrzej Bielecki, "Pytania państw Unii," *Rzeczpospolita,* July 7, 1998.

80. Artur Morka, "Janusz Steinhoff, minister gospodarki: To jest prawdziwy test dla rządu," *Rzeczpospolita*, July 6, 1998.

81. Katarzyna Jędrzejewska, "Zaciągane, potem umarzane," *Rzeczpospolita*, December 22, 2000.

82. Interview with Edward Nowak, Kraków, July 2003.

83. Interviews at the Ministry of Economy, Warsaw, July 2003, and the Ministry of State Treasury, Warsaw, July 2003, as well as with several managers at Florian Steelworks, Swiętochłowice, July 2003.

84. Barbara Cieszewska, "Czarny scenariusz," *Rzeczpospolita*, September 11, 2001.

85. Council of Ministers of the Republic of Poland, *Program Restrukturyzacji Hutnictwa Żelaza i Stali—Aktualizacja* 2001, Warsaw, accepted on June 5, 2001, 20.

86. Nowak, interview; Najwyższa Izba Kontroli, "Informacja o wynikach kontroli restrukturyzacji i przekształceń własnościowych w hutnictwie żelaza i stali" (Warsaw: NIK, February 2003), 47.

87. Commission of the European Communities, "2002 Regular Report on Poland's Progress towards Accession," SEC (2002) 1408 final (October 9, 2002), 94.

88. Agnieszka Łakoma, "Najpierw zamiana," *Rzeczpospolita*, July 12, 2002.

89. Barbara Sierszuła and Barbara Cieszewska, "Kolejny chętny na huty," *Rzeczpospolita*, September 19, 2002.

90. Barbara Cieszewska, "W oczekiwaniu na wzrost PKB," *Rzeczpospolita*, August 23, 2002; Barbara Cieszewska, "Restrukturyzacja przez likwidację," *Rzeczpospolita*, September 3, 2002.

91. Silesia's former head had since been charged with asset stripping. Witold Jajszczok, "Prześwietlanie Silesii," *Nowy Przemysł*, no. 2, 2004; *Rzeczpospolita*, "Stalowe interesy śląskiego towarzystwa," January 17, 2005.

92. Total permissible state aid to the sector over these years was to equal PLN 3,364,375,075, of which 96 percent (PLN 3,242,864,497) was to be granted over 2002–2003.

93. Interview with a journalist covering the steel sector at a major daily, Katowice, June 2003.

94. Katarzyna Jaźwińska, "PHS wciąż bez inwestora," *Puls Biznesu*, June 30, 2003.

95. Ibid.

96. Interview with a civil servant closely involved in the privatization process at the Ministry of State Treasury, Warsaw, July 2003.

97. Katarzyna Jaźwińska, "Albo dla dwóch, albo dla nikogo," *Puls Biznesu*, July 14, 2003; Kazimierz Krupa, "Bez prywatyzacji jeszcze większa kompromitacja," *Puls Biznesu*, July 15, 2003.

98. *Stal*, "Podpis wart 7,7 mld," October 31, 2003. The contract stipulated that in 2007 the State Treasury would sell the remaining 25% of the shares for the total price of PLN 66 million—that is, greatly underpriced given the enterprise's value.

99. Jan Dziadul, "Stalowa wola," *Polityka*, May 18, 2002, 68–69.

100. Witold Gadomski, "Co się wydarzyło w Ostrowcu," *Gazeta Wyborcza*, August 8, 2003.

101. One of the interested investors was the Spanish Celsa, which became the eventual buyer.

102. For example, Jerzy Jaskiernia, the leader of the SLD parliamentary group at the time, suggested that the Stalexport representatives "not obstruct the municipal government in realizing its strategy" during the upcoming shareholders' meeting" Józef Oleksy, the former prime minister and one of top SLD figures also lobbied on Ostrowiec's behalf. Both politicians had highly lucrative lectureship appointments at the local business school, owned by a politically connected businessman. Initially, the local government was supposed to purchase the steelworks for PLN 133,000 and pay back its PLN 80 million debt to Stalexport in installments from 2004 until 2009. Stalexport never obtained any of the stipulated payments. Marcin Krek et al., "Operacja prawie się udała," *Rzeczpospolita*, August 7, 2003; Witold Gadomski, "Co się wydarzyło w Ostrowcu," *Gazeta Wyborcza*, August 8, 2003.

103. Ibid.

104. *Gazeta Ostrowiecka*, "Były prezes i były członek zarządu przed sądem," January 13, 2004.

105. As part of the transaction, Celsa agreed to pay PLN 53 million for the steelworks, as well as promised a serious investment program. ZOHO claimed that it would protect the workplaces while investing in the enterprise. The Ostrowiec unions were not supportive of Celsa's entry, trusting instead that the local government would preserve employment. Celsa's negotiations with the unions had been long and difficult, but the resulting agreement contained provisions that were superior to those of the labor code. Moreover, the subsequent layoffs were voluntary and accompanied by severance packages. *Gazeta Ostrowiecka*, "Jest układ między Celsą a związkowcami," March 24, 2004; A. Śledzińska, "Znów dobrowolne odejścia? Celsa kusi odprawami," *Świętokrzyskie Wiadomości*, April 25, 2005. See also Krek et al., "Operacja prawie się udała."

106. Najwyższa Izba Kontroli, "Informacja o wynikach kontroli," 49; Barbara Cieszewska et al., "Nie tylko zmiana nazwy," *Rzeczpospolita*, October 7, 2002.

107. Maria Trepińska, "CZH spieszy z pomocą dla Huty Częstochowa," *Puls Biznesu*, June 3, 2003; Barbara Cieszewska, "Bankrut na bankrucie," *Rzeczpospolita*, April 7, 2003.

108. *Polish Press Agency*, "Nieprawidłowy przetarg na Hutę Częstochowa," *Gazeta Wyborcza*, May 9, 2004; Konrad Niklewicz, "Dlaczego Donbas przegrał?" *Gazeta Wyborcza*, March 4, 2004.

109. AFX News Limited, "Poland Gives Green Light to Donbas Takeover of Huta Czestochowa," October 4, 2005.

110. Rafał Towalski, "Donbas in Talks with Huta Częstochowa," *European Industrial Relations Observatory Online*, June 7, 2005, http://eurofound.europa.eu/observatories/eurwork/articles/donbas-in-talks-with-huta-czstochowa (accessed November 15, 2014).

111. European Commission press release, "State Aid: Commission Concludes No Aid Involved in Restructuring of Polish Steel Company Huta Czestochowa, but Orders Recovery of €4 million Restructuring Aid." IP/05/842, July 6 2005.

112. Ibid.

113. Commission of the European Communities, European Commission Staff Working Document, *Programme for Restructuring of the Romanian Steel Industry*, Final Assessment, COM (2005) 140 final (April 14, 2005), 9.

114. Janine R. Wedel, *Collision and Collusion: The Strange Case of Western Aid to Eastern Europe, 1989–1998* (New York: Palgrave, 2001), 40.

115. World Bank, "Czech Republic to 'Graduate' from World Bank Borrowing," Washington, DC: World Bank, April 18, 2005.

116. Hutní Projekt Ostrava and Vysoká Škola Báňská, *Studie Plánu Restrukturalizace Českého Ocelářského Průmyslu*. Ostrava, 1999. See also Anna Bortlíčková, "Naše hutě by si na vzájem neměly konkurovat, ale spolupracovat," *Právo*, May 31,1999; Ivo Žádný, "Hutnictví stale čeká na restrukturalizaci," *Hospodářské noviny*, March 13, 2000.

117. Michal Uryč-Gazda, "Neprodáme-li Vítkovice Steel, končíme," *Týden*, March 25, 2002.

118. Sabina Kunešová and Luděk Piaseczny, "Noví lidé v Čele Vítkovic," *Svět hospodářství*, July 3, 2001.

119. Jitka Znamenáčková, "Hutě má ozdravit nové vedení," *Hospodářské noviny*, June 26, 2001.

120. Interview with Jiří Havel, Prague, March 2004; interview with representatives at the Steel Federation, Inc. (Hutnictví Železa, a.s.), Prague, November 2003; *Nová Huť*, "Nová Huť očekává ztrátu za rok 1999 ve výši 2.9 mld. Kč," March 14, 2000; Libuše Bautzová, "NH: Zablokovaná partie," *Ekonom*, August 3, 2000.

121. Interview with a consultant at the Roland Berger Strategy Consultants, Prague, March 2004.

122. Ministerstvo průmyslu a obchodu České republiky, "Pro Schůzi Vlády, Věc: Postup Programu restrukturalizace ocelařského průmyslu v České republice, Příloha č.1 k Předkládací zprávě" (photocopy), Prague, January 14, 2002.

123. *Hospodářské noviny*, "Mertlík se dohodl s IFC na rychlé privatizaci Nové huti," April 18, 2000.

124. Ibid.

125. *Hospodářské noviny*, "Zeman podmiňuje pomoc Nové huti odchodem neúspěšného vedení," April 21, 2000; *Hospodářské noviny*, "Vláda koupí Petrcíle a pomůže tak NH," April 28, 2000.

126. See an interview with the National Property Fund chairman, Jiří Havel: Eva Sovová, "Nečekám, že by mě ministr Mertlík odvolal," *Právo*, November 21, 2000; *Hospodářské noviny*, "Prodej Nové huti ještě může zkomplikovat soudní spor," June 6, 2002; *Lidové noviny*, "Stát nebude firmě Petrcíle platit miliardy," April 12, 2005.

127. Libuše Bautzová and Petr Němec, "Nová Huť si kleknout nehodlá," *Ekonom*, No. 31, 2000, 25.

128. Bautzová, "NH: Zablokovaná partie."

129. For a discussion of the clash between the minister of industry and trade and the minister of finance, see Martin Myant, *The Rise and Fall of Czech Capitalism* (Northampton, MA: Edward Elgar, 2003), 202–204.

130. Interview with Jiří Havel, former FNM chairman, Prague, March, 2004; interview with a former top management member of the Revitalization Agency (Revitalizační Agentura, 2000–2001), Prague, June 2008.

131. Havel, interview; see also Pavel Páral and István Lékó, "S rychlostí tygra," *Euro*, February 11, 2002.

132. Páral and Lékó, "S rychlostí tygra."

133. The other two options were letting the market forces roam free and a one-time state bailout. Eurostrategy, moreover, doubted that suitable investors would immediately be found and saw individual sales as being more risky for steelworks, which were supposedly facing greater threat of postpurchase closures than would a consolidated company. Ministry of Industry and Trade of the Czech Republic and Eurostrategy Consultants, "Restructuring of the Steel Industry in the Czech Republic—Industry Restructuring Plan," Prague, May 2001, 9–14.

134. Pavel Páral and Marcela Šperkerová, "Všichni do Třince," *Euro*, May 7, 2001.

135. Interview with civil servants at the Ministry of Industry and Trade, Prague, December 2003; representatives at the Steel Federation, Inc., interview; interview with top management members #1 and #2 at North Moravia steelworks A, April 2004; consultant at the Roland Berger Strategy Consultants, interview; interview with top management member at North Moravia steelworks B, March 2004; Zuzana Kubátová, "Velká hra Třineckých železáren," *Týden*, May 7, 2001; Pavla Nováková and Vladan Gallistl, "Třinecká cesta privatizace hutí padla," *Lidové noviny*, July 24, 2001.

136. Hana Čápová, "Všechny cesty vedou do Třince," *Respekt*, May 21, 2001; Marek Pražák, "Ocelárny pohltí další miliardy," *Mladá fronta Dnes*, May 17, 2001; Václav Žák, "Byl Brusel pákou lobbistů?" *Hospodářské noviny*, December 3, 2001.

137. Civil servants at the Ministry of Industry and Trade, interview; top management member #2 at North Moravia steelworks A, interview. See also Myant, *Rise and Fall*, 208.

138. Marek Pražák, "Ocelárny pohltí další miliardy"; Pavel Šmíd, "Studie navrhuje spojit domácí hutě," *Hospodářské noviny*, May 17, 2001; Miloš Hrabě, "Odbory odmítají sloučení hutí," *Hospodářské noviny*, June 20, 2001.

139. Top management member #1 at North Moravia steelworks A, interview.

140. Consultant at the Roland Berger Strategy Consultants, interview.

141. Páral and Šperkerová, "Všichni do Třince"; Kubátová, "Velká hra."

142. Government Resolution Nr. 840 of August 29, 2001. The state's expenditure under the new plan was supposed to be far less than the nearly $1.28 billion (CZK 50 billion) total estimated by Eurostrategy, as it was to involve $35.8 million (CZK 1.4 billion) in financial restructuring, in addition to more than $512 million (CZK 20 billion) in social and environmental expenditures, and research and development. By contrast, the strategic investor was to pay nearly $282 million (CZK 11 billion) for the consolidated company and nearly $128 million (CZK 5 billion) was to come in the form of EU aid. The remaining $1 billion (CZK 41 billion) was to be invested from the enterprises' own means. *Moravskoslezský den*, "Hutě jsou proti," July 30, 2001; Libuše Bautzová, "Restrukturalizace před branami EU," *Ekonom*, No. 36, 2001.

143. *Hospodářské noviny*, "Brusel kritizuje plán na oživení ocelářství," October 17, 2001; *Hospodářské noviny*, "Evropská komise odmítla Grégrův plán restrukturalizace ocelářství," October 17, 2001; Alena Adámková, "Čeští oceláři dostali sprchu z Bruselu," *Pražské Slovo*, October 26, 2001. The plan was criticized by the new managements of Nová Huť and Vítkovice Steel (not to mention Třinecké), which demanded that the state take a more active role in the financial restructuring of the sector. The unions, on the other hand, agreed to the program during the August

meeting of the tripartite commission, claiming that holding a tender should help to select the strongest investor to enter the sector. See Vlastimil Rádl, "Šéfové hutí se přou s odbory o restrukturalizaci," *Právo*, August 20, 2001; Pavel Šmíd, "Poslední šance ocelářství," *Hospodářské noviny*, August 29, 2001; *Mladá fronta Dnes*, "Šéfové hutí nesouhlasí," July 31, 2001.

144. Michal Mocek, "Grégr: Problémy ocelářství může řešit i stát," *Mladá fronta Dnes*, November 29, 2001.

145. František Bouc, "EC Questions Steel Merger," *The Prague Post*, October 24, 2001; interview with a European Commission civil servant, Czech Republic Team, Directorate-General for Enlargement, Prague, November 2003.

146. Commission of the European Communities, "2001 Regular Report on the Czech Republic's Progress towards Accession," SEC (2001) 1746 final (November 13, 2001), 75.

147. Ministerstvo průmyslu a obchodu České republiky, "Pro Schůzi Vlády"; Bouc, "EC Questions Steel Merger."

148. Ministerstvo průmyslu a obchodu České republiky, "Pro Schůzi Vlády"; Pavel Páral, "Viditelná ruka FNM," *Euro*, January 21, 2002.

149. Dušan Šrámek, "Vlastník českých hutí se hledá zbytečně zložitě," *Profit*, January 28, 2002.

150. Former top management member of the Revitalization Agency, interview.

151. Interview with Nová Huť union leader, Ostrava, March 2004; Blanka Falcníková, "Odboráři: Nová varianta může zlikvidovat Novou huť," *Moravskoslezský deník*, January 17, 2001; Pavel Šmíd, "Odboráři Nové huti vyhlásili protestní demonstrace," *Hospodářské noviny*, February 1, 2002.

152. Pavel Šmíd, "V ostravské Nové huti odbory zvažují stávkovou pohotovost," *Hospodářské noviny*, January 21, 2002; Alena Adámková, "Hutníci hrozí vládě stávkou," *Večerník Praha*, January 22, 2002.

153. Slogans that appeared during the steelworkers' protests in Prague included the likes of: "No to Thieves from the FNM" and "We Will Defend Nova Huť against Tunnelers."

154. Vadim Fojtík and Pavel Šmíd, "Státní společnost Osinek má zachránit ocelářství," *Hospodářské noviny*, January 16, 2002 (translated by the author).

155. Consultant at the Roland Berger Strategy Consultants, interview; interview with a consultant at the Wood and Company, Prague, March 2004.

156. Jaroslav Spurný, "Tejemství šifri D47," *Respekt*, January 27, 2003; Sabina Slonková and Jan Cizner, "D47: Podivná od začátku," *Mladá fronta Dnes*, April 3, 2004; Jindřich Šídlo and Sabina Slonková, "D47: Policie jde po účtech," *Mladá fronta Dnes*, December 10, 2002; Josef Kopecký, "ČSSD má sponzora spjatého s investorem D47," *Hospodářské noviny*, March 25, 2003; *RFE/RL Newsline*, "Czech Police Want to Probe Cabinet Ministers over Highway Affair," June 30, 2003; *Respekt*, "Za vším hledej Shiran," December 10, 2001.

157. *American Metal Market*, "Three of Six Suitors Walk Away from Vitkovice Steel Auction," March 28, 2002.

158. Marek Pražák and Pavla Nováková, "Vítkovice Steel zřejmě získa stat," *Mladá fronta Dnes*, March 28, 2002; *Moravskoslezský deník*, "Osinek má být kupcem Vítkovice Steel," March 29, 2002.

159. Government Resolution Nr. 131 of February 4, 2002. See *Usnesení Vlády*

České Republiky ze dne 4. února 2002, č.131 o postupu Programu restrukturalizace ocelářského průmyslu v České republice.

160. This was according to the Ministry of Industry and Trade spokeswoman. See Vadim Fojtík, "Sloučení hutí ušetří miliardy," *Hospodářské noviny*, April 15, 2002.

161. Milena Geussová, "Pojďte do globální skupiny!" *Ekonom*, March 21, 2002 (translated by the author).

162. Civil servants at the Ministry of Industry and Trade, interview; top management member at North Moravia steelworks B, interview.

163. Sam Beckwith, "Nová huť sale stirs questions," *The Prague Post*, June 5, 2002; Alena Adámková, "Předem daný prodělek," *Večerník Praha*, May 13, 2002; Tomáš Němeček, "Sázka na Mittala," *Respekt*, June 3, 2002.

164. František Bouc, "Steel's time running short," *The Prague Post*, April 10, 2002.

165. Němeček, "Sázka na Mittala"; Beckwith, "Nová huť Sale Stirs Questions"; former top management member of the Revitalization Agency, interview.

166. Interview with Jan Mládek, former deputy minister of finance, Prague, December 2003.

167. Government Resolution Nr. 565/02 of May 29, 2002, and Nr. 587/2002 of June 4, 2002, in which the government made the decision to sell the Nová Huť shares to LNM.

168. Former top management member of the Revitalization Agency, interview.

169. Ibid; Tomáš Němeček, "Sázka na Mittala"; Alena Adámková, "Předem daný prodělek," *Večerník Praha*, May 13, 2002.

170. Consultant at the Roland Berger Strategy Consultants, interview; representatives at the Steel Federation, Inc., interview. See also Vadim Fojtík and Pavel Šmíd, "Miliarda za Vítkovice Steel je pro stat málo," *Hospodářské noviny*, October 22, 2002.

171. This privatization, too, was controversial as the FNM repeatedly refused to admit Mittal Steel (formerly LNM Holdings) into the bidding process on the grounds that Mittal Steel had a dispute with Osinek over the pricing of pig iron produced by Vysoké Pece Ostrava and sold to Vítkovice.

CHAPTER 7

1. *The Economist*, "The Challengers: A New Breed of Multinational Company Has Emerged," January 10, 2008.

2. Ibid.

3. Herbert Kitschelt et al., *Post-Communist Party Systems* (Cambridge: Cambridge University Press, 1999).

4. For a discussion of the various responses of labor to market reforms, see Maria Victoria Murillo, *Labor Unions, Partisan Coalitions, and Market Reforms in Latin America* (Cambridge: Cambridge University Press, 2001).

5. See Appendix B.

6. Milada Anna Vachudova, *Europe Undivided* (Oxford: Oxford University Press, 2005).

7. For a critical discussion of the hierarchical portrayal of the relationship between the Commission and the accession states, see Laszló Bruszt and Julia Langbein, "Anticipatory Integration and Orchestration: The Evolving EU Gover-

nance of Economic and Regulatory Integration during the Eastern Enlargement" (paper presented at the Annual Meeting of the American Political Science Association, Washington, DC, August 2014).

8. According to press accounts, the November 2002 decision to sell Dunaferr was a surprise, because during the election campaign, the Hungarian Socialist Party candidate for prime minister had promised not to privatize the plant if elected. László Neumann and András Tóth, "Dispute over Wage Cuts and Privatisation at Dunaferr Steel Mill," *European Industrial Relations Observatory On-Line*, December 4, 2002, http://www.eurofound.europa.eu/eiro/2002/12/inbrief/hu0212103n.htm (accessed May 30, 2013).

9. In 2005, the majority shareholder in Finmetals Holding sold the enterprise to Global Steel Holdings, controlled by Pramod Mittal. However, after the money borrowed to cover enterprise debts and investments was simply used to service the sale transaction, the enterprise found itself in bankruptcy and liquidation. Alex Bivol, "Kremikovtzi: 10 years after privatization," *The Sofia Echo*, August 27, 2009; Elizabeth Konstantinova, "Bulgaria Sells Bankrupt Kremikovtzi's Mill Production Assets," *Bloomberg News*, April 12, 2011; Alex Bivol, "Former Kremikovtzi CEO Vows to Sue Former Bulgarian Economy Minister," *The Sofia Echo*, March 4, 2010.

10. Sanjay Kathuria, *Western Balkan Integration and the EU* (Washington, DC: World Bank, 2008).

11. Andreas Nölke, and Arjan Vliegenthart, "Enlarging the Varieties of Capitalism: The Emergence of Dependent Market Economies in East Central Europe," *World Politics* 61 (2009): 670–702; Jadwiga Staniszkis, *Postkomunizm—Próba Opisu* (Gdańsk: Wydawnictwo Słowo/Obraz Terytoria, 2001); Lawrence King and Aleksandra Sznajder, "The State-Led Transition to Liberal Capitalism: Neoclassical, Organizational, World Systems, and Social Structural Explanations of Poland's Economic Success," *American Journal of Sociology* 112 (2006): 751–801.

12. European Bank for Reconstruction and Development, *Transition Report 2009* (London: EBRD), 73.

13. See, for example, Peter Reddaway and Dmitri Glinski, *The Tragedy of Russia's Reforms: Market Bolshevism against Democracy* (Washington, DC: United States Institute of Peace Press, 2001); Vadim Volkov, *Violent Entrepreneurs* (Ithaca, NY: Cornell University Press, 2002); Juliet Johnson, "Russia's Emerging Financial-Industrial Groups," *Post-Soviet Affairs* 13 (1997): 333–365; Anders Åslund, *How Capitalism Was Built* (Cambridge: Cambridge University Press, 2007).

14. Forbes.com, "The World's Billionaires," March 10, 2010, http://www.forbes.com/lists/2010/10/billionaires-2010_The-Worlds-Billionaires_Rank.html (accessed October 16, 2014). Compared with its ECE counterpart, the 2010 Russian list looks impressive. In all of the new EU member states from ECE, the only man who could compare to the Russian oligarchs in terms of wealth was Czech entrepreneur, Petr Kellner, who ranked in eighty-ninth place. He was followed distantly by Romania's Dinu Patriciu, in 437th place, and by Poland's Jan Kulczyk (463rd place) and Zygmunt Solorz-Żak (488th place). As of 2014, there were eight Russians among the world's top 100 billionaires, and fifteen among the top 150. Petr Kellner fell to 129th place, Jan Kulczyk moved up to 409th, and Zygmunt Solorz-Żak moved down to 498th place, while, in 2012, Dinu Patriciu moved down to 854th place (he died in 2014 as the richest man in Romania). Forbes.com,

"The World's Billionaires," http://www.forbes.com/billionaires/list/2/#tab:overall (accessed October 17, 2014).

15. Forbes.com, "#39 Alisher Usmanov," http://www.forbes.com/profile/alisher-usmanov (accessed October 17, 2014).

16. Oxford Analytica Daily Brief Service, "Russia: Government Support Props Up Steel Sector," August 18, 2009.

17. See, for example, Peter Marsh, "Novolipetsk and Evraz look to LSE STEEL," *Financial Times*, October 4, 2005.

18. Peter Marsh, "Arcelor Succumbs to Mittal," *Financial Times*, June 26, 2006; Peter Marsh, "Arcelor Suitors Await Decision—Will It be Mittal or Severstal? *Financial Times*, June 10, 2006; Paul Betts, "Men of steel," *Financial Times*, June 9, 2006.

19. Alexei Mordashov, "Why My Money Is on a Global Expansion in Steel," *Financial Times* (London), June 16, 2006, 15.

20. Evraz Group website, available at http://www.evraz.com/about/history (accessed October 17, 2014).

21. Peter Marsh and Arkady Ostrovsky, "Corus Talks with Russia's Evraz Could Herald Merger," *Financial Times*, March 21, 2006; Dan Roberts, "A Corus of Approval for More Steel Consolidation," *Financial Times*, June 2, 2006.

22. Jason Singer, Paul Glader, and Guy Chazan, "Old European Ways Meet New in Steel Standoff," *Wall Street Journal*, June 24, 2006.

23. Ibid.

24. BBC News, "Arcelor to Merge with Severstal," http://news.bbc.co.uk/go/pr/fr/-/2/hi/business/5018682.stm (accessed August 28, 2010).

25. Mordashov, "Why My Money."

26. Guy Chazan and Jason Singer, "Arcelor Hastens a Flurry of Russian Steel Activity," *Wall Street Journal*, June 2, 2006.

27. Forbes.com, "#39 Alisher Usmanov."

28. For a profile of Alisher Usmanov that summarizes the discussions in Britain, see Ian Cobain, "The Colourful Life of Football's Latest Oligarch: The Uzbek Billionaire Stalking Arsenal Has Long Been a Controversial Figure," *The Guardian*, November 19, 2007.

29. See Usmanov's 2010 profile on Forbes's list of the world's billionaires. Forbes.com, "#100 Alisher Usmanov," March 10, 2010, http://www.forbes.com/lists/2010/10/billionaires-2010_Alisher-Usmanov_GIPI.html (accessed August 29, 2010).

30. BBC Monitoring Former Soviet Union, "Gazprom Will Have Nothing to Do with Kommersant—Spokesman," August 31, 2009. Retrieved August 29, 2010, from ABI/INFORM Trade & Industry.

31. John Helmer, "Stealing Steel in Russia," *Asia Times*, January 14, 2005.

32. Ibid.

33. Henry Meyer, "Medvedev Tells Authorities to Stop Scaring Business (Update 3)," *Bloomberg News*, July 31, 2008.

34. Oxford Analytica Daily Brief Service, "Russia: Government Support Props Up Steel Sector," August 18, 2009.

35. See Mechel's press releases at http://www.mechel.com/press/press?rid=25062&apage=1 (accessed October 17, 2014).

36. *Steel Orbis*, "Metalloinvest Holding Redeems Bulk of a State-Guaranteed Loan," June 29, 2010, http://www.steelorbis.com/steel-news/latest-news/metallo-invest-holding-redeems-bulk-of-a-state_guaranteed-loan-540445.htm   (accessed June 15, 2013).

37. See the companies' annual reports, available on their websites. See also *Reuters*, "Russia Metalloinvest plans IPO," February 17, 2010.

38. *The Russia Journal*, "Novolipetsk Share Test Fall [*sic*] Short of Target," December 20, 2005, http://www.metal.com.ru/analytics/publication.php?id=391 (accessed June 15, 2013).

39. Ibid.; Elena Anankina, "Rating Russian Steel Companies," *Metal Bulletin Monthly*, June 2005, 61.

40. Polina Deveitt and Alfred Kueppers, "Mechel Sees 'Substantial' Doubt as Going Concern," *Reuters*, June 25, 2009.

41. *The Economist*, "The Challengers: A New Breed of Multinational Company Has Emerged," January 10, 2008.

42. *Financial Times*, "An Exemplary Start: Mittal Purchase Marks New Beginning for Ukraine's Economy," October 26, 2005.

43. Ukrinform (via the web-portal of the Ukrainian government), "It's a Long Way to Huta Czestochowa's Privatization, Marek Belka Says," March 4, 2005,   http://www.kmu.gov.ua/control/en/publish/article?art_id=13402574&cat_id=32598 (accessed June 28, 2013).

44. *Kyiv Post*, "Russian Group Buys Ukraine's Industrial Union of Donbass, Nation's Leading Steel Group," January 14, 2010.

45. Anders Åslund, *How Ukraine Became a Market Economy and Democracy* (Washington, DC: Peterson Institute for International Economics, 2009), 252–253.

46. Both takeover efforts were nontransparent in nature, using offshore companies that seemed to be a front for the more powerful Russian market players. *Kommersant-Ukraine*, "Mariupol Ilyich Iron and Steel in the Hands of Evraz and Severstal," May 27, 2010; John Marone, "Russian Investors Trying to Muscle in on Ukraine's Best Steel Mills, *Kyiv Post*, June 3, 2010.

47. Aleksandra Sznajder, "Effects of EU Accession on the Politics of Privatization: The Steel Sector in Comparative Perspective," in *Das Erbe des Beitritts. Mittel- und osteuropäische Gesellschaften nach dem Beitritt zur EU*, ed. Amelie Kutter and Vera Trappmann (Baden-Baden: Nomos Verlag, 2006), 215–216.

48. Riad A. Ajami et al., *International Business: Theory and Practice* (Armonk, NY: M. E. Sharpe, 2006).

49. Oxford Analytica Daily Brief Service, "European Union: Steel Prospects," October 31, 1999.

50. Russel Muir and Joseph P. Sabat, "Improving State Enterprise Performance" (World Bank Technical Paper Number 306, Washington, DC: International Bank for Reconstruction and Development, 1995), 110–115.

51. Ajami et al., *International Business*, 488.

52. Stefan Wagstyl, "Capacity for Change," *Financial Times*, March 20, 1997.

53. For information about the merger, see British Steel and Koninklijke Hoogovens, "A New Force in the Metals Industry: Background to the Proposed Merger," June 7, 1999, http://www.tatasteeleurope.com/static_files/StaticFiles/Functions/Media/Merger_Manifesto.pdf. (accessed October 18, 2014).

54. Singer, Glader, and Chazan, "Old European Ways Meet New."

55. Tarun Khanna, "Tata-Corus: India's New Steel Giant," *Economic Times / India Times*, February 14, 2007.

56. In policy areas where EU rules introduce certain restrictions or bestow benefits—the Common Agricultural Policy being the case in point—such regulations apply equally across issue areas in all member states, though the policies' implications obviously differ depending on the issue's salience in a given member state.

57. Max Lienemeyer, "The Restructuring of Huta Czestochowa," *Competition Policy Newsletter* (Spring 2006): 100–101, http://ec.europa.eu/competition/publications/cpn/cpn2006_1.pdf (accessed April 13, 2013).

58. Commission of the European Communities, "State Aid: Commission Launches Probe into Czech Steel Company Třinecké železárny," press release, IP/04/1475, December 14, 2004; Commission of the European Communities, "State Aid: Commission Concludes Czech Steel Producer Třinecké železárny Received No Illegal Aid," press release IP/06/1532, November 09, 2006.

59. Andra Timu and Irina Savu, "EU Probes ArcelorMittal, Vimetco Romania Units on Energy," *Bloomberg News*, October 21, 2011; European Commission, State Aid—Romania. State Aid SA.33623 (12/C) (ex 11/NN) "Alleged Preferential Electricity Tariffs for ArcelorMittal Galați SA," June 29, 2012, O.J. (C 189/3).

60. FHZZ-affiliated union leader at ArcelorMittal Poland, interview, Kraków, Poland, July 2010.

61. Petru Dandea, Cartel Alfa vice president, interview, Bucharest, Romania, July 2010.

62. Anna Grzymała-Busse shows that the transformation of state institutions during the transition period has been greatly affected by the quality of political competition. At the same time, Venelin Ganev's exploration of state transformation in Bulgaria shows entrenchment of groups that is unlikely to be easily dislodged. See Anna Grzymała-Busse, *Rebuilding Leviathan: Party Competition and State Exploitation in Post-Communist Democracies* (Cambridge: Cambridge University Press, 2007); Venelin Ganev, *Preying on the State: The Transformation of Bulgaria after 1989* (Ithaca, NY: Cornell University Press, 2007).

References

NEWSPAPERS, JOURNALS, AND WIRE SERVICES

American Metal Market
Asia Times
BBC News
Bloomberg News
The Economic Times/India Times
The Economist
European Industrial Relations Observatory On-Line
Financial Times
Forbes.com
The Guardian
Kommersant-Ukraine
Kyiv Post
Metal Bulletin Monthly
Oxford Analytica Daily Brief service
Reuters Company News
RFE/RL Newsline
The Russia Journal
The Sofia Echo
Steel Orbis
Ukrinform
Wall Street Journal

Czech Republic

Ekonom
Euro
Hospodářské noviny
Hutnické listy
Kovák
Lidové noviny
Magazin Vítkovice

270   *References*

Mladá fronta Dnes
Moravskoslezský den
Moravskoslezský deník
Nová Huť
Práce
The Prague Post
Pražské Noviny
Pražské Slovo
Právo
Profit
Respekt
Severočeské noviny
Svět hospodářství
Týden
Večerník Praha

Poland

Gazeta Ostrowiecka
Gazeta Wyborcza
Magazyn Huty L.W.
Nowe Życie Gospodarcze
Nowy Przemysł
Parkiet
Polish Press Agency
Polityka
Puls Biznesu
Rzeczpospolita
Stal
Świętokrzyskie Wiadomości

Romania

Adevărul
Evenimentul Zilei
InfoMetal
Mediafax (Romanian press agency)
Viaţa Liberă
Ziarul Financiar
Ziua

Slovakia

Hospodarské Noviny
Podbrezovan
Priority
Profit
The Slovak Spectator
Slovenský Profit
Trend

BOOKS AND ARTICLES

Ajami, Riad A., Karel Cool, G. Jason Goddard, and Dara Khambata. *International Business: Theory and Practice*. Armonk, NY: M. E. Sharpe, 2006.

Almond, Gabriel. "The Development of Political Development." In *Understanding Political* Development, edited by Myron Weiner and Samuel P. Huntington, 437–478. Boston: Little, Brown, 1987.

Amsden, Alice, Jacek Kochanowicz, and Lance Taylor. *The Market Meets Its Match*. Cambridge, MA: Harvard University Press, 1994.

Appel, Hilary. "The Ideological Determinants of Liberal Economic Reform: The Case of Privatization." *World Politics* 52 (2000): 520–549.

Appel, Hilary, and John Gould. "Identity Politics and Economic Reform: Examining Industrial-State Relations in the Czech and Slovak Republics." *Europe-Asia Studies* 52 (2000): 111–131.

Åslund, Anders. *How Capitalism Was Built: The Transformation of Central and Eastern Europe, Russia, and Central Asia*. Cambridge: Cambridge University Press, 2007.

Åslund, Anders. *How Ukraine Became a Market Economy and Democracy*. Washington, DC: Peterson Institute for International Economics, 2009.

Balcerowicz, Leszek. *Socialism, Capitalism, Transformation*. Budapest: Central European University Press, 1995.

Bałtowski, Maciej. *Przekształcenia Własnościowe Przedsiębiorstw Państwowych*. Warsaw: Wydawnictwo Naukowe PWN, 2002.

Bałtowski, Maciej, and Tomasz Mickiewicz. "Privatization in Poland: Ten Years After." *Post-Communist Economies* 12 (2000): 425–443.

Bandelj, Nina. *From Communists to Foreign Capitalists: The Social Foundations of Foreign Direct Investment in Postsocialist Europe*. Princeton, NJ: Princeton University Press, 2008.

Berg, Andrew, and Olivier Jean Blanchard. "Stabilization and Transition: Poland, 1990–91." In *The Transition in Eastern Europe*, vol. 1, edited by Oliver Jean Blanchard, Kenneth Froot, and Jeffrey Sachs, 51–92. Chicago: University of Chicago Press, 1994.

Bohle, Dorothee, and Béla Greskovits. *Capitalist Diversity on Europe's Periphery*. Ithaca, NY: Cornell University Press, 2012.

Bohle, Dorothee, and Béla Greskovits. "Neoliberalism, Embedded Neoliberalism and Neocorporatism: Towards Transnational Capitalism in Central-Eastern Europe." *West European Politics* 30 (2007): 443–466.

Bohle, Dorothee, and Béla Greskovits. "The State, Internationalization, and Capitalist Diversity in Eastern Europe." *Competition & Change* 11 (2007): 89–115.

Bohle, Dorothee, and Béla Greskovits. "Varieties of Capitalism and Capitalism 'tout court.'" *European Journal of Sociology* 50 (2009): 355–386.

Bruszt, Laszló, and Julia Langbein. "Anticipatory Integration and Orchestration: The Evolving EU Governance of Economic and Regulatory Integration during the Eastern Enlargement." Paper presented at the Annual Meeting of the American Political Science Association, Washington, DC, August 2014.

Cameron, David, R. "The Challenges of EU Accession." *East European Politics and Societies* 17 (2003): 24–41.

Chivu, Luminiţa. "Representativeness of the European Social Partner Organisations: Steel Industry—Romania." EIRO Briefing Paper, September 2009. http://www.eurofound.europa.eu/eiro/studies/tn0811027s/ro0811029q.htm.

Commission of the European Communities. State Aid—Romania. State Aid No. SA 33623 (12/C) (ex 11/NN)—"Alleged Preferential Electricity Tariffs for ArcelorMittal Galați SA," June 29, 2012, O.J. (C 189/3).

Commission of the European Communities. European Commission Staff Working Document. "Programme for Restructuring of the Romanian Steel Industry, Final Assessment," COM (2005) 140 final (April 14, 2005).

Commission of the European Communities. Industrial Forum on Enlargement. *Industrial Restructuring Report*, September, 2000.

Commission of the European Communities. *Steel Industry Country Fiches* (June 23, 2004).

Commission of the European Communities. *2001 Regular Report on the Czech Republic's Progress towards Accession*, SEC (2001) 1746 final (November 13, 2001).

Commission of the European Communities. *2002 Regular Report on Poland's Progress towards Accession*, SEC (2002) 1408 final (October 9, 2002).

Council of Ministers of the Republic of Poland. *Program Restrukturyzacji Hutnictwa Żelaza i Stali—Aktualizacja* [Program of iron and steel industry restructuring—an update], Warsaw, accepted on June 5, 2001.

Crowley, Stephen, and David Ost. *Workers after Workers' States: Labor and Politics in Postcommunist Eastern Europe*. Lanham, MD: Rowman and Littlefield, 2001.

Daianu, Daniel. "Macro-Economic Stabilization in Post-Communist Romania." In *Romania in Transition*, edited by Lavinia Stan, 93–125. Brookfield, VT: Dartmouth, 1997.

Daly T., and E. Dalli. "Central and Eastern European Steel after EU Accession." *Steel Times International* (November–December 2005): 41–58.

Dangerfield, Martin. "Ideology and the Czech Transformation: Neoliberal Rhetoric or Neoliberal Reality?" *East European Politics and Societies* 11 (1997): 436–469.

Desai, Raj. "Organizing Markets: Property Rights, Governance, and the Politics of Industrial Privatization in a Post-Communist Economy." PhD diss., Harvard University, 1996.

Djankov, Simeon. "The Enterprise Isolation Program in Romania." *Journal of Comparative Economics* 27 (1999): 281–293.

"Documents concerning the accession of the Czech Republic, the Republic of Estonia, the Republic of Cyprus, the Republic of Latvia, the Republic of Lithuania, the Republic of Hungary, the Republic of Malta, the Republic of Poland, the Republic of Slovenia and the Slovak Republic to the European Union." Annex XIV: List referred to in Article 24 of the Act of Accession: Slovakia, September 23, 2003, 2003 O.J. (L 236), 918.

Dostál, Vladimír, and Jaroslav Purš. *Dějiny hutnictví železa v Československu*. Vol. 3. Prague: Academia, 1988.

Drahokoupil, Jan. "After Transition: Varieties of Political-Economic Development in Eastern Europe and the Former Soviet Union." *Comparative European Politics* 7 (2009): 279–298.

Drahokoupil, Jan. *Globalization and the State in Central and Eastern Europe: The Politics of Foreign Direct Investment*. Routledge Series on Russian and East European Studies. London: Routledge, 2008.

Drahokoupil, Jan, and Martin Myant. "Putting Comparative Capitalism Research in Its Place: Varieties of Capitalism in Transition Economies." In *New Directions in Critical Comparative Capitalisms Research*, edited by Matthias Ebenau, Ian Bruff, and Christian May, 155–171. London: Palgrave Macmillan, 2015.

Dudek, Antoni. *Pierwsze Lata III Rzeczypospolitej*. Kraków: Arcana, 2002.

Earle, John S., and Dana Săpătoru. "Privatization in a Hypercentralized Economy: Case of Romania." In *Privatization in the Transition to a Market Economy*, edited by John S. Earle, Roman Frydman, and Andrzej Rapanczyski, 147–170. London: Pinter, 1993.

Easter, Gerald. *Capital, Coercion, and Postcommunist States*. Ithaca, NY: Cornell University Press, 2012.

Economist Intelligence Unit. "Romania: Country Briefing." *The Economist* (London), February 3, 2004.

Economist Intelligence Unit. "Romania Industry: Survey of the Metals Sector." *The Economist* (London), August 31, 2001.

Eke, Burcu, and Ali M. Kutan. "IMF-Supported Programmes in Transition Economies: Are They Effective?" *Comparative Economic Studies* 47 (2005): 23–40.

Ekiert, Grzegorz. "The State after State Socialism: Poland in Comparative Perspective." In *Nation-State in Question*, edited by T. V. Paul, G. John Ikenberry, and John Hall, 291–320. Princeton, NJ: Princeton University Press, 2003.

Ekiert, Grzegorz, and Jan Kubik. "Contentious Politics in New Democracies: East Germany, Hungary, Poland, and Slovakia, 1989–93." *World Politics* 50 (1998): 547–581.

Ekiert, Grzegorz, and Stephen E. Hanson, eds. *Capitalism and Democracy in Central and Eastern Europe: Assessing the Legacy of Communist Rule*. Cambridge: Cambridge University Press, 2003.

Ellison, David L. "Divide and Conquer: The European Union Enlargement's Successful Conclusion?" *International Studies Review* 8 (2006): 150–165.

Epstein, Rachel. *In Pursuit of Liberalism: International Institutions in Postcommunist Europe*. Baltimore: Johns Hopkins University Press, 2007.

Epstein, Rachel, and Wade Jacoby. "Eastern Enlargement Ten Years On: Transcending the East-West Divide?" *Journal of Common Market Studies* 52 (2014): 1–16.

Estrin, Saul, Kirsty Hughes, and Sarah Todd. *Foreign Direct Investment in Central and Eastern Europe: Multinationals in Transition*. London: Royal Institute of International Affairs, 1997.

"Europe Agreement establishing an association between the European Communities and their Member States, of the one part, and the Czech Republic, of the other part," December 31, 1994, 1994 O.J. (L 360).

"Europe Agreement establishing an association between the European Communities and their Member States, of the one part, and the Republic of Poland, of the other part," December 31, 1993, 1993 O.J. (L 348).

"Europe Agreement establishing an association between the European Communities and their Member States, of the one part, and Romania, of the other part," December 31, 1994, 1994 O.J. (L 357).

"Europe Agreement establishing an association between the European Communities and their Member States, of the one part, and the Slovak Republic, of the other part," December 31, 1994, 1994 O.J. (L 359).

European Bank for Reconstruction and Development. *Transition Report*. London: EBRD, 1994–2010.

Evans, Peter. *Embedded Autonomy: States and Industrial Transformation*. Princeton, NJ: Princeton University Press, 1995.

Evans, Peter B., Dietrich Rueschemeyer, and Theda Skocpol. *Bringing the State Back In*. New York: Cambridge University Press, 1985.

Eyal, Gil, Iván Szelényi, and Eleanor Townsley. *Making Capitalism without Capitalists: The New Ruling Elites in Eastern Europe*. New York: Verso, 2000.

Federaţia Naţională Sindicală Metarom. "Raport de Activitate, Conferinţa a XIII-a, Băile Herculane, 22–24 October, 2003."

Fortin, Jessica. "Is There a Necessary Condition for Democracy? The Role of State Capacity in Postcommunist Countries." *Comparative Political Studies* 45 (2012): 903–930.

Fortin, Jessica. "A Tool to Evaluate State Capacity in Post-Communist States, 1989–2006." *European Journal of Political Research* 49 (2010): 654–686.

Frydman, Roman, and Andrzej Rapaczynski. *Privatization in Eastern Europe: Is the State Withering Away?* Budapest: Central European University Press, 1994.

Gallagher, Tom. *Modern Romania*. New York: New York University Press, 2005.

Ganev, Venelin. *Preying on the State: The Transformation of Bulgaria after 1989*. Ithaca, NY: Cornell University Press, 2007.

Gardawski, Juliusz. *Związki zawodowe na rozdrożu*. Warsaw: Instytut Spraw Publicznych, 2001.

Gardawski, Juliusz, Barbara Gąciarz, Andrzej Mokrzyszewski, and Włodzimierz Pańków. *Rozpad bastionu? Związki zawodowe w gospodarce prywatyzowanej*. Warsaw: Instytut Spraw Publicznych, 1999.

Gaughan, Patrick A. *Mergers, Acquisitions, and Corporate Restructurings*. New York: Wiley, 1996.

Geiger, Keri. "Waiting for the Dust to Settle." *Euromoney* 377 (2000): 214–224.

Gilejko, Leszek. *Związki Zawodowe a Restrukturyzacja—Bariery czy Kompromis?* Warsaw: Oficyna Wydawnicza SGH, 2003.

Gilejko, Leszek, Paweł Gieorgica, and Paweł Ruszkowski. *Społeczni Aktorzy Restrukturyzacji*. Warsaw: Centrum Partnerstwa Społecznego 'Dialog,' 1997.

Government of the Czech Republic. *Usnesení Vlády České Republiky ze dne 4. února 2002, č.131 o postupu Programu restrukturalizace ocelářského průmyslu v České republice*. Prague: Government of the Czech Republic, 2002.

Granovetter, Mark. "Economic Action and Social Structure: The Problem of Embeddedness." *American Journal of Sociology* 91 (1985): 481–510.

Granovetter, Mark. "Problems of Explanation in Economic Sociology." In *Networks and Organizations*, edited by Nitin Nohria and Robert Eccles, 25–56. Boston: Harvard Business School Press, 1992.

Greskovits, Béla. "Leading Sectors and the Variety of Capitalism in Eastern Europe." *Actes du Gerpisa* 39 (2005): 113–128.

Greskovits, Béla. *The Political Economy of Protest and Patience*. Budapest: Central European University Press, 1998.

Grosfeld, Irena, and Gérard Roland. "Defensive and Strategic Restructuring in Central European Enterprises." *Emergo* 3 (1996): 21–46.

Grzymała-Busse, Anna. *Rebuilding Leviathan: Party Competition and State Exploitation in Post-Communist Democracies*. Cambridge: Cambridge University Press, 2007.

Grzymala-Busse, Anna, and Pauline Jones Luong. "Reconceptualizing the State: Lessons from Postcommunism." *Politics and Society* 30 (2002): 529–554.

Guvenul României. "Ordonanţă de Urgenţă [nr.146] privind unele măsuri de acompaniament social pentru salaraţii din industria metalurgică ale căror contracte individuale de muncă vor fi desfăcute ca urmare a concedierilor collective." Monitorul Oficial al României, Partea I, Nr. 281/5/X/1999.

Hall, Christophe. *Steel Phoenix: The Fall and Rise of the U.S. Steel Industry*. New York: St. Martin's, 1997.

Hall, Peter A., and David Soskice. *Varieties of Capitalism: The Institutional Foundations of Comparative Advantage*. Oxford: Oxford University Press, 2001.

Hancké, Bob, Martin Rhodes, and Mark Thatcher. *Beyond Varieties of Capitalism: Conflict, Contradictions, and Complementarities in the European Economy*. Oxford: Oxford University Press, 2007.

Hatch Associates et al. *Raport Końcowy*, June 1992.

Hellman, Joel. "Winners Take All: The Politics of Partial Reform in Postcommunist Transitions." *World Politics* 50 (1998): 203–234.

Hellman, Joel S., Geraint Jones, and Daniel Kaufmann. "Seize the State, Seize the Day: State Capture, Corruption, and Influence in Transition." Policy Research Working Paper 2444, World Bank, Washington, DC, 2000. Accessed October 10, 2014. http://elibrary.worldbank.org/doi/book/ 10.1596/1813–9450–2444.

Hunya, Gábor. "Recent Impacts of FDI on Growth and Restructuring in Central and Eastern European Transition Countries." WIIW Research Report No. 284, The Vienna Institute for International Economic Studies, 2002.

Hutní Projekt Ostrava, and Vysoká Škola Báňská. *Studie Plánu Restrukturalizace Českého Ocelářského Průmyslu*. Ostrava, 1999.

Hutnické listy. "Restrukturalizace československého hutnictví," no. 7–8 (1992): 84–89; and no. 9 (1992): 43–45.

Hutnictví Železa. "Úroveň celních tarifů s EU a ostatními zeměmi." Unpublished manuscript, photocopy, Prague, 2004.

Hutnictví Železa. *Hutnická Ročenka 1994*. Prague: Ocelot, 1994.

Iankova, Elena. *Eastern European Capitalism in the Making*. Cambridge: Cambridge University Press, 2002.

ICEM, IPROMET, IPROLAM, IEI, IEM. "Strategia Sectorială în Industria Siderurgică din România. Sinteză." Bucharest, October, 1998.

Innes, Abby. *Czechoslovakia: The Short Goodbye*. New Haven, CT: Yale University Press, 2001.

International Centre for Settlement of Investment Disputes, Washington, DC. Proceedings between Noble Ventures, Inc. (claimant) and Romania (respondent), ICSID Case No. ARB/01/11, October 12, 2005.

International Finance Corporation. *1997 Annual Report*. Washington, DC: International Finance Corporation, 1998.

International Iron and Steel Institute. *Steel Statistical Yearbook*. Brussels: International Iron and Steel Institute, 1991, 1992, 1993, 1994, 2005, 2009.

International Monetary Fund. *IMF Country Report No. 04/319*, 2004.

Jacoby, Wade. *The Enlargement of the European Union and NATO: Ordering from the Menu in Central Europe*. New York: Cambridge University Press, 2004.

Jacoby, Wade. "Managing Globalization by Managing Central and Eastern Europe: The EU's Backyard as Threat and Opportunity." *Journal of European Public Policy* 17 (2010): 416–432.

Jarosz, Maria. "Oblicza prywatyzacji." In *Manowce Polskiej Prywatyzacji*, edited by Maria Jarosz, 11–25. Warsaw: Wydawnictwo Naukowe PWN, 2001.

Jarosz, Maria. "Rady nadzorcze w kleszczach interesów partyjnych i grupowych." In *Manowce Polskiej Prywatyzacji*, edited by Maria Jarosz, 44–71. Warsaw: Wydawnictwo Naukowe PWN, 2001.

Johnson, Juliet. "Russia's Emerging Financial-Industrial Groups." *Post-Soviet Affairs* 13 (1997): 333–365.

Kathuria, Sanjay. *Western Balkan Integration and the EU*. Washington, DC: World Bank, 2008.

Kennedy, Robert E. "A Tale of Two Economies: Economic Restructuring in Post-Socialist Poland." *World Development* 25 (1997): 841–865.

Kideckel, David A. "Winning the Battles, Losing the War: Contradictions of Romanian Labor in the Postcommunist Transformation." In *Workers after Workers' States: Labor and Politics in Postcommunist Eastern Europe*, edited by Stephen Crowley and David Ost, 97–120. Lanham, MD: Rowman and Littlefield, 2001.

King, Lawrence, and Aleksandra Sznajder. "The State-Led Transition to Liberal Capitalism: Neoclassical, Organizational, World Systems, and Social Structural Explanations of Poland's Economic Success." *American Journal of Sociology* 112 (2006): 751–801.

King, Lawrence, and Iván Szelényi. "The New Capitalism of Eastern Europe: Towards a Comparative Political Economy of Post-Communism." In *Handbook of Economic Sociology*, edited by Neil J. Smelser and Richard Swedberg, 205–229. Princeton, NJ: Princeton University Press, 2005.

Kitschelt, Herbert, Zdenka Mansfeldová, Radosław Markowski, and Gábor Tóka. *Post-Communist Party Systems: Competition, Representation and Inter-Party Cooperation*. Cambridge: Cambridge University Press, 1999.

Kopanic, Michael. "Stealing the Eastern Slovak Steelworks." *Central Europe Review* 2 (2000), January 10, 2000. Accessed October 7, 2014. http://www.ce-review.org/00/1/kopanic1_steel.html.

Kornai, János. *Economics of Shortage*. Amsterdam: North-Holland, 1980.

Kuroń, Jacek, and Jacek Żakowski. *Siedmiolatka Czyli Kto Ukradł Polskę?* Wrocław: Wydawnictwo Dolnośląskie, 1997.

Kusý, Miroslav. "Slovak Exceptionalism." In *The End of Czechoslovakia*, edited by Jiří Musil. Budapest: Central European University Press, 1997.

Lane, David, and Martin Myant. *Varieties of Capitalism in Post-Communist Countries*. Basingstoke, UK: Palgrave Macmillan, 2007.

Leško, Marian. *Mečiar a Mečiarizmus*. Bratislava: VMV, 1994.

Lienemeyer, Max. "The Restructuring of Huta Czestochowa: The Commission's Decision Finding Some Compliance with Private Creditor Test but Ordering Recovery of Some Previously Granted Aid." *Competition Policy Newsletter* Spring (2006): 100–101. Accessed April 13, 2013. http://ec.europa.eu/competition/publications/cpn/cpn2006_1.pdf.

Lienemeyer, Max. "State Aid for Restructuring the Steel Industry in the New Member States." *Competition Policy Newsletter* Spring (2005): 94–102. Accessed April 13, 2013. http://ec.europa.eu/competition/publications/cpn/2005_1_94.pdf.

Lukas, Zdeněk. "Slovakia: Challenges on the path towards integration." WIIW Research Report No. 261, The Vienna Institute for International Economic Studies, 1999.

Macieja, Jan. "Restrukturyzacja hutnictwa żelaza i stali." In *Studia nad restrukturyzacją sektorów przemysłowych w Polsce*, edited by Adam Lipowski and Jan Macieja. Warsaw: Polish Academy of Sciences, Institute of Economics, 1998.

Magenheim, Ellen B., and Dennis C. Mueller. "Are Acquiring-Firm Shareholders Better Off after an Acquisition?" In *Knights, Raiders, and Targets. The Impact of the Hostile Takeover*, edited by John C. Coffee, Jr., Louis Lowenstein, and Susan Rose-Ackerman, 171–193. New York: Oxford University Press, 1988.

Mann, Michael. "The Autonomous Power of the State: Its Origins, Mechanisms and Results." *European Journal of Sociology* 25 (1984): 185–213.

Manzetti, Luigi. *Privatization South American Style*. Oxford: Oxford University Press, 2000.

Marcinčin, Anton. "Reštrukturalizácia podnikov." In *Hospodárska Politika na Slovensku 1990–1999*, edited by Anton Marcinčin and Miroslav Beblavý, 309–345. Bratislava: Slovak Foreign Policy Association, 2000.

Maxfield, Sylvia, and Ben Ross Schneider, eds. *Business and the State in Developing Countries*. Ithaca, NY: Cornell University Press, 1997.

McDermott, Gerald A. *Embedded Politics: Industrial Networks and Institutional Change in Postcommunism*. Ann Arbor: University of Michigan Press, 2002.

Meaney, Connie Squires. "Foreign Experts, Capitalists, and Competing Agendas: Privatization in Poland, the Czech Republic, and Hungary." In *Liberalization and Leninist Legacies*, edited by Beverly Crawford and Arend Lijphart, 91–125. Berkeley: University of California Press, 1997.

Medve-Bálint, Gergö. "The Role of the EU in Shaping FDI Flows to East Central Europe." *Journal of Common Market Studies* 52 (2014): 35–51.

Mény, Yves, and Vincent Wright. "State and Steel in Western Europe." In *The Politics of Steel: Western Europe and the Steel Industry in the Crisis Years*, edited by Yves Mény and Vincent Wright, 1–110. Berlin: Walter de Gruyter, 1986.

Mesaros, Serghei. "The Evolving Structure of Collective Bargaining in Europe 1990–2004. National Report. Romania." University of Florence / European Commission, 2005.

Mesežnikov, Grigorij. "Political Parties as Actors of Reforms in Slovakia." In *Slovakia: Ten Years of Independence and a Year of Reforms*, edited by Grigorij Mesežnikov and Oľga Gyárfášová, 59–72. Bratislava: Institute for Public Affairs, 2004.

Mikloš, Ivan. "Economic Transition and the Emergence of Clientalist Structures in Slovakia." In *Slovakia: Problems of Democratic Consolidation*, edited by Soňa Szomolányi and John A. Gould, 57–93. Bratislava: Slovak Political Science Association, 1997.

Ministère de l'Industrie—Roumanie, Sofres Conseil, Sofresid, Cedres. "Programme de Restructuration de la Siderurgie Roumaine." Bucharest, 1993.

Ministerstvo průmyslu a obchodu České republiky. "Pro Schůzi Vlády, Věc: Postup Programu restrukturalizace ocelařského průmyslu v České republice, Příloha č.1 k Předkládací zprávě" (photocopy). Prague, January 14, 2002.

Ministerul Economiei şi Comerţului. "Strategia de Restructurare a Industriei Siderurgice din Romania Pentru Perioada 2004–2010—Actualizare 2004—Sinteză." Bucharest, 2004.

Ministerul Industriilor, Departamentul Industriei Metalurgice. "Strategia de Restructurare a Siderurgiei Româneşti, Sinteză." Bucharest, November, 1993.

Ministry of Industry and Trade of the Czech Republic and Eurostrategy Consultants. "Restructuring of the Steel Industry in the Czech Republic—Industry Restructuring Plan" (photocopy), Prague, May 2001.

Muir, Russel, and Joseph P. Sabat. "Improving State Enterprise Performance: The Role of Internal and External Incentives." World Bank Technical Paper No. 306, International Bank for Reconstruction and Development, Washington, DC, 1995.

Muńko, Andrzej. "Ocena i Stanowisko Unii Europejskiej w Odniesieniu do Procesów Restrukturyzacji Przedsiębiorstw w Polskim Przemyśle Hutnictwa Żelaza i Stali (1990–1998)." INE PAN Working Paper No. 13, Institute of Economics of the Polish Academy of Sciences, Warsaw, 1999.

Murillo, Maria Victoria. *Labor Unions, Partisan Coalitions, and Market Reforms in Latin America.* Cambridge: Cambridge University Press, 2001.

Musil, Jiří, ed. *The End of Czechoslovakia.* Budapest: Central European University Press, 1997.

Myant, Martin. *The Rise and Fall of Czech Capitalism.* Northampton, MA: Edward Elgar, 2003.

Najwyższa Izba Kontroli. "Informacja o wynikach kontroli restrukturyzacji i przekształceń własnościowych w hutnictwie żelaza i stali." Warsaw: NIK, February, 2003.

Negrescu, Dragos. "A Decade of Privatization in Romania." In *Economic Transition in Romania: Past Present and Future; Proceedings of the Conference "Romania 2000— 10 Years of Transition,"* edited by Christof Rühl and Daniel Daianu, World Bank and Romanian Center for Economic Policies, Bucharest, 1999.

Nelson, Joan, ed. *Fragile Coalitions: The Politics of Economic Adjustment.* New Brunswick, NJ: Transaction, 1989.

Nölke, Andreas, and Arjan Vliegenthart. "Enlarging the Varieties of Capitalism: The Emergence of Dependent Market Economies in East Central Europe." *World Politics* 61 (2009): 670–702.

O'Dwyer, Conor. *Runaway State Building: Patronage Politics and Democratic Development.* Baltimore: Johns Hopkins University Press, 2006.

Olson, Mancur. *The Rise and Decline of Nations: Economic Growth, Stagflation, and Social Rigidities.* New Haven, CT: Yale University Press, 1982.

Orenstein, Mitchell. *Out of the Red: Building Capitalism and Democracy in Postcommunist Europe.* Ann Arbor: University of Michigan Press, 2001.

Orenstein, Mitchell, Stephen Bloom, and Nicole Lindstrom. *Transnational Actors in Central and East European Transitions.* Pittsburgh: University of Pittsburgh Press, 2008.

Orenstein, Mitchell, and Lisa E. Hale. "Corporatist Renaissance in Post-communist Central Europe?" In *The Politics of Labor in a Global Age: Continuity and Change in Late-Industrializing and Post-Socialist Economies,* edited by Christopher Candland and Rudra Sil, 258–282. Oxford: Oxford University Press, 2001.

Ost, David. "Illusory Corporatism in Eastern Europe: Neoliberal Tripartism and Postcommunist Class Identities." *Politics and Society* 28 (2000): 503–530.

Parlamentul României, Lege nr. 145 din 3 aprilie 2001 pentru aprobarea Ordonanţei de urgenţă a Guvernului nr. 146/1999.

Pasti, Vladimir. *The Challenges of Transition: Romania in Transition*. Boulder, CO: East European Monographs, 1997.

Pinto, Brian, Marek Belka, and Stefan Krajewski. "Transforming State Enterprises in Poland: Evidence on Adjustment by Manufacturing Firms." *Brookings Papers on Economic Activity* 24 (1993): 213–270.

Pollert, Anna. *Transformation at Work in the New Market Economies of Eastern Europe*. London: Sage, 1999.

Pytel, Barbara. "Sector Study of the Iron and Steel Industry in Poland." *Emergo* 2 (1995): 16–32.

Rączka, Eugeniusz. *Hutnictwo w Polsce na początku XXI wieku*. Katowice: SITPH, 2003.

Reczyński, Andrzej. "Omówienie i ocena studium restrukturyzacji hutnictwa wykonanego przez konsorcjum kanadyjskie." *Problemy Projektowe* 40 (January–March 1993): 1–11.

Reddaway, Peter, and Dmitri Glinski. *The Tragedy of Russia's Reforms: Market Bolshevism against Democracy*. Washington, DC: United States Institute of Peace Press, 2001.

Reed, Quentin. "Corruption in Czech Privatization: The Dangers of 'Neo-Liberal' Privatization." In *Political Corruption in Transition: A Sceptic's Handbook*, edited by Stephen Kotkin and András Sajó, 261–285. Budapest: Central European University Press, 2002.

Roll, Richard. "The Hubris Hypothesis of Corporate Takeovers." *Journal of Business* 59 (1986): 197–216.

Sachs, Jeffrey. *Poland's Jump to the Market Economy*. Cambridge, MA: MIT Press, 1993.

Schamis, Hector. *Re-Forming the State: The Politics of Privatization in Latin America and Europe*. Ann Arbor: University of Michigan Press, 2002.

Schneider, Ben Ross. *Business and the State in Twentieth Century Latin America*. Cambridge: Cambridge University Press, 2004.

Shafer, D. Michael. *Winners and Losers: How Sectors Shape the Developmental Prospects of States*. Ithaca, NY: Cornell University Press, 1994.

Shafir, Michael. *Romania: Politics, Economics and Society*. London: Pinter, 1985.

Slovak Government Information Service. *Analysis of the Inherited State of the Economy and Society*. Bratislava, 1999.

Soifer, Hillel. "State Infrastructural Power." *Studies in Comparative International Development* 43 (2008): 231–251.

Stan, Lavinia. "Romanian Privatization Program: Catching up with the East." In *Romania in Transition*, edited by Lavinia Stan, 127–161. Brookfield, VT: Dartmouth, 1997.

Staniszkis, Jadwiga. "Political Capitalism in Poland." *East European Politics and Societies* 5 (1991): 127–141.

Staniszkis, Jadwiga. *Postkomunizm—Próba Opisu*. Gdańsk: Wydawnictwo Słowo/Obraz Terytoria, 2001.

Stark, David, and László Bruszt. *Postsocialist Pathways: Transforming Politics and Property in East Central Europe*. Cambridge: Cambridge University Press, 1998.

Steinfeld, Edward S. *Forging Reform in China: The Fate of State-Owned Industry*. Cambridge: Cambridge University Press, 1998.

"Strategia de Restructurare a Industriei Siderurgice din România," approved by H.G. nr. 655/29.04.2004, published in M.O. nr. 425/12.05.2004.

Szanyi, Miklós. "Experiences with Foreign Direct Investment in Eastern Europe." *Eastern European Economics* 36 (1998): 28–48.

Sznajder, Aleksandra. "Effects of EU Accession on the Politics of Privatization: The Steel Sector in Comparative Perspective." In *Das Erbe des Beitritts. Mittel- und osteuropäische Gesellschaften nach dem Beitritt zur EU*, edited by Amelie Kutter and Vera Trappmann, 209–231. Baden-Baden: Nomos Verlag, 2006.

Sznajder, Aleksandra. "Still Restructuring? The Politics of the Steel Sector Restructuring in Poland and the Effects of EU Accession." Paper presented at the Annual Meeting of the American Political Science Association, Philadelphia, PA, August 2003.

Sznajder Lee, Aleksandra. "Between Apprehension and Support: Social Dialogue, Democracy, and Industrial Restructuring in Central and Eastern Europe." *Studies in Comparative International Development* 45 (2010): 30–56.

Sznajder Lee, Aleksandra, and Vera Trappmann. "Overcoming Postcommunist Labour Weakness: Attritional and Enabling Effects of MNCs in Central and Eastern Europe." *European Journal of Industrial Relations* 20 (2014): 111–127.

Sznajder Lee, Aleksandra, and Vera Trappmann. "Von der Avantgarde zu den Verlierern der Transformation: Gewerkschaften im Prozess der Restrukturierung der Stahlindustrie in Mittel- und Osteuropa." *Industrielle Beziehungen* 17 (2010): 170–191.

Sztobryn, Jan. "Hutnictwo Żelaza i Stali w Polsce Niepodległej." Unpublished manuscript, photocopy. Warsaw: Agencja Rozwoju Przemysłu, 2001.

Tanase, Stelian. "Reform and Counter-Reform in Romanian Political Life." *Romanian Journal of Sociology* 8, no. 2 (1997).

Tismaneanu, Vladimir. *Stalinism for All Seasons: A Political History of Romanian Communism*. Berkeley: University of California Press, 2003.

Towalski, Rafał. "Związki zawodowe w procesie restrukturyzacji sektora hutnictwa żelaza i stali." In *Związki Zawodowe a Restrukturyzacja—Bariery czy Kompromis*, edited by Leszek Gilejko, 123–143. Warsaw: SGH, 2003.

Trappmann, Vera. *Fallen Heroes in Global Capitalism: Workers and the Restructuring of the Polish Steel Industry*. Houndsmill, UK: Palgrave Macmillan, 2013.

Usinor Consultants. "Asistenţa Tehnică Pentru Restructurarea Industriei Siderurgice Româneşti, Raportul Final Nr. 0, Sinteza." Bucharest, July 2000.

Vachudova, Milada Anna. *Europe Undivided: Democracy, Leverage and Integration after Communism*. Oxford: Oxford University Press, 2005.

Volkov, Vadim. *Violent Entrepreneurs: The Use of Force in the Making of Russian Capitalism*. Ithaca, NY: Cornell University Press, 2002.

Wedel, Janine R. *Collision and Collusion: The Strange Case of Western Aid to Eastern Europe, 1989–1998*. New York: Palgrave, 2001.

Weiner, Myron. Introduction to *Understanding Political Development*, edited by Myron Weiner and Samuel P. Huntington, xiii–xxviii. Boston: Little, Brown, 1987.

Weiner, Myron, and Samuel P. Huntington, eds. *Understanding Political Development*. Boston: Little, Brown, 1987.

WIIW Industrial Database on Central and Eastern Europe. Vienna: Vienna Institute for International Economic Studies, 2005.

Williamson, John, ed. *The Political Economy of Policy Reform*. Washington, DC: Institute for International Economics, 1994.

Wolchik, Sharon L. "The Politics of Ethnicity in Post-Communist Czechoslovakia." *East European Politics and Societies* 8 (1994): 153–188.

Woo-Cumings, Meredith. *The Developmental State*. Ithaca, NY: Cornell University Press, 1999.

World Bank. *Business Environment and Enterprise Performance Survey*. Washington, DC: World Bank, 1999. Accessed October 10, 2014. http://data.worldbank.org/data-catalog/BEEPS.

World Bank. *Implementation Completion Report on a Loan in the Amount of Euro 200 Million to the Slovak Republic for an Enterprise and Financial Sector Adjustment Loan*. Report No. 31073. World Bank, Washington, DC, 2004.

World Bank. *Memorandum of the President of the International Bank for Reconstruction and Development and the International Finance Corporation to the Executive Directors on a Country Assistance Strategy for the Republic of Poland*. Report No. 24783-POL. World Bank, Washington, DC, November 13, 2002.

World Bank. *Memorandum of the President of the International Bank for Reconstruction and Development and the International Finance Corporation to the Executive Directors on a Country Assistance Strategy for Romania*. Report No. 22180-RO. World Bank, Washington, DC, 2001.

World Bank. *Romania Country Assistance Evaluation*. Report No. 32452. Washington, DC: World Bank, 2005.

World Bank. *Romania Country Brief 2004*. Washington, DC: World Bank, 2004.

World Bank. *Slovak Republic: Enterprise and Financial Sector Adjustment Loan Project*. Washington, DC: World Bank, 2000. Accessed June 13, 2013. http://documents.worldbank.org/curated/en/2000/08/443658/slovak-republic-enterprise-financial-sector-adjustment-loan-project.

World Bank. *World Development Indicators*. Washington, DC: World Bank, 2009.

World Bank. *World Development Report: The State in a Changing World*. New York: Oxford University Press, 1997.

Wypych, Andrzej et al. "Modernizacja i restrukturyzacja huty 'Częstochowa': Stan obecny i kierunki dalszych przemian." *Problemy Projektowe* 3 (1996): 66–71.

Index

Odborový Svaz KOVO (OS KOVO),
136, 137
ODS (Občanská Demokratická Strana),
58, 61–62, 65, 229n70, 253n129
Odvětvový Svaz Hutnictví Železa (Iron
and Steel Federation), 84
OFZ, a.s., Istebné, 45
Ogólnopolskie Porozumienie Związków
Zawodowych (OPZZ), 131, 132, 141,
253n142
oil crisis (1973), 193
Oleksy, Józef, 260n102
Olson, Mancur, 36, 225n62
open-hearth furnaces vs. continuous
casting technology, 21, 109, 221n18,
226n5
OPZZ (Ogólnopolskie Porozumienie
Związków Zawodowych), 131, 132,
141, 253n142
Organizația Patronală Hefaistos (Hefaistos Employers' Association), 116
organizational restructuring: challenges
of, 20–22; Czechoslovakia initial proposals, 49; Poland initial proposals, 50;
Romania initial proposals, 48–49
organized labor. *See* unions
Osinek: enabling of incomplete reform,
176; tooling system participation, 108,
110; Vítkovice operation, 172, 173,
175, 264n171
OS KOVO (Odborový Svaz KOVO),
136, 137
Ostrowiec Steelworks: financial difficulties of, 82, 83; location of, 225n10;
privatization, 104, 163–64, 259n101,
260n102, 260n105
Oțelu Roșu Steelworks. *See* Socomet
Oțelu Roșu
OZ KOVO, 128–29, 131

parastatal enterprises: in Czech Republic, 108, 110, 176; in Poland, 105, 160,
176; role in incomplete reform, 10,
176
partial vs. incomplete reform, 221n22.
See also incomplete reform; initial
policy choice and state capacity
Partidul Democrației Sociale din România. *See* PDSR

Partidul Social Democrat (PSD) (Romania), 149–50, 246n2, 255n32
partisanship argument, 180–81
Party of Civic Understanding (Slovakia),
157
Party of Democratic Left (Slovakia), 157
Party of Social Democracy in Romania.
See PDSR
Patriciu, Dinu, 265n14
patrimonial communism, 26, 27, 37
patrimonialism: in Romania, 26, 110;
in Slovakia, 27, 66, 128–29; union
autonomy and, 35, 37, 141
patronage: in Poland, 54; as reward in
initial policy choice, 23
PAV (Public Against Violence), 64–65
PD (Democratic Party) (Romania), 95
PDSR (Partidul Democrației Sociale
din România): privatization concerns,
95; PSD formation, 149, 255n32; rent
seeking of, 73, 90
Penta Group, 156
Petrcíle, s.r.o., 62, 168
Petrotub Roman: merger proposal,
226n19; privatization efforts, 73–74,
92, *93*, 94–95, *118*, *119*; sale to LNM
Holdings, 151–52, 250n73, 256n48;
social dialogue involvement, 120–21,
127
PHARE program (Poland and Hungary
Assistance for the Restructuring of the
Economy): Czechoslovakia study, 49,
226n23; Czech Republic study, 167;
EC financing of, 48, 226n14; Romania
study, 78–79, 147–48
PHS (Polskie Huty Stali). *See* Polish
Steelworks
Pinchuk, Viktor, 190
Podbrezová Steelworks, 44–45, 100–101
Poland: autonomy of unions in, 113–14,
131–36, 139; blurring of unions and
management, 132; communist legacy
in, 26–27; continuous casting use, 21,
109; direct pressure exerted by EU,
10, 13, 40; employment restructuring
in, 37, 104, 105, 134, 139, 244n153;
end of communism in, 18; enterprise
list for, 212*t*; EU accession negotiations, 6, 134–35, 159–60, 161, 162–63,